Chaos and Complexity

Scientific Perspectives on Divine Action

A Series on "Scientific Perspectives on Divine Action"

First Volume
Quantum Cosmology and the Laws of Nature:
Scientific Perspectives on Divine Action
Edited by Robert John Russell, Nancey Murphy, and C. J. Isham

Second Volume
Chaos and Complexity:
Scientific Perspectives on Divine Action
Edited by Robert John Russell, Nancey Murphy, and Arthur R. Peacocke

Future Scientific Topics
Evolutionary and Molecular Biology
Neurobiology and Brain Research
Quantum Physics and Quantum Field Theory

Jointly published by the Vatican Observatory and
The Center for Theology and the Natural Sciences

Robert John Russell, General Editor of the Series

Supported in part by a grant from
the Wayne and Gladys Valley Foundation.

Chaos and Complexity
Scientific Perspectives on Divine Action

Robert John Russell
Nancey Murphy
Arthur R. Peacocke

Editors

Vatican Observatory
Publications,
Vatican City State

The Center for Theology
and the Natural Sciences,
Berkeley, California

SECOND EDITION
1997

Robert John Russell (General Editor) is Professor of Theology and Science in Residence, Graduate Theological Union, Berkeley, California, and Founder and Director of the Center for Theology and the Natural Sciences.

Nancey Murphy is Associate Professor of Christian Philosophy, Fuller Theological Seminary, Pasadena, California.

Arthur R. Peacocke is currently Director of the Ian Ramsey Centre, Oxford, England, Warden Emeritus of the Society of Ordained Scientists, and formerly Dean of Clare College, Cambridge, England.

About the cover: the image on the cover of this volume is a graphic depiction of a Rössler attractor: the simplest case for a chaotic attractor. The image is taken from "Chaos" by James P. Crutchfield, et al., reprinted in this volume by permission of *Scientific American*.

SECOND EDITION

Distributed (except in Italy and Vatican City State) by
The University of Notre Dame Press
Notre Dame, Indiana 46556
U.S.A.

Distributed in Italy and Vatican City State by
Libreria Editrice Vaticana
V-00120 Citta del Vaticano
Vatican City State

ISBN 0-268-00812-4 (pbk.)

ACKNOWLEDGMENTS

The editors with to express their gratitude to the Vatican Observatory and the Center for Theology and the Natural Sciences for co-sponsoring this research. Particular appreciation goes to George Coyne and Bill Stoeger, whose leadership and vision made this series of conferences possible.

Editing for this volume began with an initial circulation of papers for critical responses before the conference in 1993, and continued with extensive interactions between editors and authors after the conference. The editors want to express their gratitude to all the participants for their written responses to pre-conference drafts and for their enthusiastic discussions during the conferences.

Regarding this volume, special thanks goes to Greg Maslowe, CTNS Editing Coordinator, who devoted meticulous attention and long hours to every stage in the preparation of the camera-ready manuscript of this volume. His effort ensured the quality of the final manuscript and made it possible for us to meet our deadline for publication in 1995. Thanks also goes to George Coyne for overseeing printing, jacket design, and distribution, and to Karen Cheatham and Lisa Dahlen for their help in hosting the conference at the GTU and their preliminary work on this manuscript. One of us (RJR) wishes to express special personal thanks to Nancey Murphy for the enormous effort she put into editing this volume, and to Wesley Wildman for the pleasure of co-authoring with him.

The Center for Theology and the Natural Sciences wishes to acknowledge the generous support of the Wayne and Gladys Valley Foundation. The recent renewal of their initial grant continues to make our on-going collaboration with the Vatican Observatory possible. A second grant from the Valley Foundation, also recently renewed, has supported the production of this volume.

The editors wish to thank the following for their generous permission to reprint articles in this volume:

Chaos by James P. Cruthfield, J. Doyne Farmer, Norman H. Packard, and Robert S. Shaw: From *Scientific American* 225, no. 6 (December 1986): 46-57. Reprinted by permission.

Understanding Complexity by Bernd-Olaf Küppers: From *Emergence or Reduction?*, ed. Ansgar Beckermann, Hans Flohr, and Jaegwon Kim (Berlin: Walter de Gruyter, 1992), 241-256. Reprinted by permission of the author.

Chance and Law in Irreversible Thermodynamics, Theoretical Biology, and Theology by Arthur R. Peacocke: From *Philosophy in Science* 4 (1990): 145-180. © 1990 by the Pachart Foundation dba Pachart Publishing House and reprinted by permission.

ACKNOWLEDGMENTS

TABLE OF CONTENTS

IV. ALTERNATIVE APPROACHES TO DIVINE ACTION

INTRODUCTION

Robert John Russell

1 *Background to the Volume*

In August, 1993, twenty scholars, with cross-disciplinary expertise in physics, cosmology, biology, philosophy of religion, philosophy of science, philosophical and systematic theology, history of theology, and history of science, met for a week-long conference at the Center for Theology and the Natural Sciences (CTNS) in Berkeley, California. The purpose of the conference was to explore the implications of chaos and complexity in physical, chemical, and biological systems for philosophical and theological issues surrounding the topic of divine action. The resulting papers form the contents of this volume. The conference was the second of a series of five such research conferences planned for the decade of the 1990s on theology, philosophy, and the natural sciences. The overarching goal of these conferences is to contribute to constructive theological research as it engages current research in the natural sciences, and to identify and critique the philosophical and theological elements that may be present in ongoing research in the natural sciences.

The conferences are jointly sponsored by the Vatican Observatory and CTNS. The Vatican Observatory is housed in the Papal Palace in the picturesque town of Castel Gandolfo, poised overlooking Lake Albano thirty miles southeast of Rome. Since 1935 it has been the site of basic research in both observational and theoretical astronomy. It is also here that Pope John Paul II resides during the summer. CTNS is an affiliate of the Graduate Theological Union (GTU) in Berkeley, California. CTNS sponsors and conducts a variety of research, teaching, and public service programs in the interdisciplinary field of theology and science. Facilities for the conference were provided by the GTU, a consortium of nine Roman Catholic and Protestant seminaries and a graduate school of religion located just north of the campus of the University of California, Berkeley.

This series of conferences grew out of the initiative of Pope John Paul II, who, in 1979, called for an interdisciplinary collaboration of scholars to seek a "fruitful concord between science and faith, between the Church and the world . . . (which) honors the truth of faith and of science. . . ."[1] Responding to this call, the Vatican Observatory held a number of events culminating in a major international conference at the Observatory in September, 1987. The resulting publication, *Physics, Philosophy and Theology: A Common Quest for Understand-*

---

[1]*Discourses of the Popes from Pius XI to John Paul II to the Pontifical Academy of Sciences* (Vatican City State: Pontificia Accademia Scientiarum, 1986), Scripta Varia 66, 73-84.

ing,[2] includes a message by the Pope on the relations between the church and the scientific communities. This was the first major Pontifical statement on science and religion in three decades. It was reprinted, together with nineteen responses by scientists, philosophers, and theologians, in *John Paul II on Science and Religion: Reflections on the New View From Rome*.[3]

Based on this work, George Coyne, Director of the Vatican Observatory, proposed a major new initiative: a series of five conferences for the decade of the 1990s. The goal would be to expand upon the research agenda begun in 1987 by moving into additional areas in the physical and biological sciences. Coyne convened a meeting at the Specola in June, 1990, to plan the overall direction of the research, out of which a long-term steering committee was formed with Nancey Murphy, Associate Professor of Christian Philosophy, Fuller Theological Seminary, Bill Stoeger, Staff Astrophysicist and Adjunct Associate Professor, Vatican Observatory Research Group, Steward Observatory, and myself as members. Coyne then invited CTNS to co-sponsor the decade of research. CTNS was able to accept this offer through the generous support of the Wayne and Gladys Valley Foundation.

The first conference in the series was held in September, 1991 at the Vatican Observatory in Castel Gandolfo. It resulted in the publication of *Quantum Cosmology and the Laws of Nature: Scientific Perspectives on Divine Action*.[4] Future conferences will take up issues in evolutionary and molecular biology, including the study of self-organizing systems; the mind-brain problem; and topics in quantum physics and quantum field theory.

2 *Guiding Theme: Scientific Perspectives on Divine Action*

A major issue in the way research is carried out in the field of theology and science regards the role science ought to play. Too often science tends to set the agenda for the research with little if any initiative taken by theology. From the beginning it was the clear intention of the steering committee that our research expand beyond this format to insure a two-way interaction between scientific and theological research programs. In order to achieve this goal we decided on a two-fold strategy. First, we searched for an overarching theological topic to thematize the entire series of conferences. The topic of divine action, or God's action in and interaction with the world, was quickly singled out as a promising candidate. Clearly it permeates the discussions of theology and science in both philosophical and systematic contexts, and it allows a variety of particular theological and philosophical issues to be pursued under a general umbrella.

---

[2]Robert John Russell, William R. Stoeger, and George V. Coyne, eds. (Vatican City State: Vatican Observatory, 1988).

[3]Russell, Stoeger, and Coyne, eds. (Vatican City State: Vatican Observatory, 1990).

[4]Russell, Nancey Murphy, and C.J. Isham, eds. (Vatican City State: Vatican Observatory, 1993; Berkeley, CA: Center for Theology and the Natural Sciences, 1993).

Next, we organized a series of individual conferences around specific areas in the natural sciences. We chose quantum cosmology, the origin and status of the laws of nature, and foundational issues in quantum physics and quantum field theory to build on areas of research begun in *Physics, Philosophy and Theology*. We also chose chaos and complexity, biological evolution, molecular biology, genetics, creative self-organization, and the mind-brain problem to extend our research into areas treated less directly in *Physics, Philosophy and Theology*.

The overall research methodology for the series of conferences was considered as well. It was agreed that papers for each conference would be circulated several times in advance of the conference for critical written responses from all participants. Revisions would be read in advance of the conference to maximize the critical discussion of the papers during the event. Post-conference revisions would be carefully reviewed by the volume editors in light of these discussions. We agreed to hold regional pre-conferences in Berkeley and Rome to provide an introduction for participants to relevant technical issues in science, philosophy, and theology and to foster joint research and collaboration among participants prior to the conference. An organizing committee would guide the preparation for, procedures of, and editorial process following each conference.

Since the topic of God's action in the world was chosen as the guiding theological theme for the conferences, a brief introduction to the topic is in order here. Portions of the following are drawn from the Introduction published in *Quantum Cosmology and the Laws of Nature*.

3 *Overview of Divine Action: Historical and Contemporary Perspectives*

3.1 *Biblical and Traditional Perspectives*

The notion of God's acting in the world is central to the biblical witness. From the call of Abraham and the Exodus from Egypt to the birth, ministry, death and raising of Jesus and the founding of the church at Pentecost, God is represented as making new things happen. Through these "mighty acts," God creates and saves; the themes of creation and redemption are intimately linked in Biblical theologies.[5]

---

[5]The ordering of the relation between redemption and creation theologies in the history of Israel is the subject of debate. It has long been held that the Exodus experience served as the basis for Hebrew faith in God the creator, a position developed by Gerhard von Rad (*Old Testament Theology*, 2 vols. [New York: Harper & Row, 1957-65], 1:138; and *idem, The Problem of the Hexateuch* [New York: McGraw-Hill, 1966], 131-43). This view was frequently incorporated in standard treatments of the doctrine of creation. See, e.g., Langdon Gilkey, *Maker of Heaven and Earth* (Garden City, NY: Doubleday, 1959), 70. Recent scholarship, however, has questioned this view, raising the possibility that creation theology actually permeates the historical development of Israel. See, e.g., Claus Westerman, *Creation* (Philadelphia: Fortress Press, 1974); Bernhard Anderson, ed., *Creation in the Old Testament* (Philadelphia: Fortress Press, 1984); and R.J. Clifford, "Creation in the Hebrew Bible," in

Rather than seeing divine acts as occasional events in what are otherwise entirely natural and historical processes, both the Hebrews and the early Christians conceived of God as the creator of the world and of divine action as the continuing basis of all that happens in nature and in history.[6]

The view that God works in and through all the processes of the world continued to be held throughout Patristic and Medieval times, as even a cursory reading of such theologians as Augustine and Thomas Aquinas demonstrates. Here God was understood as the first cause of all events; all natural causes are secondary or instrumental causes through which God works. In addition, God was thought to act immediately in the world through miracles, without using or by surpassing finite causes. This view of divine action led to such problematic issues as double agency—can a single and unified event issue simultaneously from two free agents (e.g., God and human agents) each of which is sufficient to accomplish the event? How does an infinite agent (God) preserve the finite freedom of a creaturely agent when they act together? Finally it lead to the problem of theodicy—if God acts in all events and God is good, why is there so much evil in the world?

The conviction that God acts universally in all events, and that God and free human agency can act together in specific events, was maintained by the Protestant Reformers and the ensuing Protestant orthodoxy. John Calvin, for example, argued that God is in absolute control over the world and at the same time maintained that humans are responsible for evil deeds.[7] In general, the topic of divine agency was treated as part of the doctrine of providence and formulated in terms of divine preservation, concurrence, and government. Questions about human freedom and the reality of evil were seen more as problems requiring serious theological attention than as reasons for abandoning belief in God's universal agency.

The rise of modern science and modern philosophy in the seventeenth century led many to reject the traditional views of divine action, especially belief in miracles. Although Isaac Newton argued for the essential role of God in relation to the metaphysical underpinnings of his mechanical system, and in this way defended the sovereignty of God in relation to nature,[8] Newtonian mechanics

---

Physics, Philosophy and Theology, 151-70. For a recent theological discussion of the relation between creation, redemption, and natural science, see Ted Peters, "Cosmos as Creation," in *Cosmos as Creation*, ed. Ted Peters (Nashville: Abingdon Press, 1989), 45-113.

[6]See, e.g., Gen. 45:5; Job 38:22-39:30; Ps. 148:8-10; Is. 26:12; Phil. 2:12-13; 1 Cor. 12:6; 2 Cor. 3:5.

[7]See, e.g., John Calvin, *The Institutes*, II, iv, 2.

[8]Newton's mechanics and his system of the world led to profound philosophical issues through his introduction of absolute space and absolute time to ground the distinction between uniform and accelerated motion, as well as to important theological reconstructions of the relation of God to nature in terms of the *divine sensorium* and the design of the universe. See E.A. Burtt, *The Metaphysical Foundations of Modern Science* (Garden City, NY: Doubleday, 1954). Michael Buckley has argued that the reliance on Newtonian science as a foundation for theology, and the abandonment of the "God hypothesis" thereafter, were

seemed to depict a causally closed universe with little, if any, room for God's *special* action in specific events. With the ascendancy of deism in the eighteenth century, the scope of divine agency was limited to an initial act of creation. Moreover, David Hume and Immanuel Kant raised fundamental questions concerning the project of natural theology, challenged belief in miracles, undercut metaphysical speculation about causality and design, and restricted religion to the sphere of morality.

Given this background, as well as the rise of the historical-critical approach to the Bible, theology in the nineteenth century underwent a fundamental questioning not only of its contents and structure, but even of its method. The response of Friedrich Schleiermacher (1768-1834) was to understand religion as neither a knowing (the activity of pure reason) nor a doing (the activity of practical reason) but as grounded in an entirely separate domain, the "feeling of absolute dependence." Theological assertions can only have as their basis the *immediate* assertions of religious self-consciousness. Schleiermacher understood God's relation to the world in terms of universal divine immanence. By miracle we mean ". . . simply the religious name for event. Every event, even the most natural and usual, becomes a miracle, as soon as the religious view of it can be the dominant."[9]

3.2 *The "Travail" of Divine Action in the Twentieth Century*

Protestant theology in the first half of the twentieth century was largely shaped by Karl Barth. In his rejection of nineteenth-century liberal theology, Barth turned to the sovereignty of the God who is "wholly other" and stressed God's initiating action in our redemption. "The Gospel is . . . not an event, nor an experience, nor an emotion—however delicate! . . . [I]t is a communication which presumes faith in the living God, and which creates that which it presumes."[10] But do Barthian

---

principal causes of the rise of atheism in the West. See Michael J. Buckley, *At the Origins of Modern Atheism* (New Haven, CT: Yale University Press, 1987). For a brief historical account see Ian G. Barbour, *Issues in Science and Religion* (New York: Harper Torchbook, 1966), chap. 3. One can also argue that the concept of inertia played an important role in deflecting attention away from the view of God as acting ubiquitously to sustain nature in being. See Wolfhart Pannenberg, "Theological Questions to Scientists," in *The Sciences and Theology in the Twentieth Century*, ed. A.R. Peacocke (Notre Dame, IN: University of Notre Dame Press, 1981), 3-16, esp. 5-6.

[9]Friedrich Schleiermacher, *On Religion: Speeches to its Cultured Despisers* (New York: Harper Torchbook, 1958), 88. In a long discussion in *The Christian Faith* (Edinburgh: T&T Clark, 1968), he wrote: "[A]s regards the miraculous . . . we should abandon the idea of the absolutely supernatural because no single instance of it can be known by us. . . ."(#47.3, 183) For an excellent analysis of Schleiermacher and other important developments in the nineteenth century see Claude Welch, *Protestant Thought in the Nineteenth Century*, 2 vol. (New Haven, CT: Yale University Press, 1972).

[10]Karl Barth, *The Epistle to the Romans*, 6th ed. (London: Oxford University Press, 1968), 28.

neo-orthodoxy and the ensuing "biblical theology" movement of the 1940s and 1950s succeed in producing a credible interpretation of God's acts?

In a well-known article written in 1961, Langdon Gilkey forcefully argued that they do not.[11] According to Gilkey, neo-orthodoxy is an unhappy composite of biblical/orthodox language and liberal/modern cosmology. It attempts to distance itself from liberal theology by retaining biblical language about God acting through wondrous events and by viewing revelation as an objective act, not just a subjective perception. Yet, like liberalism, it accepts the modern premise that nature is a closed causal continuum, as suggested by classical physics. The result is that, whereas orthodoxy used language univocally, neo-orthodoxy uses language at best analogically. In fact, since its language has been emptied of any concrete content, its analogies devolve into equivocations. "Thus the Bible is a book descriptive not of the acts of God but of Hebrew religion. . . . [It] is a book of the acts Hebrews believed God might have done and the words he [*sic*] might have said had he done and said them—but of course we recognize he did not."[12]

Thus, a "two-language" strategy sets in: neo-orthodox theologians use biblical language to speak confessionally about God's acts, but they use secular language when speaking historically or scientifically about "what actually happened." Similarly, the insistence that revelation presupposes faith founders when one challenges the possibility of God acting in the event in which faith originates. Thus, neo-orthodoxy, and with it much of contemporary theology, involves a contradiction between orthodox language and liberal cosmology.

3.3 *Current Approaches to Divine Action*

In the wake of these problems, several approaches to divine action are being explored in current literature. The following lists a few of the more prominent ones as suggested in part by a major anthology on divine action edited by Owen Thomas in 1983.[13]

---

[11]Langdon B. Gilkey, "Cosmology, Ontology, and the Travail of Biblical Language," *The Journal of Religion* 41 (1961): 194-205.

[12]Ibid., 198.

[13]See Owen C. Thomas, ed., *God's Activity in the World: The Contemporary Problem* (Chico, CA: Scholars Press, 1983). According to Thomas, only process and neo-Thomistic theologies offer full-blown metaphysical theories. They are therefore superior to the personal action approach with its limited analogies, and these are all preferable to liberal, uniform action and two-perspectives theologies, which offer neither theories nor analogies. In a more recent publication, Ian Barbour offers a lucid description and creative comparison of the problem of divine action in classical theism, process theism, and their alternatives, including several types of personal agency models. See Barbour, *Religion in an Age of Science*, The Gifford Lectures 1989-1991, vol. 1 (San Francisco: HarperSanFrancisco, 1990), chap. 9. See also Peacocke, *Theology for a Scientific Age: Being and Becoming—Natural and Divine* (Oxford: Basil Blackwell, 1990; and second enlarged edition, London: SCM Press, 1993 and reprinted, Minneapolis: Fortress Press, 1993), chap. 9; and Thomas F. Tracy, ed., *The*

3.3.1 ***Neo-Thomism*** Here traditional Thomistic distinctions between primary and secondary causality, characteristic of both Roman Catholicism and Protestant orthodoxy, have been modified in light of the work of Kant. Advocates include Bernard Lonergan, Joseph Mareschal, Jacques Maritain, and Karl Rahner. Related works are those of Austin Farrer and Eric Mascall. Recent contributors to theology and science from a neo-Thomistic perspective include W. Norris Clarke, Ernan McMullin, and Bill Stoeger.[14]

3.3.2 ***Process Theology*** Process philosophy represents a fundamental shift away from Aristotelian metaphysics. Based on the metaphysics of Alfred North Whitehead, process theology rejects the traditional conception of divine action in terms of primary and secondary causes. Instead, God is seen as acting persuasively in all events, though never exclusively determining their character, since each "actual occasion" also includes an irreducible element of genuine novelty as well as its response to the influence to the past. Representatives include Charles Birch, John Cobb, Jr., David Griffin, Charles Hartshorne, and Schubert Ogden. Ian Barbour endorses a process perspective in a nuanced way, calling for a "social or interpersonal analogy" of divine action, and provides a balanced assessment of the strengths and weaknesses of other approaches.[15]

3.3.3 ***Uniform Action*** Here God is thought of as acting uniformly in all events in the world. Distinctions in meaning and significance are due entirely to human interpretation. This view is found in the writings of Gordon Kaufman and Maurice Wiles.

3.3.4 ***Personal Agent Models***

3.3.4.1 ***Literal Divine Action*** Some philosophers of religion, notably William Alston,[16] have questioned the assumption that the causal determinism of all natural events by other natural events prevents us from speaking of God, literally, as a personal agent who brings about particular states of affairs at particular times and places. First, he holds that there is no sufficient reason to adopt strict naturalistic causal determinism. Second, he argues that deterministic laws hold only for closed

---

God Who Acts: Philosophical and Theological Explorations (University Park, PA: Pennsylvania State University, 1994).

[14]See their articles in *Physics, Philosophy and Theology.*

[15]Barbour echoes Thomas Tracy's objections to viewing "the world as the body of God" since the universe lacks bodily unity and an environment (see section 3.3.4.2 below). He also believes that embodiment would fail to provide adequately for the independence of God and the world. See Barbour, *Religion in an Age of Science*, 259.

[16]William P. Alston, *Divine Nature and Human Language* (Ithaca: Cornell University Press, 1989). See also, *idem*, "Divine Action, Human Freedom, and the Laws of Nature," in *Quantum Cosmology*, 185-207.

systems and we need not suppose that all systems are closed to outside influences (including divine ones).

3.3.4.2 ***Embodiment and Immanentist Models*** The analogy of God: world:: mind:body has been explored in differing ways by such authors as Arthur Peacocke,[17] Grace Jantzen,[18] and Sallie McFague.[19] Peacocke stresses the immanence of God in the world and views the universe as dynamic and interconnected, suggesting biological and feminine analogies for divine agency. These analogies counteract the tendency of traditional language both to stress God's transcendence and to neglect God's immanence with respect to creation. Peacocke adopts a panentheistic approach which combines the language of immanence with transcendence in speaking about God's relation to creation.[20] Jantzen proposes that the entire universe is God's body. God is immediately aware of all events in nature, and acts both universally throughout nature and particularly in unique events. Moreover, all of God's acts are basic or direct acts, analogous to the direct acts we perform when we move our own bodies. She recognizes that the embodiment model raises several problems, including the problem of evil, the relation of God's action to the laws of nature, and the significance for the divine life if the universe has a finite past and/or future, but suggests important responses to each of these. McFague employs embodiment models to view "God as mother, lover and friend. . . ."[21] She admits, however, that the embodiment model suggests that God is at risk, being vulnerable to the sufferings of nature.

3.3.4.3 ***Non-Embodiment*** Thomas Tracy rejects both the claim that personal agency requires embodiment and that the world is like an organism. Instead he develops a conception of God as a non-embodied agent.[22] Tracy sees himself as combining aspects of classical and process theism.

---

[17]Peacocke, *Creation and the World of Science* (Oxford: Clarendon, 1979), 142ff., 207; and *idem, Intimations of Reality* (Notre Dame, IN: University of Notre Dame Press, 1984), 63ff., 76. See also section 4.2 of his paper, "God's Interaction with the World," in this volume.

[18]Grace Jantzen, *God's World, God's Body* (Philadelphia: Westminster Press, 1984).

[19]Sallie McFague, *Models of God: In an Ecological Nuclear Age* (Philadelphia: Fortress Press, 1987), 69-78; and *idem, The Body of God: An Ecological Theology* (Minneapolis: Fortress Press, 1993). See also, idem, "Models of God for an Ecological, Evolutionary Era: God as Mother of the Universe," in *Physics, Philosophy and Theology*, 249-271.

[20]According to panentheism, "the world is regarded as being . . . 'within' God, but the being of God is regarded as not exhausted by, or subsumed within, the world" (Peacocke, *Creation and the World of Science*, 207). See also *Theology for a Scientific Age*, n. 75, p. 370-2.

[21]McFague, *Models of God*, 71.

[22]Tracy, *God, Action and Embodiment* (Grand Rapids, MI: Eerdmans, 1984).

3.3.4.4 ***Interaction Models*** Many of the objections to embodiment models cited above have been used by supporters of one or another form of an interaction model. John Polkinghorne argues that a clue to God's interaction with the world can come from examining how humans interact with the world. He turns to quantum physics and chaos theory to suggest how the combination of lawlike behavior with openness and flexibility in nature makes human and, in some preliminary way, divine agency conceivable.[23]

3.3.5 ***Assessment*** Owen Thomas returned to the problem in 1991, with this rather caustic comment:

> Theologians continue to talk a great deal about God's activity in the world, and there continue to be only a very few who pause to consider some of the many problems involved in such talk.[24]

According to Thomas, the question of double agency still remains "the key issue in the general problem. . . ."[25] How can we assert coherently that both divine and creaturely agents are fully active in one unified event? After evaluating the current state of the discussion, Thomas's position is that one must either follow the primary/secondary path of traditional theism or the process theology approach. He asserts that if there *is* another solution to the problem, he has not heard of it, and concludes that this question should be a major focus of future discussions.[26]

3.4 *Working Typology of Theological Positions on Divine Action in Light of Science*

The following working typology (see Figure 1) presents a correlation of various theological views with the types of claims their proponents tend to make about divine action, including new developments in light of science. It is meant primarily as a framework to guide the reader in interpreting the varieties of positions taken in the papers in this volume. It is presented here in its current form as work in progress. Its heuristic nature should be underscored, as well as the fact that, like any typology, it represents an abstraction from the extant literature; actual authors may hold differing positions at various places in their writings.[27]

The upper portion of the table includes five positions frequently found in historical and contemporary theology regarding divine action. It serves as a

---

[23]John Polkinghorne, *Science and Providence* (Boston: Shambhala, 1989). See also, *idem*, "The Laws of Nature and the Laws of Physics," in *Quantum Cosmology*, 437-448. See also *Reason and Reality: The relationship between science and theology* (Philadelphia: Trinity Press International, 1991), chap. 3.

[24]Thomas, "Recent Thought on Divine Agency," in *Divine Action*, ed. Brian Hebblethwaite and Edward Henderson (Edinburgh: T&T Clark, 1990), 35-50.

[25]Ibid., 46.

[26]Ibid., 50.

[27]Since its inception in 1991, the typology has been developed further through discussion with a number of scholars including Nancey Murphy and Thomas Tracy.

backdrop to the current discussions in theology and science, as found in the lower portion of the table.

For completeness, the table begins with the possibility that no events are God's acts; this is obviously the position of atheism which denies God's very existence.

A second position is that God acts only "in the beginning,"[28] creating the universe and the laws of nature. After this, the universe runs entirely on its own according to the laws of nature. This is the deist option. If science could show that there was no beginning, the deist view would be seriously undermined.

According to a third view, it is the existence of the universe as such, and not primarily its beginning, that requires explanation in terms of God's action. Thus, God acts not only in the beginning to create the universe (if indeed there was a beginning); more importantly, God acts uniformly in all events throughout the history of the universe to sustain its existence moment by moment. This means that without God's continuous action the universe would simply cease to exist, since matter does not contain its own sufficient principle of existence. Whether the laws of nature prescribe or merely describe the regularities of natural processes (i.e., whether causal efficacy is contained in the laws governing nature or in nature itself), God is required as the ground of natural (secondary) causality and the primary cause of every event in the universe. Still, on this view, no events are *special* acts of God in any sense. We are calling this view "garden-variety theism," although pantheism could also fall into this category.

A fourth view is held by those who agree with theists that God acts both in the beginning (if there was one) and that God acts uniformly in all events. However, they add the further claim that certain events in nature and history can be said to be special acts of God, but only in a subjective sense: they are *seen as* revealing something special about the character or intentions of God even though God does not really act differently in an *objective* way in these events. We call this the "liberal" view. In recent work in which a much more dynamic conception of the universe has come to prevail over the older, static cosmology, God's action in sustaining the universe has often been seen as God's "continuous creation" of the universe. Those who view the universe in these more dynamic terms and speak about continuous creation are often eager to attribute special significance to what appears to be the occasional appearance of genuine novelty, even if they agree that all events are in fact uniformly caused by God through the unfolding realizations of the potentialities of nature represented by the laws of nature.

Finally, it is possible to hold that there are events that in some *objective* sense are special acts of God. When we call these events "special" we do so as a response to God's initiative, an acknowledgment of what God is doing to bring them about. It has most often taken the form of *intervention*: God performs such acts by intervening in or suspending the laws of nature. We will call this the "traditional"

---

[28]Assuming there was one, such as Big Bang cosmology supports! For a detailed discussion of the scientific, philosophical, and theological complexity of this assumption, see *Quantum Cosmology*.

	Creation	Uniform Divine Action		Objectively Special Divine Action		
		Sustenance	Subjectively Special	Interventionist	Non-Interventionist	
					Apparent with religious presuppositions	Apparent without religious presuppositions
Atheism						
Deism	X					
'Theism'	X	X				
'Liberal'	X	X	X			
'Traditional'	X	X	X	X		
Theology and Science:						
(1) Top-Down or Whole-Part (e.g., mind/brain, non-linear thermodynamics)	X	X	X		X	?
(2) Bottom-Up (e.g., quantum indeterminacy)	X	X	X		X	?
(3) Lateral Amplification (e.g., chaos)	X	X	X		X	?
(4) Primary/ Secondary	X	X	X		?	?

Figure 1: Working Typology of Divine Action

view. Note that the traditional view usually *includes* the theist's claim that the existence of the universe, both its continued sustaining in being and its beginning (if it had one), is a product of God's action.

Much of the *current* discussion in the field of theology and science regarding divine action now turns on the question whether there are *objectively* special divine acts that are *neither* interventions nor suspensions of the laws of nature. We will call these "non-interventionist" views. Among non-interventionists, it is still an open question whether these events are objectively perceptible as special, that is, whether they would be seen as God's acts by anyone present regardless of their prior religious presuppositions, or whether it is precisely these presuppositions which allow observers to recognize that God has objectively acted in a special way in the event in question.

Among those who opt for objectively special divine action, some authors distinguish between direct and indirect divine acts, or equivalently between basic and non-basic divine acts. The argument here is that if God acts indirectly through secondary causes to achieve an objectively special act, there has to be at least one direct or basic act somewhere that initiates the chain of events the outcome of which we call God's (indirect, objectively special) act. This is a logical point, and not a theological point: if an agent is to do something indirectly, at some point the agent has to do something directly which eventuates in the indirect act. This issue is related to the problem of the "causal joint."

We should note that it is possible to affirm objectively special divine acts without deciding whether these acts are themselves direct acts or are the indirect result of other, perhaps hidden, direct acts of God. On the one hand, the special event in question may be seen as the direct act of God. On the other hand, most of the current arguments tend to see it as an indirect act of God, that is, as the product of secondary causes stemming from a direct act of God located elsewhere. But if this is the case, then the question in turn becomes: Where is this "elsewhere," that is, where is the real domain of God's direct or basic act?

In surveying the continuing conversations with scholars in theology and science, it has become clear that there are four distinct *non-interventionist* approaches to this question, though combinations of them are also viable: (1) "top-down" or "whole-part" causality; (2) "bottom-up" causality; (3) what we might call "lateral" causality, and (4) "primary/secondary" causality. First of all, then, some describe divine action in terms of "top-down" or "whole-part" causality. Here, a localized, special event in the world is viewed as the indirect result of God acting directly in one of two ways: either in a top-down way from a higher level in nature (using such analogies as "mind/brain"), or in a whole-part way starting either at the physical boundaries or environment of the system (an analogy here is the formation of vortices in a liquid heated in a container), or, ultimately, at the boundary of the universe as a whole. The second, or "bottom-up," approach views a special event in the macroscopic world as the indirect result of a direct act of God at the quantum mechanical level, amplified by a stream of secondary causes linked in a bottom-up way. This view presupposes that quantum uncertainty be given an ontological interpretation, namely that of indeterminism.

Some authors in theology and science have pointed to a third non-interventionist option, which was of particular importance at the 1993 conference. They stress the "supple" and "subtle" nature of chaotic and complex systems in physics, meteorology, biology, and so on, focusing on their vulnerability to minute changes in their environment and their ability to amplify these effects (the analogy here is the "butterfly effect"). Since this view is entirely restricted to the classical level,[29] involving a chain of events within a group of phenomena encompassed within a single epistemic level (e.g., classical physics), we can call this view "lateral" causality. Perhaps, then, chaos and complexity would offer a new, non-interventionist understanding of God's objectively special action in nature, or at least suggest that new, holistic laws of nature might be indicated, which in turn could lead to new insights into divine action.

Fourth, some authors are committed to accounts of divine action that work strictly within the distinctions between primary and secondary causality. These authors see no need to speak in terms of objectively special divine acts, with the possible exception of miracles.

4 *Chaos and Complexity: Brief Introduction to the Scientific Themes of This Volume*

Among the many facets of our intellectual landscape inherited from the Enlightenment, the "received view" from physics is that of nature as a machine. The universe is governed by a set of deterministic equations discovered by Newton at the close of the seventeenth century. By the eighteenth century, most scientists were convinced that these equations would allow the exact prediction of the future state of the world given precise knowledge of its present state and all relevant forces. As Pierre Simon Marquis de Laplace put it succinctly, to one of sufficient intelligence "nothing would be uncertain and the future, as the past, would be present to its eyes."[30] In such a world, there would be little need—or possibility—for God to act in special events, as the biblical witness recounts. Notice that Laplace used predictability as a criterion for evaluating the applicability of Newtonian physics to the world and its confirmation in his own work on the stability of planetary orbits against perturbations then served as a warrant in support of a metaphysics of causal determinism.

We are now, however, faced with a fundamental challenge to this "equation" between determinism and predictability. Though severe problems in predicting the actual course of events were known to exist from the outset—even the three-body

---

[29]It is possible to appeal to quantum mechanics in order to introduce indeterminism into the problem in terms of the initial conditions which are then amplified by chaotic dynamics, but this move (1) is challenged, in turn, by the apparent lack of "quantum chaos," and (2) effectively changes this approach into the previous one, viz., "bottom-up" causality. For comments on "quantum chaos" see the papers by Crutchfield, et al.; and Wesley Wildman and Robert Russell, both in this volume.

[30]Pierre Simon Marquis de Laplace, *A Philosophical Essay on Probabilities*, 6th ed., trans. F.W. Truscott and F.L. Emory (New York: Dover, 1961), 4.

problem in Newtonian gravity defied an analytic solution—recent discoveries in so-called "chaotic dynamic systems" have complicated the problem of prediction to the extreme. These chaotic systems now seem ubiquitous in nature; they are known to occur throughout hydrodynamics and plasma physics, and to pervade the domains of physics, meteorology, chemistry, biology, evolution, and cosmology. Chaotic systems are typically governed by simple, non-linear, deterministic equations. Thus, their future states are in principle entirely determined by their present state and the forces acting on them, and in this sense they are predictable. Nevertheless in many cases the future states *appear* to be random. Moreover, even if we start with a chaotic equation in hopes of modeling the system's behavior, the unusual properties of these equations and our inability to specify their initial states with infinite precision means that we will eventually lose the ability to predict this behavior. Chaotic systems thus embody elements of both determinism, predictability, *and* unpredictability, even when they are treated entirely within the domain of classical physics. What, then, are the philosophical and metaphysical implications of chaos, and do chaotic systems, in turn, shed any light on the topic of divine action?

5 *The Present Publication: Summary*

These broad scientific, philosophical, and theological themes surface in nuanced and interwoven patterns in the essays included in this volume. Section I introduces some of the basic mathematical and scientific issues in the subjects of chaos and complexity. Section II deals with related issues in the philosophy of nature. Section III focuses on the implications of chaos and complexity for the problem of divine action. The authors in Section IV develop alternative proposals regarding divine action, most of which were pursued after the authors concluded that chaos theory is not helpful to the problem.

5.1 *Scientific Background: Chaos and Complexity*

We begin with a previously published paper by **James P. Crutchfield, J. Doyne Farmer, Norman H. Packard,** and **Robert S. Shaw**. Their paper is reprinted here to give a broad introduction and background to the science of chaos and complexity.

Until recently scientists assumed that natural phenomena such as the weather or the roll of the dice could in principle be predictable given sufficient information about them. Now we know that this is impossible. "Simple deterministic systems with only a few elements can generate random behavior." Though the future is determined by the past, small uncertainties are amplified so radically by chaotic systems that, in practice, their behavior rapidly becomes unpredictable. Still there is "order in chaos," since elegant geometrical forms underlie and generate chaotic behavior. The result is "a new paradigm in scientific modeling" which both limits predictability in a fundamental way and yet extends the domain in which nature can be at least partially predictable.

The article acknowledges the challenge to Laplacian determinism posed by quantum mechanics for subatomic phenomena like radioactive decay, but stresses that large-scale chaotic behavior, which focuses instead on macroscopic phenomena like the trajectory of a baseball or the flow of water, "has nothing to do with quantum mechanics." In fact many chaotic systems display both predictable and unpredictable behavior, like fluid motion which can be laminar or turbulent, even though they are governed by the same equations of motion. As early as 1903, Henri Poincaré suggested that the explanation lay in the exponential amplification of small perturbations.

Chaos is an example of a broad class of phenomena called dynamical systems. Such systems can be described in terms of their state, including all relevant information about them at a particular time, and an equation, or dynamic, that governs the evolution of the state in time. The motion of the state, in turn, can be represented by a moving point following an orbit in what is called state space. The orbits of non-chaotic systems are simple curves in state space. For example, the orbit of a simple pendulum in state space is a spiral ending at a point when the pendulum comes to rest. A pendulum clock describes a cyclic, or periodic, orbit, as does the human heart. Other systems move on the surface of a torus in state space. Each of these structures characterizing the long-term behavior of the system in state space—the point, the cycle, the torus—is called an attractor since the system, if nudged, will tend to return to this structure as it continues to move in time. Such systems are said to be predictable.

In 1963, Edward Lorenz of MIT discovered a chaotic system in meteorology which showed exponential spreading of its previously nearby orbits in state space. The spreading effect is due to the fact that the surface on which its orbits lie is folded in state space. Such a surface is called a strange attractor, and it has proven, in fact, to be a fractal. The shape of a strange attractor resembles dough as it is mixed, stretched, and folded by a baker. With this discovery we see that "random behavior comes from more than just the amplification of errors and the loss of the ability to predict; it is due to the complex orbits generated by stretching and folding."

The essay closes with some profound questions about scientific method. If predictability is limited in chaotic systems, how can the theory describing them be verified? Clearly this will involve "relying on statistical and geometric properties rather than on detailed prediction." What about the assumption of reductionism in simple physical systems? Chaotic systems display a level of behavioral complexity which frequently cannot be deduced from a knowledge of the behavior of their parts. Finally, the amplification of small fluctuations may be one way in which nature gains "access to novelty" and may be related to our experience of consciousness and free will.

In contrast to the general introduction given by Crutchfield, et. al., **Wesley J. Wildman** and **Robert John Russell's** article surveys the mathematical details of a single equation, the logistic equation, which has become a hallmark of this field, at least within the circles of "theology and science." The logistic equation displays many of the generic features of chaotic dynamical systems: the transition from regular to apparently random behavior, the presence of period-doubling bifurcation cascades, the influence of attractors, and the underlying characteristics of a fractal.

They then raise philosophical questions based on the mathematical analysis and conclude with possible theological implications.

The logistic equation is a simple, quadratic equation or "map," $x_{n+1} = kx_n(1-x_n)$, which iteratively generates a sequences of states of the system represented by the variable x. The tuning constant k represents the influence of the environment on the system. One starts from an initial state x_0 and a specified value for the tuning constant k to generate x_1. Substituting x_1 back into the map generates x_2, and so on. Although incredibly simple at face value, the logistic map actually displays remarkably complex behavior, much of which is still the focus of active scientific research.

The behavior of the iterated sequence produced by the logistic map can be divided into five regimes. The constant k determines which regime the sequence occupies as well as much of the behavior within that regime. In Regime I, the sequence converges to 0. In Regime II, the sequence converges on a single positive limit which depends on k. In Regime III, bifurcations set in and increase in powers of two as k increases. Moreover, the initial conditions have a significant permanent effect on the system in the form of "phase shifts." Chaos sets in in Regime IV. Here chaotic sequences are separated by densely packed bifurcation regions and there is maximal dependence on initial conditions. For most values of k, the sequences seem to fluctuate at random and the periodic points found in previous regimes appear to be absent. Nevertheless, for almost all values of k we actually find highly intricate bifurcation structures, and the sequences fall within broad bands, suggesting an underlying orderliness to the system. Finally in Regime V, chaos is found on the Cantor subset of x.

There is no universally accepted mathematical definition of chaos capturing all cases of interest. Defining chaos simply as randomness proves too vague because this term acquires new and more precise shades of meaning in the mathematics of chaos theory. Defining chaos in terms of sensitive dependence on initial conditions (the butterfly effect) results in the inclusion of many maps that otherwise display no chaotic behavior. The definition adopted here requires a chaotic map to meet three conditions: mixing (the effect of repeated stretching and folding), density of periodic points (a condition suggesting orderliness), and sensitive dependence. Interestingly, in the case of the logistic map and many similar chaotic maps, mixing is the fundamental condition, as it entails the other two.

The paper also addresses the question of the predictability of chaotic systems. On the one hand, a chaotic system such as the logistic map is predictable in principle, since the sequence of iterations is generated by a strict governing equation. On the other hand, chaotic systems are "eventually unpredictable" in practice, since most values of the initial conditions cannot be specified precisely, and even if they could, the information necessary to specify them cannot be stored physically. Yet these systems are also "temporarily predictable" in practice, since one can predict the amount of time which will elapse before mathematical calculations will cease to match the state of the system. This leads to a definition of 'chaotic randomness' as a *tertium quid* between strict randomness (as in one common interpretation of quantum physics), and the complete absence of randomness.

What implications does mathematical chaos have for a philosophy of nature? It is superficial to say that the mathematical determinism of chaotic equations requires metaphysical determinism in nature, because of complexities in the experimental testing of the mathematical models used in chaos theory. In particular, it may be very difficult to distinguish phenomenologically between chaos, sufficiently complicated periodicity, and strict randomness, even though these are entirely distinct mathematically. There are additional practical limitations to the testing of chaotic models of natural systems, including sensitivity to the effects of the environment (such as heat noise or long-range interactions), and the fact that the development of the physical system eventually out paces even the fastest calculations.

Two philosophical conclusions are drawn from this. On the one hand, the causal whole-part relations between environment and system, the causal connnectedness implied in the butterfly effect, and the fact that much of the apparent randomness of nature can now be brought under the umbrella of chaos, are best seen as supporting evidence for the hypothesis of metaphysical determinism. On the other hand, however, there are profound epistemic and explanatory limitations on the testing of chaos theory due to the peculiar nature of chaotic randomness. In this sense, chaos theory places a fundamental and unexpected new limit on how well the hypothesis of metaphysical determinism can be supported.

On the basis of these philosophical conclusions, what relevance does chaos theory have for theology? On the one hand, it will be "bad news" to those who simply assume that nature is open to the free actions of God and people, and particularly bad news to those who mistakenly appeal to chaos theory to establish this. On the other hand, chaos theory will be irrelevant to theologians operating with a supervening solution to the problem of divine action, such as Kant's, that is able to affirm human freedom and divine action even in the presence of strict metaphysical determinism. At still another level chaos theory is "good news" to the theological project and "bad news" for "polemical determinists." Due to the fundamental, new limitation in the testability of chaos theory, one can never fully exclude the possibility that classical physics as we now have it, including chaos theory, will be replaced by a better model of the world at the *classical* level which allows for divine causality in some way. This "opens a window of hope for speaking intelligibly about special, natural-law-conforming divine acts, and it is a window that seems to be impossible in principle to close."

The article includes an extended bibliography of textbooks, key technical articles, experimental applications, useful introductions and surveys, and selected works on chaos theory and theology.

5.2 *Chaos, Complexity, and the Philosophy of Nature*

The three papers in this section take up issues in the philosophy of nature which relate in various ways to the topics of chaos or complexity.

According to the paper by **Bernd-Olaf Küppers** reprinted here, the reductionistic research program "is based on the central working hypothesis that all biological phenomena can be explained totally within the framework of physics and chemistry." It assumes that there is no essential difference between non-living and living matter; life arises as a "quasi-continuous" transition requiring no additional epistemic principles other than those of physics and chemistry. Restrictions in our current understanding are merely the result of the complexity of the problem and its computability. Epistemic reductionism leads to ontological reductionism in which "life is nothing but a complex interplay of a large number of atoms and molecules." Even consciousness must ultimately be reducible to physical laws.

To counter this program, some biologists and philosophers of science appeal to "emergence" and "downward causation," claiming that genuinely novel properties and processes arise in highly complex phenomena. According to this view, physics is a necessary part of the explanation but it cannot provide a sufficient explanation on its own. Küppers summarizes the claims of emergence and downward causation, respectively, as follows: "(1) The whole is more than the sum of its parts. (2) The whole determines the behavior of its parts."

Since these concepts seem "vague and mysterious" to scientists in physics and biology, Küppers focuses here on a general problem concerning the transition from the non-living to the living: can we adequately characterize the emergence of life in terms of the concept of complexity. Küppers thinks not, since non-living systems may themselves be extraordinarily complex. In addition, one may find evidence of emergence even within a field, such as within physics, and not just between fields.

In a similar way, those supporting intratheoretical reduction (e.g., reductionism within physics) frequently appeal to "bridge laws," while defenders of emergence deny their availability and their fruitfulness. Arguments such as these also apply to the question of downward causation. In Küppers' opinion, both emergence and downward causation are to be found within physics. Since no "non-physical principle" is involved, apparently, in the transition to life, Küppers concludes that "both (emergence and downward causation) must be thought of as characteristics of self-organizing matter that appear at all levels when matter unfolds its complexity by organizing itself." Still, there are examples of biological systems, such as the DNA macromolecule, which are immensely more complex than complex physical systems. Do they point to a limitation in physical method or in the reductionistic research program, or will physics undergo a paradigm shift as it seeks to encompass these phenomena within its domain?

To understand these questions better, Küppers begins by distinguishing between laws and initial or boundary conditions in physical theory. His central claim is that "the complexity of a system or a phenomenon lies in the complexity . . . of its boundary conditions." Following the analysis of Michael Polanyi, Küppers argues that in a human construction, such as a complex machine, the design, or boundary conditions, governs the physical processes but cannot be deduced from them. In this way a machine, by its structure and operation, is an emergent system, a whole which is "neither additive nor subtractive," whose properties cannot be reduced to those of its components, and whose boundary conditions represent a form of downward causation. A similar case can be made for a living organism.

Now the question becomes, what determines the boundary conditions? For a machine, the answer is a blueprint. For the living organism, however, the "blueprint" lies in the organism's genome which, in contrast to the machine, is an inherent part of the living system. Küppers then distinguishes complex from simple systems in terms of both their sensitivity to small changes in their boundary conditions and the uniqueness of these conditions, given all possible physically equivalent conditions.

The concept of boundary conditions thus becomes the key to understanding the paradigm shift that is occurring within physics regarding the problem of complex phenomena. This shift is not of the Kuhnian type, with its revolutionary change in the fundamental laws and the theoretical framework of a field. Instead it is an "internal shift of emphasis" within the given explanatory structure of the paradigm. As Küppers sees it, the shift of emphasis within the reductionistic research program consists in the move to regard the boundary conditions of complex phenomena as that which needs explanation. He calls this shift of emphasis the "paradigm of self-organization." It entails a sequence of explanations, in which boundary conditions at one level (such as the boundary conditions of the DNA molecule) are explained by those of another level (such as the random molecular structures), which themselves need explanation. In effect, the nested structures found in living matter are reflected by the nested structures of the paradigm of self-organization.

Finally, Küppers points out that biological self-organization is only possible in the context of non-equilibrium physics. Still, though the existence of specific boundary conditions can be understood within the framework of physics, their detailed physical structure cannot be deduced from physics. "The fine structure of biological boundary conditions reflects the historical uniqueness of the underlying evolutionary process" and these, by definition, transcend the powers of natural law to describe.

"The eternal mystery of the world is its comprehensibility." This is, of course, Albert Einstein's famous claim, and it serves as the point of departure for **Michael Heller**'s paper. According to Heller, this mystery is present in our prescientific cognition, but it reveals itself in full light only when one contemplates what Eugene Wigner called "the unreasonable effectiveness of mathematics in the natural sciences." It is not *a priori* self-evident that the world should be "algorithmically compressible," that is, that many of its phenomena should be captured by a few mathematical formulae.

There have been attempts to neutralize this wonder by reducing all regularities in the universe to the blind game of chance and probability. Heller briefly reviews two such attempts: the so-called chaotic gauge program and André Linde's chaotic inflation in cosmology. If complete anarchy is the only law of nature ("laws from no laws"), then the fundamental rationality of the world is lost. The problem is important from a theological point of view. At the most fundamental level, God's action in the world consists in giving the world its existence and giving it in such a way that everything that exists participates in its rationality, that is, is subject to the mathematically expressible laws of nature. If the ideology of the "pure game of chance and probability" turns out to be correct, then God's action seems to be in jeopardy.

Heller responds by arguing that such attempts to neutralize the "mystery of comprehensibility" lead us even deeper into the problem. Probability calculus is as much a mathematical theory as any other, and even if chance and probability lie at the core of everything, the important philosophical and theological problem remains of why the world is *probabilistically comprehensible*. The probabilistic compressibility of the world is a special instance of its mathematical compressibility. Heller clarifies this point by reminding us that there are two kinds of elements (in the Greek sense of this word) in the universe—the *cosmic elements*, such as integrability, analyticity, calculability, predictability; and the *chaotic elements*, such as probability, randomness, unpredictability, and various stochastic properties. The chaotic elements are in fact as "mathematical" as the cosmic ones. If the cosmic elements provoke the question of why the world is mathematical, the same is true of the chaotic elements. In this view, *cosmos* and *chaos* are not antagonistic forces but rather two components of the same *Logos* immanent in the structure of the universe.

Arthur Peacocke's topic in this reprint is the general relationship between chance and law in thermodynamics and biology, and its implications for belief in God as creator.

Chance may be the means for actualizing the possibilities of the world, but it need not be seen as a metaphysical principle opposed to order or undercutting the meaning of life. Chance actually has two quite distinct meanings: it can refer simply to our ignorance of the processes which underlie an event, or it can refer to unpredictable intersections of previously unrelated causal chains. Recently, the interplay between chance and law has come to be seen as crucial to the origin and development of life, particularly through the work of Jacques Monod in molecular biology and Ilya Prigogine in irreversible thermodynamics and theoretical biology. As Monod emphasizes, evolution depends on chance, in the sense of two independent causal chains, operating in living organisms: one is at the genetic level, including changes in the nucleotide bases of DNA; the other is at the level of the organism, including interactions between the organism expressing these changes and its environment. Chance also arises here in the sense that we cannot now (nor may we ever be able to) specify the mechanisms underlying genetic mutations.

Though agreeing with him this far, Peacocke challenges both Monod's generalization of the role of chance from the context of evolution to include all of human culture, and his subsequent conclusion to the meaningless of life. Instead, Peacocke sees chance as the means by which all possibilities for the organization of matter are explored in nature.

Peacocke then turns to irreversible thermodynamics and theoretical biology. Thermodynamics is "the *science* of the possible" which prescribes how nature can behave. Classical thermodynamics, with its focus on systems in equilibrium, centers on the second law of increasing entropy in closed systems. Through the statistical thermodynamics of Boltzmann this came to be seen as increasing disorder or randomness in closed systems. How, then, do living organism maintain themselves in a high state of organization and a low state of entropy, given the second law? The answer, as Peacocke points out, is that living systems are open to their environment. By exchanging energy and matter with it they can decrease in entropy as long as there is an increase in the net environmental entropy.

But does thermodynamics help us to understand how more complex organisms come to be in the first place? The answer comes only with the extension of classical thermodynamics, first to linear, and then to non-linear, irreversible processes involved in what are called *dissipative structures*. According to Prigogine, if fluctuations in these non-linear, non-equilibrium structures are amplified, they can change the structures and result in new, more ordered states. The answer also includes the key role played by multiple and relatively stable strata in the hierarchy of biological complexity. These intermediate strata enhance the rate of evolution of more complex organisms from very simple ones, in effect directing evolution towards increased complexity. In essence, the evolution of chemical, pre-biological, and biological complexity is seen as probable, perhaps even inevitable, although the particular path taken in nature is unpredictable. Still, detailed kinetic and dynamic requirements, as well as thermodynamic ones, must be met for evolution to occur.

Peacocke then turns briefly to theological reflections. God is creator of the world through a timeless relation to it in two ways. God is totally other than the world, its transcendent ground of being. God is also immanent in the world, continuously creating all that is through its inbuilt evolutionary processes. These processes, revealed by the natural sciences, are in fact God's action in the world, and eventually include the evolution of humanity. Thus, all-that-is is *in* God, but God is "more" than nature and humanity. The complex interplay of law and chance is itself "written into creation by the creator's intention and purpose," to emerge in time by the explorations of nature. Here Peacocke suggests the metaphor of God as a musical composer and nature as God's composition, perhaps like a rich fugue.

But does this metaphor carry deistic overtones, as H. Montefiore claims? Not according to D. J. Bartholomew, who sees chance as conducive to the production of a world in which freedom can operate purposefully. Still, the best response to the charge of deism, as Peacocke emphasizes, is to see God's action as immanent within natural processes. Moreover, as Rustum Roy points out, the interplay of chance and law in nature means that we should accept a similar interplay as characteristic of God's creativity in human life and society, and we should be critical of belief in a God who "intervenes in the natural nexus for the good or ill of individuals and societies." Peacocke concludes that just as it takes a stream to have eddies, it is the existence of the universe, flowing as it does towards overall increasing entropy, that is required if there are to be eddies of biological life.

5.3 *Chaos, Complexity, and Divine Action*

Five essays comprise this section, each focusing on issues in philosophical theology relating to the problem of divine action in light of the sciences and the philosophical arguments contained in the preceding material.

In "The Metaphysics of Divine Action," **John Polkinghorne** notes that any discussion of agency requires the adoption of a metaphysical view of the nature of reality. He claims that there is no "deductive" way of going "from epistemology to ontology," but the strategy of critical realism is to maximize the connection. This leads most physicists, he claims, to interpret Heisenberg's uncertainty principle as implying an actual indeterminacy in the physical world, rather than an ignorance of its detailed workings.

Polkinghorne is critical of physical reductionism, which makes unsubstantiated and implausible claims for the explanatory power of the idea of self-organizing systems. Moreover, it focuses strictly on the generation of large-scale structure rather than the temporal openness necessary to accommodate agency. A theological appeal to divine primary causality is too vague to yield an understanding of providential action. We need not be stymied by the problem of the "causal joint" that makes this possible. Top-down causality is a valuable idea, but it is not unproblematic and its plausibility depends upon exhibiting intrinsic gaps in the bottom-up description in order to afford it room for maneuver.

Polkinghorne believes that such gaps might originate from indeterminate quantum events. However, there are problems about amplifying their effects, and the idea also leads to an episodic account of divine agency. Polkinghorne prefers an approach based upon interpreting the unpredictabilities of chaotic dynamics (in accord with the strategy of critical realism) as indicating an ontological openness to the future whereby "active information" becomes a model for human and divine agency. He interprets sensitivity to small triggers as indicators of the vulnerability of chaotic systems to environmental factors, with the consequence that such systems have to be discussed holistically. It is *not* supposed, however, that such triggers are the local mechanism by which agency is exercised.

The resulting metaphysical conjecture Polkinghorne calls a complementary dual-aspect monism, in which mind and matter are opposite poles or phases of the single stuff of created reality. This scheme is antireductionist, stressing instead a contextualist approach in which the behavior of parts depends on the whole in which they participate. Polkinghorne then discusses some of the consequences of adopting this point of view, including the insight that divine agency has its own special characteristics and that God's knowledge of the world of becoming will be truly temporal in character.

Denis Edwards begins by pointing to a major shift in science: the old worldview is giving way to a new paradigm of an open and self-organizing universe. Similarly, in systematic theology the old concept of God as the individual Subject is giving way to a relational, dynamic, trinitarian concept of God.

The first part of Edwards' paper explores the general concept of divine action from the perspective of what many are calling a "retrieved" trinitarian theology. In the West, trinitarian theology as inherited from Augustine and Aquinas emphasized an individual and psychological model of the Trinity rather a communitarian one. It focused on divine unity rather than three persons, and on divine being rather than divine love. The newer trinitarian theology builds instead on the writings of Richard of St. Victor and Bonaventure. Edwards outlines a theology of divine action which understands the Trinity as a communion of mutual relationships which are dynamic, ecstatic, and fecund. He argues that the universe is God's trinitarian self-expression, that there are "proper" roles for the trinitarian persons in creation, and that divine interaction with creation is characterized by the vulnerability and liberating power of love.

The second part of the paper asks what this trinitarian theology of divine action has to say about *particular* divine actions, such as the incarnation, the Holy

Spirit, and divine providence. Edwards explores these questions by assessing the views of John Polkinghorne and Arthur Peacocke. He finds both significant agreements as well as some disagreements between them, particularly over the issue of whether the unpredictability of chaotic systems points towards an ontological indeterminism in nature.

Edwards' reflections can be summarized in the form of six statements: (1) The trinitarian God works in and through the processes of the universe, through laws and boundary conditions, through regularities and chance, through chaotic systems and the capacity for self-organization. (2) This trinitarian God allows for, respects, and is responsive to, the freedom of human persons and the contingency of natural processes, but is not necessarily to be denied a knowledge of future contingent events. (3) We must take into account not only the divine action of continuous creation, but also particular or special divine acts. (4) If God is acting creatively and responsively at all times and also in particular ways, then this seems to demand action at the level of the whole system as well as at the everyday level of events, and at the quantum level. (5) Particular divine acts are always experienced as mediated through created realities. (6) The unpredictability, openness, and flexibility discovered by contemporary science is significant for talk of particular divine action because it provides the basis for a worldview in which divine action and scientific explanation are understood as mutually compatible, but it is not possible or appropriate to attempt to identify the "causal joint" between divine action and created causality.

According to **Stephen Happel**, Christian theology in the thought of Thomas Aquinas has a coherent understanding of the interaction between God and creation. By developing a clear theory of transcendence and of universal instrumentality, Aquinas was able to articulate the basic ways in which inanimate, animate, and human secondary causes cooperate or conflict with the divine act of love for the universe (i.e., providence). These terms can be transposed into an historical ontology and a language of mutual mediation such that all levels of reality have their relative autonomy. Contemporary science, with its analysis of self-organizing systems, provides an understanding of the regularities and contingencies of inanimate and animate created realities. Its language permits us to understand how an open, flexible universe can provide the conditions for cooperation with one another and with divine action without conflict or violence to the integrity of creation.

Happel's analysis is basically optimistic. It is born of a religious conviction that though the cosmos (whether human or non-human) is flawed and finite, its internal logic is not vitiated, malicious, or deceptive. Images, the body and the non-verbal are no more (and no less) prone to sin than reason. Within the temporal being of "nature," self-organizing, living, self-conscious beings can engage with their environments in a cooperative way. Ultimately, Happel argues that self-conscious creatures may learn that cooperating with the ultimate environment, an unfathomable Other, will not do violence to their own complex teleonomies.

The Christian claim, however, goes further. It maintains that this mysterious enveloping environment is involved in a *mutual self-mediation* with creation. When one is in love, one mediates oneself in and through an other who is discovering, planning, negotiating his or her personal identity in and through oneself. That is mutual self-mediation. Christians claim that they are not merely projecting themselves abstractly into an alien environment to mediate themselves, but that the Other has chosen out of love to mediate the divine subjectivity in and through natural self-organization (because God is ultimately a community of mutual self-mediation). The story of the Christ could have been quite different than it was. Jesus could have mediated himself in some other fashion, but he did not. He chose to offer his life for others in self-sacrificing generosity. In this action, he operated as though neither the natural nor the human environment nor God were an enemy. In loving creation, entrusting his own life to others, even in death, faith claims that there is here a divine love. This is what Happel has called elsewhere the "double dative of presence." We are present *to* the divine who in that same movement is present *to* us. What we discover in this fragile and stumbling process of mediating ourselves and our world is an antecedent lover and friend.

In his paper, **Jürgen Moltmann** first describes five models of the God-world relation: (1) According to the Thomistic model, God is the *causa prima* of the world. God also acts through the *causae secundae* which serve as God's instruments. (2) The interaction model postulates a degree of reciprocal influence between God and the world. This model can include the Thomistic model, but not vice versa. (3) The whole-part model, taken from biological systems theory, emphasizes that the whole is more than and different from its parts. In complex and chaotic systems this difference shows up in the form of top-down causality. The whole-part model is more inclusive than the previous models and sheds light on God's indirect effect upon the world as a whole. (4) The model of life processes emphasizes the open character of biological systems. The present state of a living system is constituted by its fixed past and its open future or, more generally, by what can be called tradition and innovation. Here the world process is open to God as its transcendental future. (5) Finally, Moltmann considers two central theological models: creation and incarnation. Here God creates by a process of self-limitation (or *tzitzum*). The limitation on God's omnipresence creates a habitation for the world; the limitation on God's omniscience provides the world with an open future. God's self-limitation allows God to be present within the world without destroying it. Moltmann believes this model is the most inclusive of the five.

Moltmann next offers three comments on how these models function in current theological discussions about chaotic, complex and evolutionary systems.

(1) He is critical of the interaction model, seeing it as a *theistic* model in which God is the absolute Subject who may intervene at will in nature. In the modern period it was replaced by two even more problematic models: deism and pantheism. In their place Moltmann commends to us a trinitarian model in which "God the Father creates through the Logos/Wisdom in the power of the Holy Spirit. . . . God not only transcends the world but is also immanent in the world." According to this model God acts upon the world through God's presence in and *perichoresis* with all things.

(2) Next Moltmann discusses eschatology, the new creation of all things. For Moltmann the future is not a state of completion but a process of continuing openness, in which all finite creatures will participate in God's unending and open eternity even as God participates in their temporality. The openness of chaotic,

complex, and evolutionary systems is suggestive of this vision, and seems inconsistent with a future conceived of as completed. "The future of the world can only be imagined as the openness of all finite life systems to the abundance of eternal life. In this way they can participate in the inexhaustible sources of life and in the divine creative ground of being."

(3) Finally Moltmann asks whether the universe as a whole should be thought of as an open system. The growth of possibilities for such systems, their undetermined character, and their dependence on an influx of energy suggest that the universe itself might be open to energy. "In this case the world would be a 'system open to God' and God a 'Being open to the world.'"

Langdon Gilkey's paper considers two questions: whether nature's processes suggest the existence of a God, and if so, what sort of God. However, he emphasizes that the traces of God that may be found in nature are not the main source of religious belief; for Christians, God is encountered primarily in history.

Science, including chaos theory, provides a picture of reality that combines both order and novelty; the ascending order can well be described as an order of increasing value. Reflection on this scientific picture of reality, along with the wider data of human experience (history) leads to ontology or metaphysics—the effort to understand the structure of being *qua* being. This level of reflection is crucial for both the scientist and the theologian. For the scientist it provides the rational grounding for science itself. Metaphysics is crucial for the theologian, since "proofs" of any sort for the existence of God are always conditional upon a particular metaphysical structuring of experience.

Consequently, the aspects of nature suggested by the sciences must be represented in ontological categories. Beyond and through the abstractions of the scientific understanding of nature, nature's reality has manifested itself as *power*, as *spontaneity* or *life*, as *order*, and as implying a *redemptive principle*, a strange dialectic of sacrifice, purgation, redemption, and rebirth. In nature each of these appear as vulnerable and ambiguous as well as creative. Each of these characteristics, therefore, raises a "limit question," and thus represents a trace of God. For example, what is the deeper, more permanent power that makes possible the transitory appearance of power in nature? 'God' is the name for that ever-continuing source of power. To know God truly is to know God's presence in the power, life, order, and redemptive unity of nature; to know nature truly is to know its mystery, depth, and ultimate value—to know it as an image of the sacred.

5.4 *Alternative Approaches to Divine Action*

The six authors in this section are skeptical of the fruitfulness of chaos theory for the problem of divine action, and are pursuing alternative approaches. The first part of this section deals with general issues, the second with specific issues, including the potential relevance of other topics in science beyond those of chaos and complexity.

Willem B. Drees argues that theories of chaotic and complex systems have made it clearer than ever before that a naturalistic explanation of the world is possible, even in light of the lack of predictability of these systems. These theories have effectively closed certain gaps in our understanding of nature. He is therefore critical of Polkinghorne's suggestion that the unpredictability of natural processes provides a potential locus for divine action. Polkinghorne suggests that God brings about an input of information into the world without an input of energy. Drees claims that this is inconsistent with quantum physics and thermodynamics. In addition, Polkinghorne seems to interpret the unpredictability of chaotic systems as a sign of intrinsic openness, but this ignores the real meaning of deterministic chaos. Moreover, discarding the theory of deterministic chaos would be inconsistent with the very critical realism that Polkinghorne promotes.

However, denying any such gaps within natural processes need not foreclose all options for a religious view of reality. In fact Drees claims that science raises religious questions about nature as a whole and about the most fundamental structures of reality. To make his case, he distinguishes between two conceptions of explanation in contemporary philosophy of science. Ontic views of explanation consider an event explained if it is understood as a possible consequence of a causal mechanism. Epistemic views of explanation consider phenomena and laws explained if they are seen as part of a wider framework. Hence if one adopts an epistemic view of explanation the framework itself still requires an explanation. Along these lines, Peacocke and others have argued for divine action on the whole of natural reality: God could cause specific events in nature via "top-down" or "whole-part" causation. Drees, however, rejects the attempt to extrapolate from the context of nature as environment to the concept of God as the world's environment.

Given the various problems with attempts to envisage God's action *in* the world, Drees prefers to understand the world *as* God's action. Whatever strength explanations have, there always remain limit questions about reality and about understanding which allow us to develop a religious interpretation of "secular naturalism."

The approach to divine action taken by **William Stoeger** is to accept with critical seriousness our present and projected knowledge of reality from the sciences, philosophy and other disciplines, including theology which has already developed in response to the sciences. By "critical seriousness," Stoeger means that this knowledge, though critically assessed by the disciplines themselves, by philosophy and by the other human sciences, does indeed indicate something about the realities it talks about. Stoeger then integrates these results into a roughly sketched theory of God's action. Implicit here is the methodological problem of how the languages of science and theology are to be integrated.

Next Stoeger employs a philosophical presupposition which he calls a "weakly critical realist" stance. Included are elements of Aristotelianism and Thomism, particularly the notions of primary and secondary causality. These seem to him more adequate to both the scientific and the theological data. They also lead to fewer difficulties in explicating the essential differences between God and God's creation, the relationships between them, and the ideas of divine immanence and transcendence. Stoeger uses the term 'law' in the context of both physical processes and free human actions to mean any pattern, regularity, process, or relationship, and its description. 'Law' is thus used to describe or explain order. It does not necessarily imply determinism.

Stoeger concludes this section by saying that there are aspects of divine action which we are able to understand better by letting science and theology critically interact. There are other aspects which seem thoroughly resistant to our understanding, particularly the nexus between God and the secondary causes through which God acts, or between God and the direct effects of divine action, such as *creatio ex nihilo*. The analog of human agency is of some limited help here. However, the principal barrier is that we could only know the critical nexus if we ourselves were divine, or if God revealed such knowledge to us.

Turning to the problem of God's action, Stoeger argues that if God acts through secondary causes it would seem to require the injection of information, and therefore energy, from outside the physical system. Though we cannot rule out such injections, they have never been observed and are unattractive from many points of view. Some scholars try to evade this problem by allowing God to influence events at the quantum level. Stoeger admits that this is a solution, and may in fact be the case, but he finds it unattractive. God's working through secondary causes is almost always a function of God's invitation, or response, to persons. To locate such divine action at the quantum level removes it from the level of the personal. It is also unclear whether its intended effects can surface at the level of the complex and the personal.

Stoeger provides no answer to this issue, but he believes that the framework he has established may move us in the right direction. Either there is some injection of information and energy at the level of personal relationships, or God works within what is already given to make the recipient more receptive to what is available. Stoeger prefers the latter, though something of the former may be involved as well. The difficulty with higher regularities subsuming those at the lower-level is that we usually experience the lower level laws as constraining what can be done on a higher level while not being supplanted by them. Nevertheless there is a great deal left under determined by the lower-level constraints within which agents, including God, can function.

According to **Arthur Peacocke**, the long-established aim in science of predicting the future macroscopic states of natural systems has recently come to be recognized as unattainable in practice for those systems capable of manifesting "deterministic chaos." The possibility of prediction has also been closely associated with the conviction that there is a causal nexus which scientific procedures will unambiguously ascertain. In this paper, Peacocke surveys the applicability of these concepts with respect to relatively simple, dynamic, law-obeying systems; to statistical properties of assemblies; to Newtonian systems which are deterministic yet unpredictable; and to "chaotic" and "dissipative" systems. In doing so he also analyzes the limitations to predictability stemming from quantum theory.

Chaotic and dissipative systems prove to be unpredictable in practice, primarily because of the nature of our knowledge of the real numbers, and possibly (and more problematically) because of quantum uncertainties. The notion of causality still proves to be applicable to these systems in an unambiguous, even if only in a probabilistic, fashion. However, for many significant interconnected and interdependent complex systems the concept of causality scarcely seems applicable, since whole-part constraints operate, whereby the *state of a system-as-a-whole* influences what occurs among its constituents at the microscopic level.

Peacocke acknowledges that in the past this phenomenon has also, perhaps somewhat misleadingly, been denoted by himself and others as "downward" or "top-down" causation, in particular in relation to evolution and to the brain-body relation.

Peacocke then considers how to conceive of God's relation to the world in the light of modifications in the scientific concepts of predictability and causality which the phenomena of deterministic chaos and dissipative systems on the one hand, and of "whole-part constraints" on the other hand, have induced. Consideration of the former has to take account of the possible, and as yet unclear, effects of quantum uncertainty on chaotic and dissipative systems. Peacocke concludes that, whatever is decided about those effects, the unpredictabilities for us of non-linear chaotic and dissipative systems do not, as such, help us in the problem of articulating more coherently and intelligibly how *God* interacts with the world, illuminating as they are concerning the flexibilities built into natural processes. The discussion is based in part on the assumption that God logically cannot know the future, since it does not exist *for* God to know.

However, Peacocke argues that the notion of "whole-part constraints" in interconnected and interdependent systems *does* provide a new conceptual resource for modeling how God might be conceived of as interacting with and influencing events in the world. This is particularly true in conjunction with a prime exemplification of the whole-part constraint in the unitive relation of the human-brain-in-the-human-body—in fact, this model of personal agency is the biblical and traditional model for God's action in the world. He evokes the notion of a flow of information as illuminating this 'whole-part' interaction of God with the world, which could then be conceived of as a communication by God to that part of the world (namely, humanity) capable of discovering God's meanings.

Thomas Tracy's paper takes up a persistent modern problem in relating scientific descriptions of the world as a natural order and theological claims about divine action. Do some traditional ways of speaking about divine action require "gaps" in the causal order and therefore incompleteness in scientific explanations? This appears to be the case, for example, if we claim that God acts in the world at particular times and places to affect the unfolding course of events. Must this kind of theological claim compete with scientific descriptions of the world, so that we cannot both explain an event scientifically and affirm that it as a particular divine action?

Tracy considers three strategies which reply to these questions. The first avoids conflict between scientific and theological claims by insisting that, strictly speaking, God does not act *in* history but rather *enacts* history as a whole. Its paradigmatic modern development comes from Friedrich Schleiermacher, who holds that every event both stands in a relation of absolute dependence upon God's immediate agency and is integrated into a complete system of natural causes. On this account, particular events can be singled out as acts of God only in the sense that they especially evoke in us a recognition of God's universal activity, or play a distinctive role in advancing divine purposes "built into" the causal processes of nature. This eliminates any risk of conflict between science and theology, but it does so at the cost of imposing significant limits on the claims that can be made, for example, about the person and work of Christ, about the divine-human interaction, and about human freedom and the problem of evil. Tracy considers a contemporary and widely influential version of this strategy developed by Gordon Kaufman. Unfortunately, Kaufman's proposal, Tracy argues, leaves us with a series of questions about how God can be understood to enact history without acting in history.

The second strategy affirms that God does act in the world to affect the course of events, but holds that this does not require any gaps in the causal structures of nature. There are at least two recent proposals that take this form. Brian Hebblethwaite contends that God acts in and through the causal powers of creatures, so that the whole network of created agencies is "pliable, or flexible, to the providential hand of God" without any gaps in the natural order. This leaves the crucial puzzle unsolved, however; for if God affects the course of events once they are underway, then an explanation of those events that appeals strictly to other finite causes must be incomplete. John Compton has suggested another way to pursue this second strategy. Just as we routinely describe certain movements of the human body both as a series of physical events and as intentional action, so we can describe events in the world both as part of a causally complete natural order and as acts of God. Compton's proposal hinges on the claim that the language we use in discussing physical events, on the one hand, and intentional actions, on the other, are not interdependent. But this claim, Tracy argues, cannot be sustained even within the terms of Compton's own discussion. These two versions of the second strategy, then, are undone by internal inconsistencies.

The third strategy grants that theologically motivated talk of particular divine action carries with it a commitment to the causal incompleteness of the natural order, and then argues that this is at least consonant with contemporary physical theory. Two key issues must be addressed by any such proposal. First, a case must be made that the natural sciences now describe a world whose causal structure is "open" in certain respects. Second, it must be shown that this openness is relevant to the theological concern with divine action. Tracy argues that chaos theory, for all its power to demonstrate the limits of predictability, does not provide the needed openness, since it presupposes an unbroken causal determinism. More promising are interpretations of quantum mechanics that acknowledge the role in nature of indeterministic chance. With regard to the second question, Tracy contends that such chance (whether at the quantum level or elsewhere) will be theologically interesting if the determination of such events by God can make a macroscopic difference. If so, then God could affect the course of events without disrupting the structures of nature, since they will provide for both novelty and regularity in the world.

The essay concludes by exploring God's relation to indeterministic chance: must we say that God determines all events left undetermined by secondary causes? God might, Tracy argues, choose not to determine some events, and thereby (contra Einstein) play dice with the universe.

Nancey Murphy directs our attention away from chaos and complexity to the arena of quantum physics. In her paper, Murphy argues that the problem of divine action will be solved by nothing less than a revised metaphysical theory of the nature of matter and of natural causes. Her proposal is that we view the causal powers of created entities as inherently incomplete. No event occurs without divine participation but, apart from creation *ex nihilo*, God never acts except by means of cooperation with created agents. Her paper attempts to show how this account can be reconciled with contemporary science, focusing on divine action at the quantum level.

First Murphy proposes criteria, derived from both theology and science, which any satisfactory theory of divine action must meet. She claims that it must allow for objectively special divine acts, yet not undercut our scientific picture of the law-like regularity of many natural processes. Then she surveys changes in metaphysical views of matter and causation from ancient to modern philosophy. The historical survey is intended to put in question current metaphysical assumptions about the nature of matter and of natural causes, as a prelude to considering the consequences of recent developments in science for these metaphysical issues.

Murphy's proposal is that any adequate account of divine action must include a "bottom-up" approach: if God is to be active in all events, then God must be involved in the most basic of natural events. Current science suggests that this most basic level is quantum phenomena. It is a bonus for theology that we find a measure of indeterminacy at this level, since it allows for an account of divine action wherein God has no need to overrule natural tendencies or processes. This cooperation rather than coercion is in keeping with God's pattern of respecting the integrity of other higher-level creatures, especially human creatures.

Consequences of this proposal are spelled out regarding the character of natural laws and regarding God's action at the macroscopic level. One of these consequences is that the "laws of nature" must be descriptive, rather than prescriptive; they represent our human perceptions of the regularity of God's action. In the end, she replies to some of the objections that have been raised against theories of divine action based on quantum indeterminacy and explains how the essay's proposal meets the criteria of adequacy set out in the beginning.

In "Ordinary and Extraordinary Divine Action," **George Ellis** intends to elaborate the conclusions reached by Tracy, Murphy, and others concerning the role of quantum indeterminacy in a contemporary understanding of divine action. He claims that some account of *special* divine action is necessary if the Christian tradition is to make sense. However, there are two important constraints to be reckoned with. One is that an ideal account of divine action must not conflict with a scientific understanding of nature; the other is that some explanation must be given of why a God capable of special action would not exercise that ability regularly to oppose evil and ameliorate suffering.

Ellis' analysis focuses on the nature of bottom-up and top-down causation in hierarchical systems. It is predicated upon the assumption that chaotic dynamics does not provide the required openness in physical systems. Furthermore, his analysis of top-down causation convinces him that this concept alone does not provide for an adequate account of divine action. He distinguishes between generic top-down causation, in which boundary conditions produce a global effect upon all the entities in a system, and specific top-down causation, which involves local interactions with elements of the lower-level system. Special divine actions would seem to entail the latter. However, specific top-down causation seems to require, in turn, that there be an intrinsic openness or indeterminacy at the very lowest level of the hierarchy of complexity. Thus, a study of the possibilities for divine action via top-down causation leads inevitably to a consideration of divine action at the quantum level.

Ellis takes God's action to be largely *through* the ordinary created processes. God initiates the laws of physics, establishes the initial conditions for the universe, and sustains the universe and its processes, which in turn result in the emergence of higher levels of order, including, finally, free human beings. Special divine action focuses on providing to human beings intimations of God's will for their social lives. Thus, the problem of the mode of divine action is largely a question of how God might communicate directly with those who are open to revelation. Ellis speculates that quantum events in the brain (directed by God) might be amplified to produce revelatory thoughts, images, and emotions. If it is supposed that God has adequate reason to restrict divine action to a combination of ordinary action (in and through natural processes) and revelation (such as the Resurrection of Christ) then the problem of evil does not take on the same dimensions as it does when it is assumed that God might freely intervene in any sort of process at any time.

Finally, Ellis addresses the question of support for his view. He claims that while individual moves made in the paper (such as the focus on divine action at the quantum level) may not appear to be justified, the combined constraints imposed by the need to make sense of the Christian tradition and by science actually limit the possible acceptable positions quite severely; thus, the view herein presented is, in Ellis' opinion, highly credible relative to the broad range of data.

I

SCIENTIFIC BACKGROUND: CHAOS AND COMPLEXITY

CHAOS

James P. Crutchfield, J. Doyne Farmer, Norman H. Packard,
and Robert S. Shaw

There is order in chaos: randomness has an underlying geometric form. Chaos imposes fundamental limits on prediction, but it also suggests causal relationships where none were previously suspected.

The great power of science lies in the ability to relate cause and effect. On the basis of the laws of gravitation, for example, eclipses can be predicted thousands of years in advance. There are other natural phenomena that are not as predictable. Although the movements of the atmosphere obey the laws of physics just as much as the movements of the planets do, weather forecasts are still stated in terms of probabilities. The weather, the flow of a mountain stream, the roll of the dice all have unpredictable aspects. Since there is no clear relation between cause and effect, such phenomena are said to have random elements. Yet until recently there was little reason to doubt that precise predictability could in principle be achieved. It was assumed that it was only necessary to gather and process a sufficient amount of information.

Such a viewpoint has been altered by a striking discovery: simple deterministic systems with only a few elements can generate random behavior. The randomness is fundamental; gathering more information does not make it go away. Randomness generated in this way has come to be called chaos.

A seeming paradox is that chaos is deterministic, generated by fixed rules that do not themselves involve any elements of chance. In principle the future is completely determined by the past, but in practice small uncertainties are amplified, so that even though the behavior is predictable in the short term, it is unpredictable in the long term. There is order in chaos: underlying chaotic behavior there are elegant geometric forms that create randomness in the same way as a card dealer shuffles a deck of cards or a blender mixes cake batter.

The discovery of chaos has created a new paradigm in scientific modeling. On one hand, it implies new fundamental limits on the ability to make predictions. On the other hand, the determinism inherent in chaos implies that many random phenomena are more predictable than had been thought. Random-looking information gathered in the past and shelved because it was assumed to be too complicated can now be explained in terms of simple laws. Chaos allows order to be found in such diverse systems as the atmosphere, dripping faucets, and the heart. The result is a revolution that is affecting many different branches of science.

What are the origins of random behavior? Brownian motion provides a classic example of randomness. A speck of dust observed through a microscope is seen to move in a continuous and erratic jiggle. This is owing to the bombardment of the dust particle by the surrounding water molecules in thermal motion.

Because the water molecules are unseen and exist in great number, the detailed motion of the dust particle is thoroughly unpredictable. Here the web of causal influences among the subunits can become so tangled that the resulting pattern of behavior becomes quite random.

The chaos to be discussed here requires no large number of subunits or unseen influences. The existence of random behavior in very simple systems motivates a reexamination of the sources of randomness even in large systems such as weather.

What makes the motion of the atmosphere so much harder to anticipate than the motion of the solar system? Both are made up of many parts, and both are governed by Newton's second law, $F=ma$, which can be viewed as a simple prescription for predicting the future. If the forces F acting on a given mass m are known, then so is the acceleration a. It then follows from the rules of calculus that if the position and velocity of an object can be measured at a given instant, they are determined forever. This is such a powerful idea that the eighteenth-century French mathematician Pierre Simon de Laplace once boasted that given the position and velocity of every particle in the universe, he could predict the future for the rest of time. Although there are several obvious practical difficulties to achieving Laplace's goal, for more than 100 years there seemed to be no reason for his not being right, at least in principle. The literal application of Laplace's dictum to human behavior led to the philosophical conclusion that human behavior was completely predetermined: free will did not exist.

Twentieth-century science has seen the downfall of Laplacian determinism, for two very different reasons. The first reason is quantum mechanics. A central dogma of that theory is the Heisenberg uncertainty principle, which states that there is a fundamental limitation to the accuracy with which the position and velocity of a particle can be measured. Such uncertainty gives a good explanation for some random phenomena, such as radioactive decay. A nucleus is so small that the uncertainty principle puts a fundamental limit on the knowledge of its motion, and so it is impossible to gather enough information to predict when it will disintegrate.

The source of unpredictability on a large scale must be sought elsewhere, however. Some large-scale phenomena are predictable and others are not. The distinction has nothing to do with quantum mechanics. The trajectory of a baseball, for example, is inherently predictable; a fielder intuitively makes use of the fact every time he or she catches the ball. The trajectory of a flying balloon with the air rushing out of it, in contrast, is not predictable; the balloon lurches and turns erratically at times and places that are impossible to predict. The balloon obeys Newton's laws just as much as the baseball does; then why is its behavior so much harder to predict than that of the ball?

The classic example of such a dichotomy is fluid motion. Under some circumstances the motion of a fluid is laminar—even, steady, and regular—and easily predicted from equations. Under other circumstances fluid motion is turbulent—uneven, unsteady, and irregular—and difficult to predict. The transition from laminar to turbulent behavior is familiar to anyone who has been in an airplane in calm weather and then suddenly encountered a thunderstorm. What causes the essential difference between laminar and turbulent motion?

To understand fully why that is such a riddle, imagine sitting by a mountain stream. The water swirls and splashes as though it had a mind of its own, moving first one way and then another. Nevertheless, the rocks in the stream bed are firmly fixed in place, and the tributaries enter at a nearly constant rate of flow. Where, then, does the random motion of the water come from?

The late Soviet physicist Lev D. Landau is credited with an explanation of random fluid motion that held sway for many years, namely that the motion of a turbulent fluid contains many different, independent oscillations. As the fluid is made to move faster, causing it to become more turbulent, the oscillations enter the motion one at a time. Although each separate oscillation may be simple, the complicated combined motion renders the flow impossible to predict.

Landau's theory has been disproved, however. Random behavior occurs even in very simple systems, without any need for complication or indeterminacy. The French mathematician Henri Poincaré realized this at the turn of the century, when he noted that unpredictable, "fortuitous" phenomena may occur in systems where a small change in the present causes a much larger change in the future. The notion is clear if one thinks of a rock poised at the top of a hill. A tiny push one way or another is enough to send it tumbling down widely differing paths. Although the rock is sensitive to small influences only at the top of the hill, chaotic systems are sensitive at every point in their motion.

A simple example serves to illustrate just how sensitive some physical systems can be to external influences. Imagine a game of billiards, somewhat idealized so that the balls move across the table and collide with a negligible loss of energy. With a single shot the billiard player sends the collection of balls into a protracted sequence of collisions. The player naturally wants to know the effects of the shot. For how long could a player with perfect control over his or her stroke predict the cue ball's trajectory? If the player ignored an effect even as minuscule as the gravitational attraction of an electron at the edge of the galaxy, the prediction would become wrong after one minute!

The large growth in uncertainty comes about because the balls are curved, and small differences at the point of impact are amplified with each collision. The amplification is exponential: it is compounded at every collision, like the successive reproduction of bacteria with unlimited space and food. Any effect, no matter how small, quickly reaches macroscopic proportions. That is one of the basic properties of chaos.

It is the exponential amplification of errors due to chaotic dynamics that provides the second reason for Laplace's undoing. Quantum mechanics implies that initial measurements are always uncertain, and chaos ensures that the uncertainties will quickly overwhelm the ability to make predictions. Without chaos Laplace might have hoped that errors would remain bounded, or at least grow slowly enough to allow him to make predictions over a long period. With chaos, predictions are rapidly doomed to gross inaccuracy.

The larger framework that chaos emerges from is the so-called theory of dynamical systems. A dynamical system consists of two parts: the notions of a state (the essential information about a system) and a dynamic (a rule that describes how the state evolves with time). The evolution can be visualized in a state space, an abstract construct whose coordinates are the components of the state. In general the coordinates of the state space vary with the context; for a mechanical system they might be position and velocity, but for an ecological model they might be the populations of different species.

A good example of a dynamical system is found in the simple pendulum. All that is needed to determine its motion are two variables: position and velocity. The state is thus a point in a plane, whose coordinates are position and velocity. Newton's laws provide a rule, expressed mathematically as a differential equation, that describes how the state evolves. As the pendulum swings back and forth the state moves along an "orbit," or path, in the plane. In the ideal case of a frictionless pendulum the orbit is a loop; failing that, the orbit spirals to a point as the pendulum comes to rest.

A dynamical system's temporal evolution may happen in either continuous time or in discrete time. The former is called a flow, the latter a mapping. A pendulum moves continuously from one state to another, and so it is described by a continuous-time flow. The number of insects born each year in a specific area and the time interval between drops from a dripping faucet are more naturally described by a discrete-time mapping.

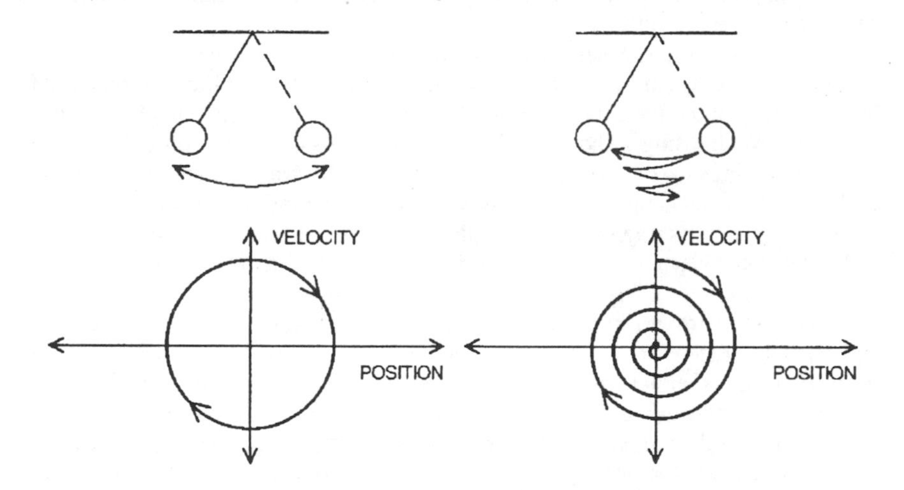

Figure 1: STATE SPACE is a useful concept for visualizing the behavior of a dynamical system. It is an abstract space whose coordinates are the degrees of freedom of the system's motion. The motion of a pendulum (*top*), for example, is completely determined by its initial position and velocity. Its state is thus a point in a plane whose coordinates are position and velocity (*bottom*). As the pendulum swings back and forth it follows an "orbit," or path, through the state space. For an ideal, frictionless pendulum the orbit is a closed curve (*bottom left*); otherwise, with friction, the orbit spirals to a point (*bottom right*).

To find how a system evolves from a given initial state one can employ the dynamic (equations of motion) to move incrementally along an orbit. This method of deducing the system's behavior requires computational effort proportional to the desired length of time to follow the orbit. For simple systems such as a frictionless pendulum the equations of motion may occasionally have a closed-form solution, which is a formula that expresses any future state in terms of the initial state. A closed-form solution provides a short cut, a simpler algorithm that needs only the initial state and the final time to predict the future without stepping through intermediate states. With such a solution the algorithmic effort required to follow the motion of the system is roughly independent of the time desired. Given the equations of planetary and lunar motion and the earth's and moon's positions and velocities, for instance, eclipses may be predicted years in advance.

Success in finding closed-form solutions for a variety of simple systems during the early development of physics led to the hope that such solutions exist for any mechanical system. Unfortunately, it is now known that this is not true in general. The unpredictable behavior of chaotic dynamical systems cannot be expressed in a closed-form solution. Consequently there are no possible short cuts to predicting their behavior.

The state space nonetheless provides a powerful tool for describing the behavior of chaotic systems. The usefulness of the state-space picture lies in the ability to represent behavior in geometric form. For example, a pendulum that moves with friction eventually comes to a halt, which in the state space means the orbit approaches a point. The point does not move—it is a fixed point—and since it attracts nearby orbits, it is known as an attractor. If the pendulum is given a small push, it returns to the same fixed-point attractor. Any system that comes to rest with the passage of time can be characterized by a fixed point in state space. This is an example of a very general phenomenon, where losses due to friction or viscosity, for example, cause orbits to be attracted to a smaller region of the state space with lower dimension. Any such region is called an attractor. Roughly speaking, an attractor is what the behavior of a system settles down to, or is attracted to.

Some systems do not come to rest in the long term but instead cycle periodically through a sequence of states. An example is the pendulum clock, in which energy lost to friction is replaced by a mainspring or weights. The pendulum repeats the same motion over and over again. In the state space such a motion corresponds to a cycle, or periodic orbit. No matter how the pendulum is set swinging, the cycle approached in the long-term limit is the same. Such attractors are therefore called limit cycles. Another familiar system with a limit-cycle attractor is the heart.

A system may have several attractors. If that is the case, different initial conditions may evolve to different attractors. The set of points that evolve to an attractor is called its basin of attraction. The pendulum clock has two such basins: small displacements of the pendulum from its rest position result in a return to rest; with large displacements, however, the clock begins to tick as the pendulum executes a stable oscillation.

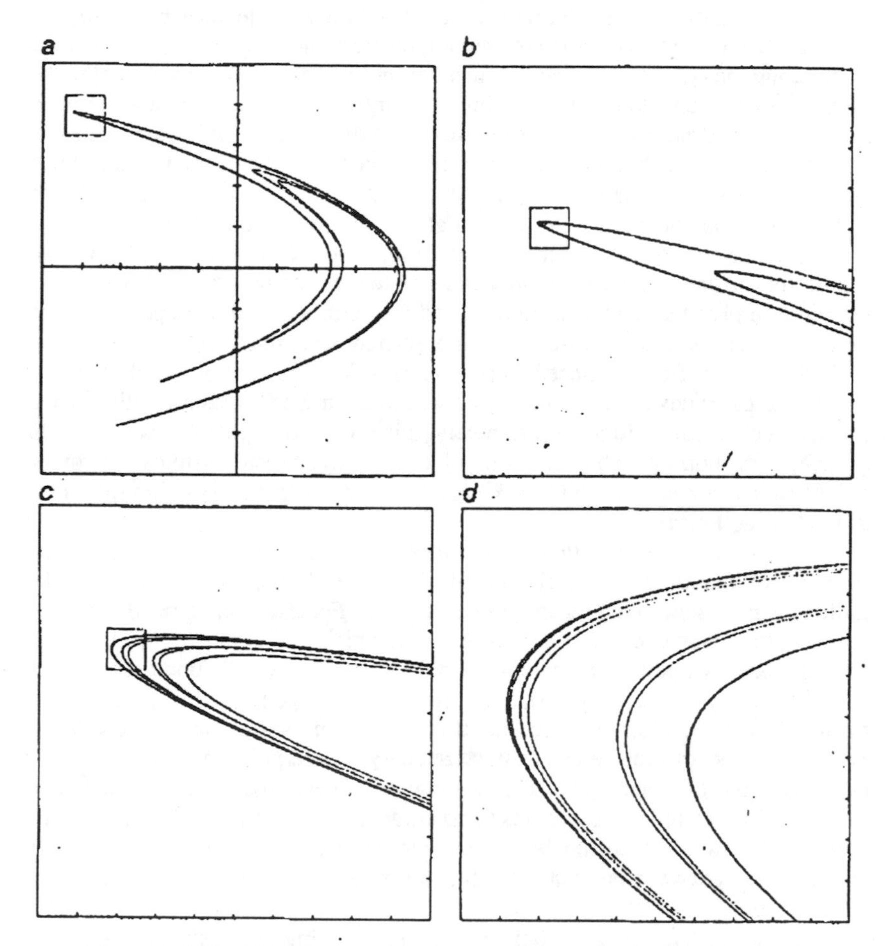

Figure 2: CHAOTIC ATTRACTORS are fractals: objects that reveal more detail as they are increasingly magnified. Chaos naturally produces fractals. As nearby trajectories expand they must eventually fold over close to one another for the motion to remain finite. This is repeated again and again, generating folds within folds, *ad infinitum*. As a result chaotic attractors have a beautiful microscopic structure. Michel Hénon of the Nice Observatory in France discovered a simple rule that stretches and folds the plane, moving each point to a new location. Starting from a single initial point, each successive point obtained by repeatedly applying Hénon's rule is plotted. The resulting geometric form (*a*) provides a simple example of a chaotic attractor. The small box is magnified by a factor of 10 in *b*. By repeating the process (*c, d*) the microscopic structure of the attractor is revealed in detail.

The next most complicated form of attractor is a torus, which resembles the surface of a doughnut. This shape describes motion made up of two independent oscillations, sometimes called quasi-periodic motion. (Physical examples can be constructed from driven electrical oscillators.) The orbit winds around the torus in state space, one frequency determined by how fast the orbit circles the doughnut in the short direction, the other regulated by how fast the orbit circles the long way around. Attractors may also be higher-dimensional tori, since they represent the combination of more than two oscillations.

The important feature of quasi-periodic motion is that in spite of its complexity it is predictable. Even though the orbit may never exactly repeat itself, if the frequencies that make up the motion have no common divisor, the motion remains regular. Orbits that start on the torus near one another remain near one another, and long-term predictability is guaranteed.

Until fairly recently, fixed points, limit cycles, and tori were the only known attractors. In 1963 Edward N. Lorenz of the Massachusetts Institute of Technology discovered a concrete example of a low-dimensional system that displayed complex behavior. Motivated by the desire to understand the unpredictability of the weather, he began with the equations of motion for fluid flow (the atmosphere can be considered a fluid), and by simplifying them he obtained a system that had just three degrees of freedom. Nevertheless the system behaved in an apparently random fashion that could not be adequately characterized by any of the three attractors then known. The attractor he observed, which is now known as the Lorenz attractor, was the first example of a chaotic, or strange, attractor.

Employing a digital computer to simulate his simple model, Lorenz elucidated the basic mechanism responsible for the randomness he observed: microscopic perturbations are amplified to affect macroscopic behavior. Two orbits with nearby initial conditions diverge exponentially fast and so stay close together for only a short time. The situation is qualitatively different for nonchaotic attractors. For these, nearby orbits stay close to one another, small errors remain bounded and the behavior is predictable.

The key to understanding chaotic behavior lies in understanding a simple stretching and folding operation, which takes place in the state space. Exponential divergence is a local feature: because attractors have finite size, two orbits on a chaotic attractor cannot diverge exponentially forever. Consequently the attractor must fold over onto itself. Although orbits diverge and follow increasingly different paths, they eventually must pass close to one another again. The orbits on a chaotic attractor are shuffled by this process, much as a deck of cards is shuffled by a dealer. The randomness of the chaotic orbits is the result of the shuffling process. The process of stretching and folding happens repeatedly, creating folds within folds *ad infinitum*. A chaotic attractor is, in other words, a fractal: an object that reveals more detail as it is increasingly magnified [*see Figure 2*].

Chaos mixes the orbits in state space in precisely the same way as a baker mixes bread dough by kneading it. One can imagine what happens to nearby trajectories on a chaotic attractor by placing a drop of blue food coloring in the dough. The kneading is a combination of two actions: rolling out the dough, in

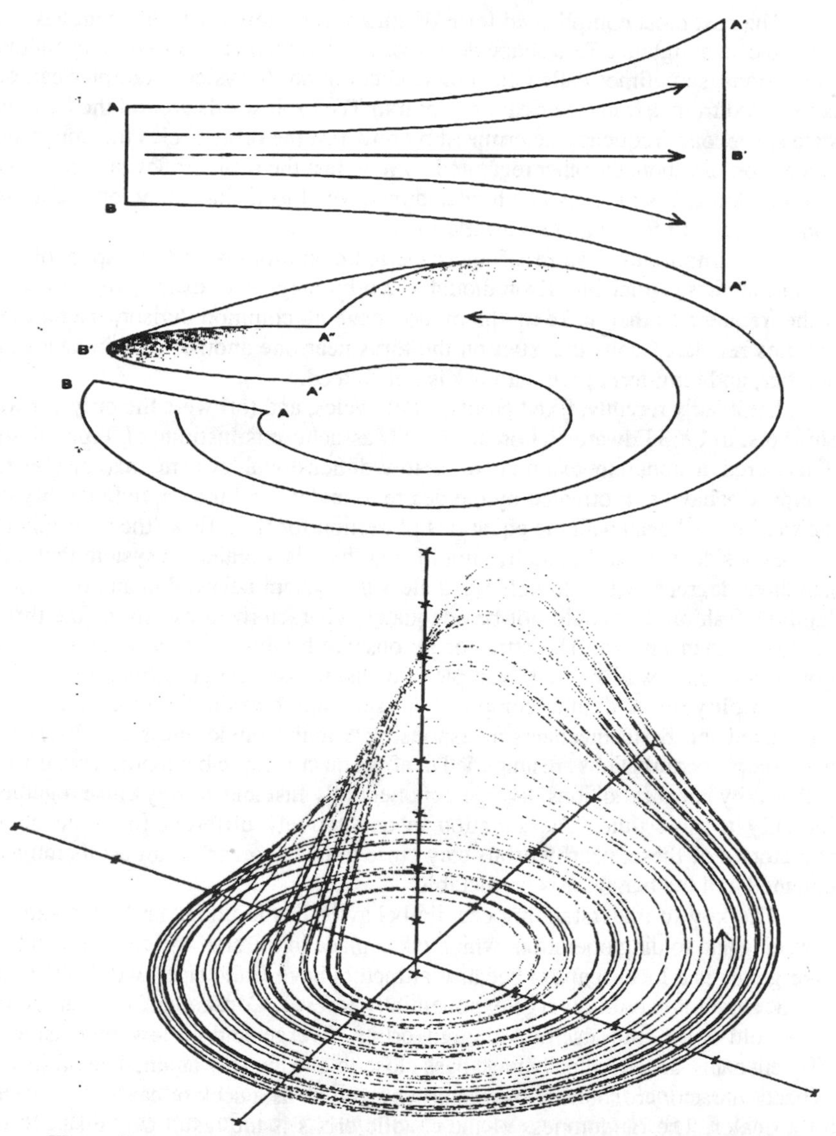

Figure 3: A CHAOTIC ATTRACTOR has a much more complicated structure than a predictable attractor such as a point, a limit cycle, or a torus. Observed at large scales, a chaotic attractor is not a smooth surface but one with folds in it. The illustration shows the steps in making a chaotic attractor for the simplest case: the Rössler attractor (*bottom*). First, nearby trajectories on the object must "stretch," or diverge, exponentially (*top*); here the distance between neighboring trajectories roughly doubles. Second, to keep the object compact, it must "fold" back onto itself (*middle*): the surface bends onto itself so that the two ends meet. The Rössler attractor has been observed in many systems, from fluid flows to chemical reactions, illustrating Einstein's maxim that nature prefers simple forms.

which the food coloring is spread out, and folding the dough over. At first the blob of food coloring simply gets longer, but eventually it is folded, and after considerable time the blob is stretched and refolded many times. On close inspection the dough consists of many layers of alternating blue and white. After only 20 steps the initial blob has been stretched to more than a million times its original length, and its thickness has shrunk to the molecular level. The blue dye is thoroughly mixed with the dough. Chaos works the same way, except that instead of mixing dough it mixes the state space. Inspired by this picture of mixing, Otto E. Rössler of the University of Tübingen created the simplest example of a chaotic attractor in a flow [*see Figure 3*].

When observations are made on a physical system, it is impossible to specify the state of the system exactly owing to the inevitable errors in measurement. Instead the state of the system is located not at a single point but rather within a small region of state space. Although quantum uncertainty sets the ultimate size of the region, in practice different kinds of noise limit measurement precision by introducing substantially larger errors. The small region specified by a measurement is analogous to the blob of blue dye in the dough. Locating the system in a small region of state space by carrying out a measurement yields a certain amount of information about the system. The more accurate the measurement is, the more knowledge an observer gains about the system's state. Conversely, the larger the region, the more uncertain the observer. Since nearby points in nonchaotic systems stay close as they evolve in time, a measurement provides a certain amount of information that is preserved with time. This is exactly the sense in which such systems are predictable: initial measurements contain information that can be used to predict future behavior. In other words, predictable dynamical systems are not particularly sensitive to measurement errors.

The stretching and folding operation of a chaotic attractor systematically removes the initial information and replaces it with new information: the stretch makes small-scale uncertainties larger, the fold brings widely separated trajectories together and erases large-scale information. Thus, chaotic attractors act as a kind of pump bringing microscopic fluctuations up to a macroscopic expression. In this light it is clear that no exact solution, no short cut to tell the future, can exist. After a brief time interval the uncertainty specified by the initial measurement covers the entire attractor and all predictive power is lost: there is simply no causal connection between past and future.

Chaotic attractors function locally as noise amplifiers. A small fluctuation due perhaps to thermal noise will cause a large deflection in the orbit position soon afterward. But there is an important sense in which chaotic attractors differ from simple noise amplifiers. Because the stretching and folding operation is assumed to be repetitive and continuous, any tiny fluctuation will eventually dominate the motion, and the qualitative behavior is independent of noise level. Hence chaotic systems cannot directly be "quieted," by lowering the temperature, for example. Chaotic systems generate randomness on their own without the need for any external random inputs. Random behavior comes from more than just the amplification of errors and the loss of the ability to predict—it is due to the complex orbits generated by stretching and folding.

It should be noted that chaotic as well as nonchaotic behavior can occur in dissipationless, energy-conserving systems. Here orbits do not relax onto an attractor but instead are confined to an energy surface. Dissipation is, however, important in many if not most real-world systems, and one can expect the concept of attractor to be generally useful.

Low-dimensional chaotic attractors open a new realm of dynamical systems theory, but the question remains of whether they are relevant to randomness observed in physical systems. The first experimental evidence supporting the hypothesis that chaotic attractors underlie random motion in fluid flow was rather indirect. The experiment was done in 1974 by Jerry P. Gollub of Haverford College and Harry L. Swinney of the University of Texas at Austin. The evidence was indirect because the investigators focused not on the attractor itself but rather on statistical properties characterizing the attractor.

The system they examined was a Couette cell, which consists of two concentric cylinders. The space between the cylinders is filled with a fluid, and one or both cylinders are rotated with a fixed angular velocity. As the angular velocity increases, the fluid shows progressively more complex flow patterns, with a complicated time dependence. Gollub and Swinney essentially measured the velocity of the fluid at a given spot. As they increased the rotation rate, they observed transitions from a velocity that is constant in time to a periodically varying velocity and finally to an aperiodically varying velocity. The transition to aperiodic motion was the focus of the experiment.

The experiment was designed to distinguish between two theoretical pictures that predicted different scenarios for the behavior of the fluid as the rotation rate of the fluid was varied. The Landau picture of random fluid motion predicted that an ever higher number of independent fluid oscillations should be excited as the rotation rate is increased. The associated attractor would be a high-dimensional torus. The Landau picture had been challenged by David Ruelle of the Institut des Hautes Études Scientifiques near Paris and Floris Takens of the University of Groningen in the Netherlands. They gave mathematical arguments suggesting that the attractor associated with the Landau picture would not be likely to occur in fluid motion. Instead their results suggested that any possible high-dimensional tori might give way to a chaotic attractor, as originally postulated by Lorenz.

Gollub and Swinney found that for low rates of rotation the flow of the fluid did not change in time: the underlying attractor was a fixed point. As the rotation was increased the water began to oscillate with one independent frequency, corresponding to a limit-cycle attractor (a periodic orbit), and as the rotation was increased still further the oscillation took on two independent frequencies, corresponding to a two-dimensional torus attractor. Landau's theory predicted that as the rotation rate was further increased the pattern would continue: more distinct frequencies would gradually appear. Instead, at a critical rotation rate a continuous range of frequencies suddenly appeared. Such an observation was consistent with Lorenz' "deterministic nonperiodic flow," lending credence to his idea that chaotic attractors underlie fluid turbulence.

Although the analysis of Gollub and Swinney bolstered the notion that chaotic attractors might underlie some random motion in fluid flow, their work was by no means conclusive. One would like to explicitly demonstrate the existence in experimental data of a simple chaotic attractor. Typically, however, an experiment does not record all facets of a system but only a few. Gollub and Swinney could not record, for example, the entire Couette flow but only the fluid velocity at a single point. The task of the investigator is to "reconstruct" the attractor from the limited data. Clearly that cannot always be done; if the attractor is too complicated, something will be lost. In some cases, however, it is possible to reconstruct the dynamics on the basis of limited data.

A technique introduced by us and put on a firm mathematical foundation by Takens made it possible to reconstruct a state space and look for chaotic attractors. The basic idea is that the evolution of any single component of a system is determined by the other components with which it interacts. Information about the relevant components is thus implicitly contained in the history of any single component. To reconstruct an "equivalent" state space, one simply looks at a single component and treats the measured values at fixed time delays (one second ago, two seconds ago and so on, for example) as though they were new dimensions.

The delayed values can be viewed as new coordinates, defining a single point in a multidimensional state space. Repeating the procedure and taking delays relative to different times generates many such points. One can then use other techniques to test whether or not these points lie on a chaotic attractor. Although this representation is in many respects arbitrary, it turns out that the important properties of an attractor are preserved by it and do not depend on the details of how the reconstruction is done.

The example we shall use to illustrate the technique has the advantage of being familiar and accessible to nearly everyone. Most people are aware of the periodic pattern of drops emerging from a dripping faucet. The time between successive drops can be quite regular, and more than one insomniac has been kept awake waiting for the next drop to fall. Less familiar is the behavior of a faucet at a somewhat higher flow rate. One can often find a regime where the drops, while still falling separately, fall in a never repeating patter, like an infinitely inventive drummer. (This is an experiment easily carried out personally; the faucets without the little screens work best.) The changes between periodic and random-seeming patterns are reminiscent of the transition between laminar and turbulent fluid flow. Could a simple chaotic attractor underlie this randomness?

The experimental study of a dripping faucet was done at the University of California at Santa Cruz by one of us (Shaw) in collaboration with Peter L. Scott, Stephen C. Pope, and Philip J. Martein. The first form of the experiment consisted in allowing the drops from an ordinary faucet to fall on a microphone and measuring the time intervals between the resulting sound pulses. By plotting the time intervals between drops in pairs, one effectively takes a cross section of the underlying attractor. In the periodic regime, for example, the meniscus where the drops are detaching is moving in a smooth, repetitive manner, which could be represented by a limit cycle in the state space. But this smooth motion is inaccessible in the actual experiment; all that is recorded is the time intervals between the breaking off of the individual drops. This is like applying a stroboscopic light to regular motion around a loop. If the timing is right, one sees only a fixed point.

The exciting result of the experiment was that chaotic attractors were indeed found in the nonperiodic regime of the dripping faucet. It could have been the case that the randomness of the drops was due to unseen influences, such as small vibrations or air currents. If that was so, there would be no particular relation between one interval and the next, and the plot of the data taken in pairs would have shown only a featureless blob. The fact that any structure at all appears in the plots shows the randomness has a deterministic underpinning. In particular, many data sets show the horseshoe like shape that is the signature of the simple stretching and folding process discussed above. The characteristic shape can be thought of as a "snapshot" of a fold in progress, for example, a cross section partway around the Rössler attractor shown in Figure 2. Other data sets seem more complicated; these may be cross sections of higher-dimensional attractors. The geometry of attractors above three dimensions is almost completely unknown at this time.

If a system is chaotic, how chaotic is it? A measure of chaos is the "entropy" of the motion, which roughly speaking is the average rate of stretching and folding, or the average rate at which information is produced. Another statistic is the "dimension" of the attractor. If a system is simple, its behavior should be described by a low-dimensional attractor in the state space, such as the examples given in this article. Several numbers may be required to specify the state of a more complicated system, and its corresponding attractor would therefore be higher-dimensional.

The technique of reconstruction, combined with measurements of entropy and dimension, makes it possible to reexamine the fluid flow originally studied by Gollub and Swinney. This was done by members of Swinney's group in collaboration with two of us (Crutchfield and Farmer). The reconstruction technique enabled us to make images of the underlying attractor. The images do not give the striking demonstration of a low-dimensional attractor that studies of other systems, such as the dripping faucet, do. Measurements of the entropy and dimension reveal, however, that irregular fluid motion near the transition in Couette flow can be described by chaotic attractors. As the rotation rate of the Couette cell increases so do the entropy and dimension of the underlying attractors.

In the past few years a growing number of systems have been shown to exhibit randomness due to a simple chaotic attractor. Among them are the convection pattern of fluid heated in a small box, oscillating concentration levels in a stirred- chemical reaction, the beating of chicken-heart cells and a large number of electrical and mechanical oscillators. In addition computer models of phenomena ranging from epidemics to the electrical activity of a nerve cell to stellar oscillations have been shown to possess this simple type of randomness. There are even experiments now under way that are searching for chaos in areas as disparate as brain waves and economics.

It should be emphasized, however, that chaos theory is far from a panacea. Many degrees of freedom can also make for complicated motions that are effectively random. Even though a given system may be known to be chaotic, the fact alone does not reveal very much. A good example is molecules bouncing off one another in a gas. Although such a system is known to be chaotic, that in itself does not make prediction of its behavior easier. So many particles are involved that all that can be hoped for is a statistical description, and the essential statistical properties can be derived without taking chaos into account.

There are other uncharted questions for which the role of chaos is unknown. What of constantly changing patterns that are spatially extended, such as the dunes of the Sahara and fully developed turbulence? It is not clear whether complex spatial patterns can be usefully described by a single attractor in a single state space. Perhaps, though, experience with the simplest attractors can serve as a guide to a more advanced picture, which may involve entire assemblages of spatially mobile deterministic forms akin to chaotic attractors.

The existence of chaos affects the scientific method itself. The classic approach to verifying a theory is to make predictions and test them against experimental data. If the phenomena are chaotic, however, long-term predictions are intrinsically impossible. This has to be taken into account in judging the merits of the theory. The process of verifying a theory thus becomes a much more delicate operation, relying on statistical and geometric properties rather than on detailed prediction.

Chaos brings a new challenge to the reductionist view that a system can be understood by breaking it down and studying each piece. This view has been prevalent in science in part because there are so many systems for which the behavior of the whole is indeed the sum of its parts. Chaos demonstrates, however, that a system can have complicated behavior that emerges as a consequence of simple, nonlinear interaction of only a few components.

The problem is becoming acute in a wide range of scientific disciplines, from describing microscopic physics to modeling macroscopic behavior of biological organisms. The ability to obtain detailed knowledge of a system's structure has undergone a tremendous advance in recent years, but the ability to integrate this knowledge has been stymied by the lack of a proper conceptual framework within which to describe qualitative behavior. For example, even with a complete map of the nervous system of a simple organism, such as the nematode studied by Sidney Brenner of the University of Cambridge, the organism's behavior cannot be deduced. Similarly, the hope that physics could be complete with an increasingly detailed understanding of fundamental physical forces and constituents is unfounded. The interaction of components on one scale can lead to complex global behavior on a larger scale that in general cannot be deduced from knowledge of the individual components.

Chaos is often seen in terms of the limitations it implies, such as lack of predictability. Nature may, however, employ chaos constructively. Through amplification of small fluctuations it can provide natural systems with access to novelty. A prey escaping a predator's attack could use chaotic flight control as an element of surprise to evade capture. Biological evolution demands genetic variability; chaos provides a means of structuring random changes, thereby providing the possibility of putting variability under evolutionary control.

Even the process of intellectual progress relies on the injection of new ideas and on new ways of connecting old ideas. Innate creativity may have an underlying chaotic process that selectively amplifies small fluctuations and molds them into macroscopic coherent mental states that are experienced as thoughts. In some cases the thoughts may be decisions, or what are perceived to be the exercise of will. In this light, chaos provides a mechanism that allows for free will within a world governed by deterministic laws.

CHAOS: A MATHEMATICAL INTRODUCTION WITH PHILOSOPHICAL REFLECTIONS

Wesley J. Wildman and Robert John Russell

1 *Introduction*

There is a simple quadratic equation often used to illustrate chaos, the so-called logistic map:

$$x_{n+1} = k\, x_n\, (1 - x_n)$$

Here, x_n is the nth number in a sequence $x_0, x_1, x_2, \ldots$, and $k>0$. The logistic map is called an iterative map, because each number in the sequence is generated by applying the map to the previous number in the sequence. So, for example, if x_0 and k are stipulated, $x_1 = kx_0(1-x_0)$, $x_2 = kx_1(1-x_1)$, and so on.

This primary aim of this essay is, using the logistic map, to explain what 'chaos' means. We focus on the logistic map for two reasons. Firstly, this map displays most of the features of the broad class of chaotic dynamical systems: the transition from regular to apparently random behavior, period-doubling bifurcation cascades, the influence of attractors, and the underlying characteristics of a fractal. A certain lack of generality is inevitable due to the discrete, one-dimensional character of the dynamical system obtained by iterating the logistic map. The mathematics of the one-dimensional case is much simpler, however, and simplicity is a compelling virtue in an introductory essay.

Secondly, the logistic map has been specifically singled out in many of the papers in previous science and religion publications that discuss the possible philosophical and theological implications of chaos (see the bibliography for examples). Our tour of the logistic map will involve showing that the behavior of the logistic map is more complex than this tradition of discussion in the science-religion specialization has sometimes supposed. This is important, we hold, because clarity about the mathematics of chaos is essential for such philosophical and theological reflections to be relevant.

While the bulk of this paper is an introduction to the mathematics of chaos, we also intend to make a small contribution to debates surrounding the philosophical and theological import of chaos theory. This subsidiary goal involves treatment of the themes of unpredictability, randomness, determinism, chaos in nature, and divine action.

1.1 *About Technical Material*

The logistic map, with its parabolic graph (see Figure 1 and Figure 2), is one of the simplest possible equations, familiar to high school students of mathematics. Yet, under iteration, it produces an extraordinarily complex chaotic dynamical system. In fact, mathematicians are intensively engaged in trying to find out more about the dynamical system this quadratic map produces. Most of this mathematical labor lies well beyond the scope of an introductory paper such as this one. The aim here is to introduce the mathematics of chaos, however, so technical material cannot be avoided entirely.

We have tried to present the mathematical results as accessibly as possible, avoiding any indication of how they are obtained, in order to minimize technical language. The technical parts that remain have been flagged in the text by means of a vertical bar in the right margin, such as the one that accompanies this paragraph. We have also divided the mathematical material into bite-sized pieces, making it easier for non-mathematicians to maintain a sense of the flow of the discussion. In the final analysis, however, non-mathematical readers will have to make their own decisions about which parts to skim, and which to read more carefully. On the other hand, mathematically informed readers will be frustrated by the absence of proofs, and so we refer them to the bibliography for more details. This is the double problem facing essays introducing technical subjects, and we have tried to steer a middle way between the diverse expectations of the science and religion audience.

1.2 *About Graphs*

There are a number of different kinds of figures included in this paper, all of which have been generated by the authors. One type of graph is called a 'cobweb diagram,' and appears in Figure 1, Figure 2, and elsewhere. It illustrates the process of iteration by overlaying the logistic map's parabolic shape with a plot of x_{n+1} against x_n for several values of n. Another type is called a 'time series plot.' It graphs the sequence of values produced by iterations of the logistic map against the number of iterations completed. In such graphs, the vertical axis records the value, x_n, and the horizontal axis the iteration, n. This type offers a way to visualize what will soon be defined as the orbits of the logistic map. Other types will be explained when they are used.

1.3 *About References*

The material in section 2 is available in a number of sources, and these sources include more on almost every topic than can be covered here. Rather than citing a source for every piece of information, therefore, we have appended a short bibliography to this paper. The papers and books listed there are organized according to relevant topics and annotated where appropriate.

2 *The Logistic Map*

Figure 1 illustrates the first few numbers in the sequence produced by iterating the logistic map for k=1, beginning from x_0=0.5. To understand the diagram, follow the line from x_0 up to the parabola, and then left to the vertical axis; the point at which the line meets the vertical axis is x_1. The diagonal line helps to locate x_1 on the horizontal axis. Now the second iteration begins, with x_2 lo-

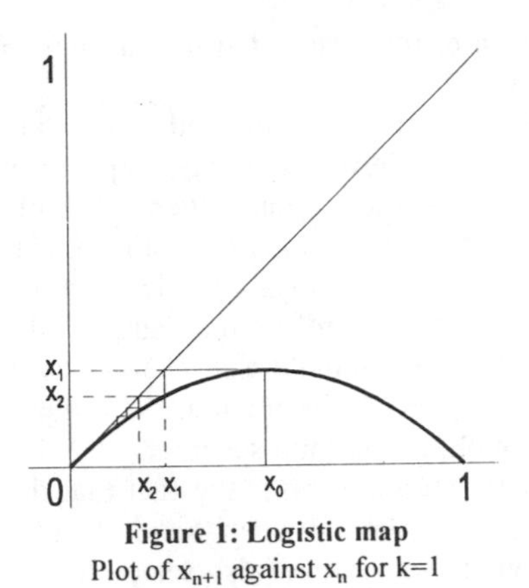

Figure 1: Logistic map
Plot of x_{n+1} against x_n for k=1

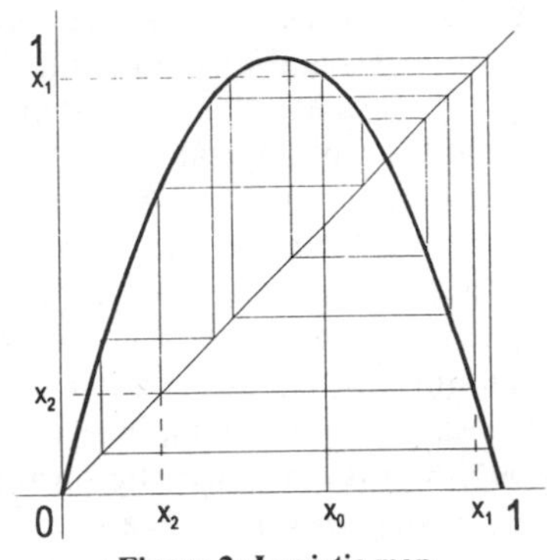

Figure 2: Logistic map
Plot of x_{n+1} against x_n for k=4

cated first on the vertical axis, and then on the horizontal axis, in the same way. Figure 2 shows a steeper parabola, representing the logistic map for k=4. A few iterations, beginning from $x_0 \cong 0.6$, are illustrated there.

It is possible to ask when the iterated sequence converges, or exhibits some other patterned behavior. It turns out that, for some values of k, the sequence of numbers appears random, and highly sensitive to the initial choice, x_0; this is the case for the sequence whose beginning is plotted in Figure 2. For other values of the constant k there is a recognizable pattern in the sequence of numbers generated by iterating the logistic map, no matter what is chosen for x_0. For example, some iterated sequences steadily decline toward 0, like the one plotted in Figure 1. Others converge to some non-zero limit. Still other sequences never converge but oscillate stably between two or more values. In other cases, though the final pattern that emerges does not depend on the initial value chosen for x_0, the *way* that pattern emerges *does* depend on x_0.

It is fascinating that a single equation such as this displays both stable and chaotic behavior, depending merely on the value of k, which is often called the tuning constant. Imagine being able to increase the metabolic rate of Dr. Jekyll by turning a single knob on some dastardly machine, eventually reaching a point at which Mr. Hyde emerges, with his characteristically unpredictable behavior. That is an apt analogy for what happens to the sequence of numbers generated by the logistic map as k is increased. Such characteristics are extremely common in mathematical dynamical systems. But many physical systems also seem to behave in the same way as this innocent looking equation, with its Jekyll and Hyde personality.

2.1 *Modeling Nature: The Moth Colony Example*

In just the past two decades, scientists have made dramatic breakthroughs by using the logistic map and similar dynamical systems to study a vast range of previously intractable or unnoticed problems in the physical, biological, and social sciences. A sampling of the areas studied with some success include financial markets, meteorology, ecology, turbulence in fluids, periodic behavior in chemical reactions, the spread of disease, the onset of war, and human physiology (see the bibliography for references).

For instance, the population of successive generations of moths in certain habitats is a biological variable that displays both predictable and apparently random behavior. It turns out that the humble logistic map is quite serviceable as a model for moth colony populations when the variations in moth population are predictable. The fact that the logistic map displays chaotic behavior under certain mathematical conditions (values of k) therefore may suggest the beginnings of an explanation for the apparent randomness that characterizes successive moth populations under certain special environmental conditions, though this is a problematic form of explanation, as we shall see later.

Thought of as such a model, the logistic map expresses the fact that the moth population of one generation (scaled to a number between 0 and 1; call it x_{n+1}), depends on the population of the previous generation (also scaled down; call it x_n) in three ways:

1. It depends directly on x_n since, if there are lots of moths this generation, there will be lots of baby moths to boost the next generation.

2. It also depends on $(1-x_n)$ because, if there are too many moths, there will be a shortage of food and extremely rapid spread of disease.

3. Finally, the population depends upon the 'tuning' constant, k, which corresponds to such effects as the general availability of resources, the level of fecundity of the moth population, and other factors that remain unchanged from generation to generation.

It may seem unlikely that such a simple model actually works, and the truth is that it does not always work that well; the complexities of moth life can quickly muddy the logistic map's clear waters. It is quite useful, nevertheless, especially when the 'tuning' of the system is low; that is, when the generationally invariant factors lead to a constant, k, that is not too high.

2.2 *Regimes of Behavior*

The unit interval, [0,1], is of great importance for understanding the logistic map. To see this, notice that the graph of $x_{n+1}=kx_n(1-x_n)$ is an inverted parabola when $k>0$, intersecting the horizontal x_n-axis at 0 and 1 (see Figure 1 for $k=1$). The value of k determines how steep the inverted parabola is, because the maximum value, which occurs at $x=0.5$, is $0.25k$. It follows that, when x_0 is in [0,1] and the constant k lies between 0 and 4, the numbers generated by iterating the logistic map never lie outside the bounds of 0 and 1, no matter what x_0 might be. Iteration in this range of x_0 and k creates a kind of feedback process, making possible such dynamics as convergence of the iterated sequence.

Of course, the logistic map can be iterated no matter what k and x_0 might be. But when $k\geq1$, iterations tend rapidly toward infinity unless the first number in the sequence, x_0, lies in the unit interval, [0,1]. For $0<k<1$, exponential

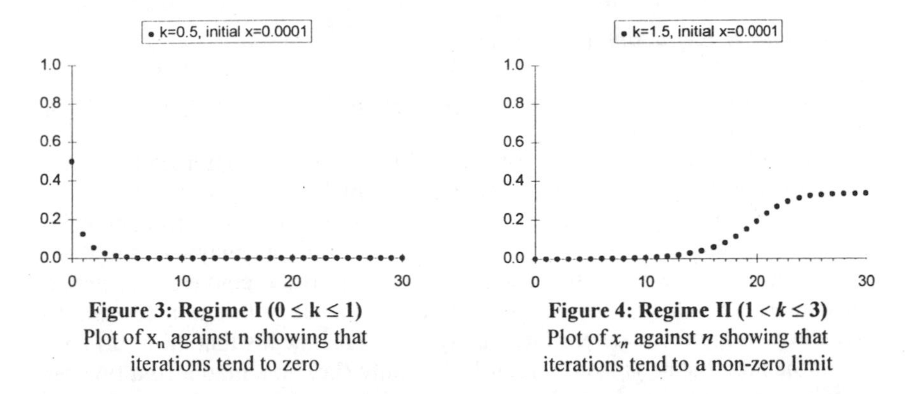

Figure 3: Regime I ($0 \leq k \leq 1$)
Plot of x_n against n showing that iterations tend to zero

Figure 4: Regime II ($1 < k \leq 3$)
Plot of x_n against n showing that iterations tend to a non-zero limit

explosion of the sequence occurs unless x_0 is in the interval $[(k-1)/k, 1/k]$. On that interval, which includes [0,1] as a subset, iterations converge to 0. Thus, the interesting dynamics of the logistic map for $0<k<4$—that is, the various behaviors of the sequences obtained by iteration—are encountered when x_0 is constrained to lie in the unit interval.

Interesting dynamics also occur when $k>4$. Though the iterated sequence explodes to infinity for almost every starting value, x_0, it remains within [0,1] if x_0 belongs to a special subset of [0,1] that depends on k. In this case too, therefore, the unit interval is crucial.

Once we limit the discussion to x_0 in the unit interval, the dynamics of the logistic map turn out to depend fundamentally on k. Just as a caterpillar passes through a number of stages in its life cycle, or a civilization takes on new forms as one era gives way to the next, so the behavior of the logistic map undergoes beautiful changes as the tuning constant, k, is increased. Here we describe the logistic map with reference to five Regimes of the tuning constant, k, in which the logistic map has distinctive dynamics.

We will discuss these Regimes beginning with $0<k\leq 1$ and ending with $k>4$. Along the way we will make excurses to consider themes of more general significance. We will sometimes describe the behavior of the logistic map using the highly suggestive moth colony population example.

To help readers keep track of the movement of the discussion through the various Regimes of the logistic map, we list them here, with a summary of the dynamics they exhibit. The number k_{crit} is roughly equal to 3.59, and its importance will be described below.

Regime	*k*-values	Summary of Dynamics
I	$0<k\leq 1$	*Extinction Regime:* Convergence to 0
II	$1<k\leq 3$	*Convergence Regime:* Convergence to non-zero limit
III	$3<k<k_{crit}$	*Bifurcation Regime:* Splitting of stable limit states
IV	$k_{crit}\leq k\leq 4$	*Chaotic Regime:* Mixture of stable periodic and chaotic behavior
V	$4<k$	*Second Chaotic Regime:* Chaotic dynamics on a subset of [0,1]

2.3 *Regime I: Extinction* ($0 < k \leq 1$)

Monotonically decreasing sequences converging to 0; no significant dependence on initial conditions.

In this Regime sequences of iterated values fall monotonically (that is, always in one direction) to zero. (See Figure 3 for k=0.5).

If k=1, or is very close to 1, the sequence descends to zero appreciably more slowly than if k is close to 0. The initial value, x_0, plays no significant role, except to give some sequences a head start on the convergence process (of course, if x_0= 0 or 1, convergence of the sequence to 0 is immediate). The rate of convergence is determined solely by k. This means that all information about the initial state is lost more or less rapidly (i.e., in a matter of a few generations, or iterations, depending if we are thinking about moths or numbers). In moth language, the population inevitably declines to extinction.

2.4 *Regime II: Convergence* ($1 < k \leq 3$)

Increasing sequences converging to single limits>0 that depend on k; no significant dependence on initial conditions.

When k is between 1 and 3, iterated sequences converge to non-zero limits (see Figure 4 for k=1.5). Whereas in Regime I the limit is always 0, the value of the limit in Regime II depends on k; it can be calculated easily, since it occurs when $x_n=x_{n+1}=kx_n(1-x_n)$. Graphically, this is the intersection of the line $y=x$ with the parabola $y=kx(1-x)$. The limit, therefore, is $(k-1)/k$. Of course, if x_0=0 or $(k-1)/k$—or one of the points that are iterated by the logistic map into 0 or $(k-1)/k$—then convergence occurs in a finite number of steps. Beyond this trivial dependence on initial conditions, the value of x_0 only gives the process of convergence more or less of a head start, as in Regime I.

The way convergence occurs, however, varies within this regime. For $1<k\leq 2$, after possible initial bumps, convergence is monotonic from below (Figure 4 for k=1.5), whereas for $2<k\leq 3$, the sequence fluctuates above and below the eventual limit, moving gradually closer with each iteration (Figure 5 for k=2.5). Figure 6 is a close up illustration of the fluctuating convergence of the iterated sequence.

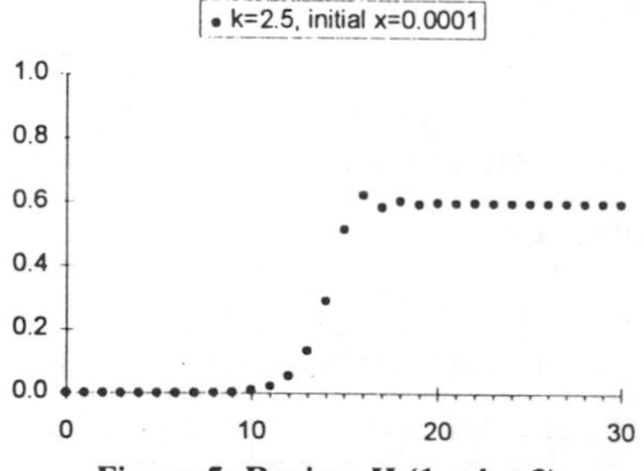

Figure 5: Regime II ($1 < k \leq 3$)
Plot of x_n against n illustrating the breakdown of monotonic convergence

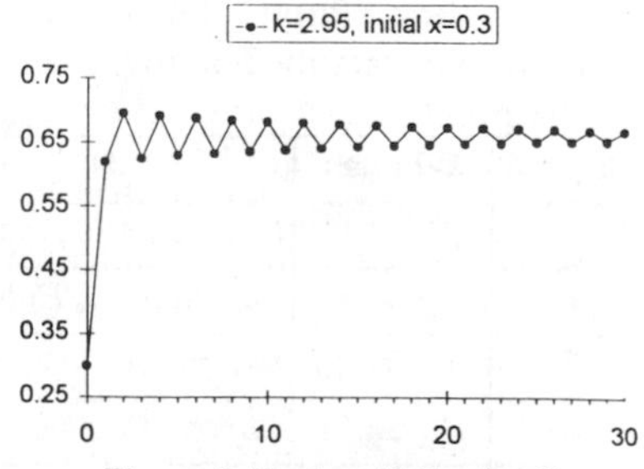

Figure 6: Regime II ($1 < k \leq 3$)
Plot of x_n against n illustrating a fluctuating population close up

2.5 *Excursus: Technical Definitions*

The following technical terms will be helpful for understanding the details of the logistic map as the more complex aspects of its behavior are introduced. Some require abstract definitions, but all are quite intuitive. Each is a specification of a concept important for studying chaotic dynamical systems.

(i) A *fixed point* of the logistic map is one that remains unchanged by iteration. For example, 0 and $(k-1)/k$ are fixed points for $k>0$.

(ii) A *periodic point of period n* is one transformed into itself by the logistic map after exactly and no fewer than n iterations. For example, fixed points are periodic of period 1, and we shall soon see that, in Regime III, the logistic map has periodic points of period 2, 4, 8, and so on. When the logistic map is chaotic, it has infinitely many periodic points.

(iii) The *forward orbit* of the logistic map for a given k and x, denoted $O^+(k,x)$, is the set of numbers generated by iterating the logistic map for k, beginning from x. So, for example, $O^+(k,(k-1)/k)=\{(k-1)/k\}$ for $k>0$. The *backward orbit* or *preimage* for a given k and x—the same thing in the reverse direction, denoted $O^-(k,x)$—is the set of numbers that the logistic map iterates into x. For example, $O^-(k,0)$ includes at least $\{0,1\}$ for $k>0$, because 0 and 1 are iterated to 0. The *orbit* for a given k and x, denoted $O(k,x)$, is just the union of the forward and backward orbits. For convenience, forward orbits will sometimes be called orbits when the abbreviation is not misleading.

(iv) Orbits only rarely consist of a finite number of points, as in these examples. In fact, orbits of the logistic map usually have infinitely many points with intriguing patterns. For example, suppose that $0<k\leq 3$, and further suppose that x_0 is in $[0,1]$, but not in the preimages of 0 or $(k-1)/k$, so that convergence is never interrupted by hitting a fixed point. Then $O(k,x_0)$ is an infinite set of points. The interesting feature of the orbit just described is that it has one and only one *limit point*: 0 for k in Regime I, and $(k-1)/k$ for k in Regime II. Note that a limit point does not have to be in the orbit itself (and usually is not). However, the *closure* of an orbit includes all of its limit points. *The time series plots in this paper are essentially diagrams of forward orbits organized in such a way as to draw attention to their limit points, if they exist.*

(v) A fixed point or a periodic point is *attracting* for a given k if it is a limit point of some orbit for k; otherwise it is *repelling*. For example, $(k-1)/k$ is an attracting fixed point for $1\leq k\leq 3$, and repelling otherwise. Orbits draw attention to attracting fixed and periodic points because they are limit points, and the points of an orbit accumulate around its limit points. Repelling fixed and periodic points are less visible, especially in the sense that they are undetectable by computer calculations of orbits.

(vi) The set of limit points for a given orbit is called the *attractor* for that orbit; for non-chaotic orbits, it is usually the set of attracting fixed and periodic points. For example, the attractor of orbits in Regimes I and II is a single point. In Regime III, as we shall shortly see, the attractor may contain 2, 4, 8, or more points, depending on k. In Regimes IV and V, the attractor may have infinitely many points, none of them attracting fixed or periodic points. Attractors for chaotic orbits usually have extremely complex and strange structures.

In dynamical systems that are described with ordinary differential equations, attractors are typically attracting, simple closed curves.

(vii) The logistic map is said to be in *equilibrium* for a given k if the attractor for $O(k,x)$ consists of the same single point for almost all x in [0,1].

2.6 *Excursus: Convergence*

The sequences in Figure 7, Figure 8, and Figure 9 illustrate how the rate of convergence slows dramatically as k nears the value 3. Note that these sequences are *oscillating* above and below their eventual limit. For instance, when k=2.95 (Figure 7), variation between successive iterations is barely discernible on the scale of that diagram after 80 iterations, whereas convergence is much slower (maximally slow, strictly speaking) for k=3 (Figure 9). This is a reminder of an important general point: when convergence is not cut short by iterations hitting a fixed or periodic point, *the occurrence of convergence, as well its limit and rate, depend only on the value of* k. By contrast, the initial value, x_0, does not have a consistent influence on convergence.

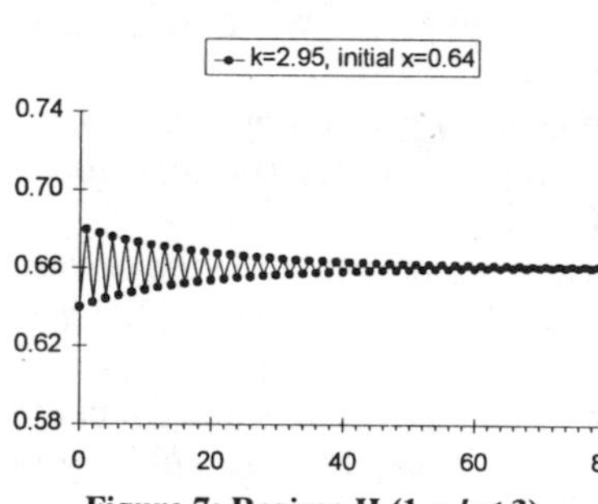

Figure 7: Regime II ($1 < k \leq 3$)
Rapid convergence for k=2.95

From a technical point of view, we must always keep in mind the (by now) familiar exceptional case that convergence only requires a finite number of iterations, which happens whenever x_0 lies in the preimage of a fixed or periodic point. For a given k, we shall call the set of values of x_0 for which convergence occurs in a finite number of iterations the *finite convergence set* of k, and denote it FC_k. The finite convergence set for a given k can be described as follows: Denote by P_k the set of fixed and periodic points, both attracting and repelling. Convergence will be achieved in a finite number of iterations whenever x_0 is in one of the preimages of the points in P_k, so $FC_k=O^-(k,P_k)$.

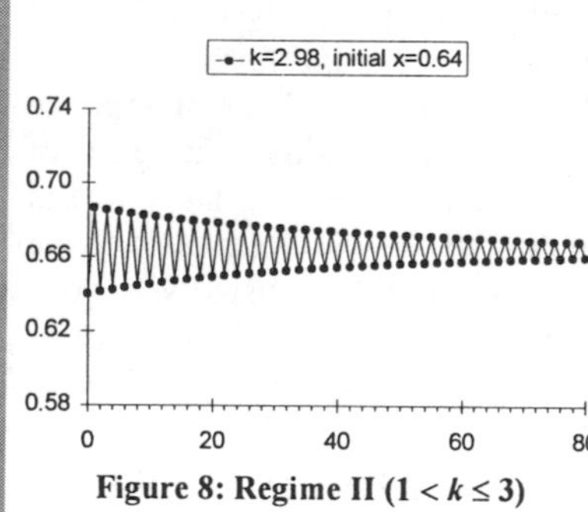

Figure 8: Regime II ($1 < k \leq 3$)
Slow convergence for k=2.98

When $0<k\leq1$, $O^-(k,P_k)=\{0,1\}$. When $1<k\leq3$, $O^-(k,P_k)$ contains infinitely many points: the preimage of the repelling fixed point 0, which consists only of 0 and 1; and the preimage of the attracting fixed point $(k-1)/k$, which is an infinite set scattered symmetrically through the unit inter-

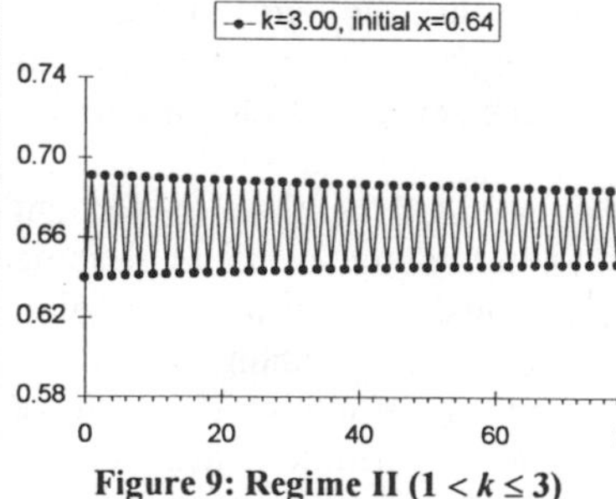

Figure 9: Regime II ($1 < k \leq 3$)
Maximally slow convergence for k=3

val. As k increases above 3, FC_k becomes larger with an extremely complex structure, but its overall size remains comparatively small in the sense that it has Lebesgue measure zero.

We can define the *infinite convergence set* of k, denoted as IC_k, to be the complement of the finite convergence set on the unit interval. That is to say, $IC_k=[0,1]-FC_k$. If x_0 lies in the infinite convergence set, then convergence, if it occurs, is never completed after a finite number of iterations, and the orbit of k and x_0, $O(k,x_0)$, contains infinitely many numbers. The infinite convergence set is much larger than the finite convergence set—though both sets contain infinitely many points, IC_k contains almost all possible starting values—but its precise size (Lebesgue measure) in some cases is a tricky problem (we shall have more to say about this presently). We will need to refer to the infinite convergence set, IC_k, repeatedly in the succeeding pages, because the general pattern of the logistic map's behavior is determined solely by k only when x_0 lies within IC_k. For convenience we will sometimes simply say "for almost all starting values, x_0" when we mean "for all x_0 in IC_k."

We can summarize these technical observations as follows. It is true that when x_0 is in the finite convergence set of k, convergence is affected by initial conditions. Apart from this interesting way—trivial though it may be for $k \leq 3$—in which convergence depends on x_0, *the value of k determines all convergence issues for almost all* x_0. In graphical terms, the steepness of the inverted parabola is the key; the choice of starting point for iteration results only in more or less of a head start on the process, for almost all x_0. This is typical of dynamical systems: the tuning variable that determines the Regime of the system also settles all convergence questions.

2.7 *Regime III: Bifurcation* ($3.0 < k < k_{crit} \cong 3.57$)

Bifurcation Regime consisting of fluctuating sequences with multiple attracting periodic points that depend on k; minor, permanent dependence on initial conditions in the form of phase shifts.

Immediately after the tuning constant is increased beyond k=3, something truly striking begins to occur: equilibrium, or convergence to a single limit, breaks down. In its place, the iterated sequences oscillate closer and closer to *multiple distinct values*: first two, then four, then eight, and so on in

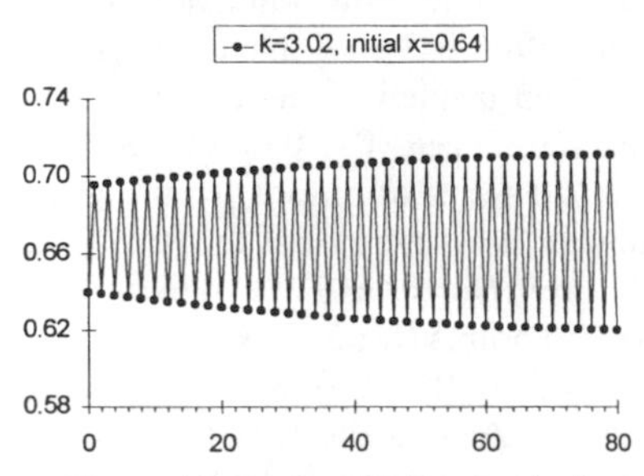

Figure 10: Regime III ($3 < k < k_{crit}$)
Slow divergence for k=3.02

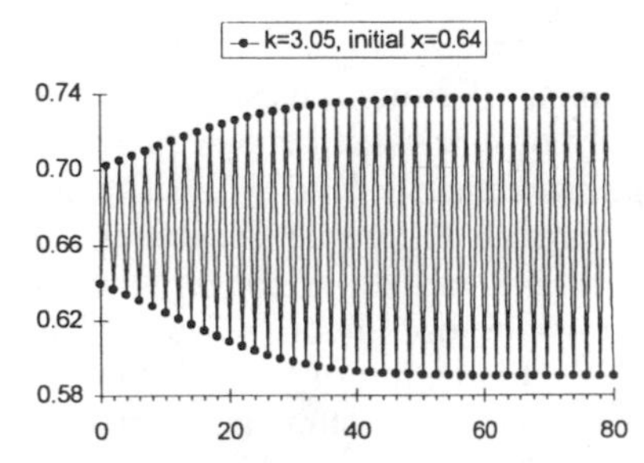

Figure 11: Regime III ($3 < k < k_{crit}$)
Rapid divergence for k=3.05

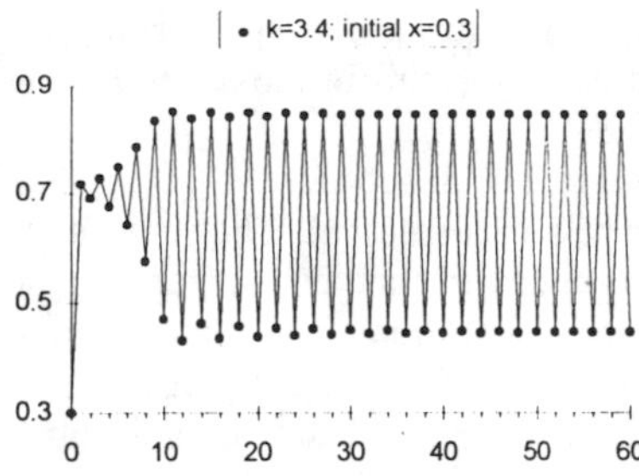

Figure 12: Regime III ($3 < k < k_{crit}$)
Limit cycle, period 2

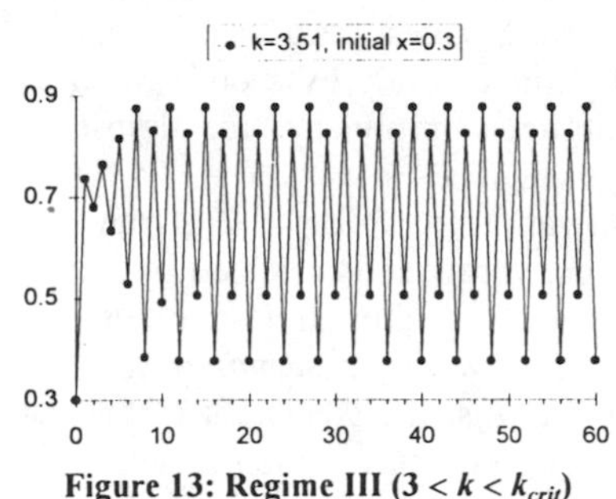

Figure 13: Regime III ($3 < k < k_{crit}$)
Limit cycle, period 4

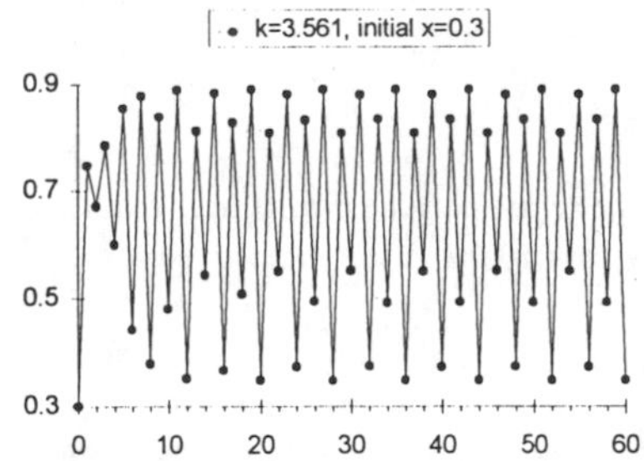

Figure 14: Regime III ($3 < k < k_{crit}$)
Limit cycle, period 8

increasing powers of 2 as k increases. Put precisely, for a given k, and for all x_0 in IC_k, the orbits for k and x_0, $O(k,x_0)$, have the same limit points, and they are all attracting periodic points of the same period, a power of 2. *The value of k alone determines the limit points and their period, for almost all* x_0. (The fixed points 0 and $(k-1)/k$ are both repelling in Regime III, and so are not limit points of any orbits).

An apparently stable oscillation between two population levels has been noticed in moth colonies, and was unexplained until the effectiveness of the logistic map as a model suggested a way. In moth language, moth colonies can be in a stable state even when they do not have a unique population: after some initial bumps, they can simply flip-flop from generation to generation between two distinct population levels. In number language, the oscillation of the sequence of iterated values we noticed in Regime III eventually overpowers the tendency to converge to a single limit. Put figuratively, alternate halves of the iterated sequence converge, even though the sequence as a whole is oscillating with each iteration, and displays no tendency toward overall convergence whatsoever. (See Figure 10 for k=3.02 and Figure 11 for k=3.05.)

The splitting of the equilibrium state of the iterated sequence into stable oscillation between attracting periodic points of period 2 is called a *bifurcation*. Whereas orbits have just one limit point before the bifurcation, after k=3 they have two limit points. Of course, these are not limits of iterated sequences as a whole, but only of intertwined sub-sequences of iterated sequences. The pair of periodic points defines a *limit cycle* of period 2. The entire series of graphs from Figure 7 through Figure 11 illustrates a bifurcation as k is increased above 3: the equilibrium state of convergence to a single limit (for $0<k\leq 3$) is transformed into stable oscillation between attracting periodic points (for $3<k<k_{crit}$).

Bifurcation of 'limits' does not stop with the split at k=3. As k increases, more and more attracting periodic points appear. At k=3 the first bifurcation occurs, producing stable oscillation of the iterated sequence between two attracting periodic points of period 2, and the formerly attracting fixed point,

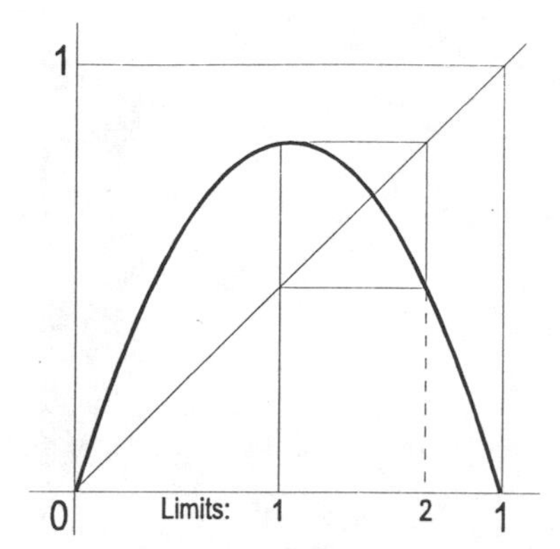

Figure 15: Limit Cycle (period 2)
x_{n+1} against x_n for k=3.4, corresponding to the stable limit state of Figure 12

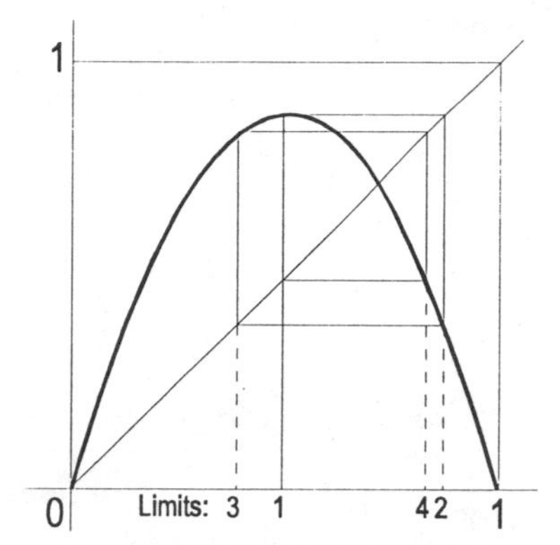

Figure 16: Limit Cycle (period 4)
x_{n+1} against x_n for k=3.51 corresponding to the stable limit state of Figure 13

$(k-1)/k$, becomes a repelling fixed point. At $k \cong 3.449$, the second bifurcation occurs, producing four attracting periodic points of period 4, with the periodic points of period 2 becoming repelling instead of attracting. Figure 12 (for k=3.4) illustrates stable oscillation between two clearly defined attracting periodic points of period 2. Compare that to Figure 13 (for k=3.51) where each of these limit points of the orbit has, in turn, given rise to two new limit points each, all periodic points of period 4.

Successive bifurcations occur at $k \cong 3.544$ (8 periodic points of period 8; see Figure 14 for k=3.561), $k \cong 3.564$ (16 periodic points of period 16), $k \cong 3.569$ (32 periodic points of period 32), and so on. Because the bifurcations occur for values of k that are packed increasingly closely together, the usefulness of the logistic map as a model for physical systems such as moth colony populations breaks down; the real world is not well enough behaved to maintain fecundity and food supplies (corresponding to a precise value of the tuning constant, k) perfectly invariant.

Each one of these limit cycle situations is stable, in the sense that the values for k and x_0 can be changed a little and, so long as x_0 is in IC_k (i.e., for almost all x_0), the number of attracting periodic points in the perturbed orbit remains the same as that of the unchanged orbit. Stability can be mathematically quantified by measuring the effects of perturbations of this kind. Not surprisingly, it turns out that stability is weakest in the vicinity of bifurcations, and greatest between bifurcations.

It is possible to illustrate geometrically the way iteration of the logistic map produces sequences that oscillate stably among attracting periodic points: Figure 15 (for k=3.4) shows 2 attracting periodic points of period 2, and Figure 16 (for k=3.51) displays 4 attracting periodic points of period 4. The intersection of the diagonal line and the parabola in these two figures locates the fixed point at $(k-1)/k$. No orbit will have this point as a limit point, however, because it is repelling for $k>3$, though all of the points in its preimage terminate iteration there.

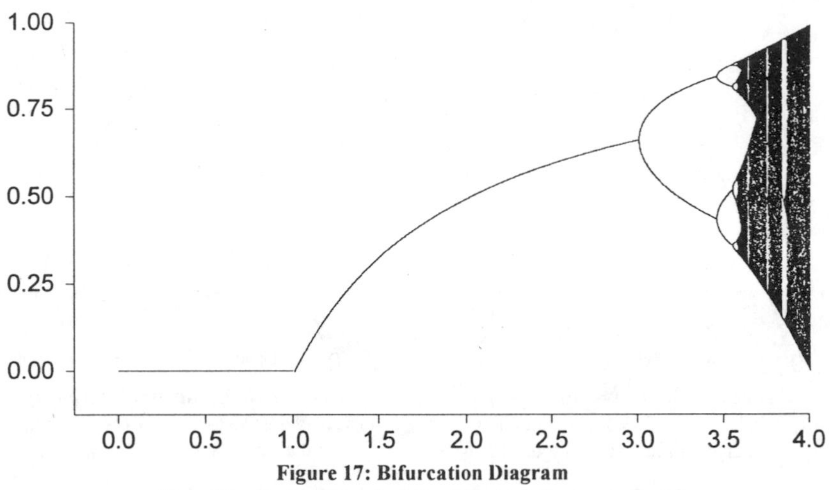

Figure 17: Bifurcation Diagram
Plot of the limit points of orbits of the logistic map against *k*

The change from being an attracting to a repelling fixed point can be pictured graphically as being caused by the slope of the tangent to the parabola at (k-1)/k becoming progressively steeper. As it passes 45 degrees to the horizontal axis at k=3, it changes from attracting to repelling. In a similar way, periodic points of period 2 are present in Figure 16, but they can never be the limit points of any orbit because, as k moves through the bifurcation at k=3.5, they change permanently from attracting to repelling.

2.8 *Excursus: The Bifurcation Diagram*

The limits of orbits of the logistic map can be plotted against values of k in what is called a bifurcation diagram (Figure 17). For a given k, the orbits $O(k,x_0)$, have the same limit points, for almost all x_0. The vertical slice of the bifurcation diagram at that value of k is effectively a plot of the limit points (the attractor) for those orbits.

The bifurcation diagram graphically illustrates how the limit points split successively and without ceasing in Regime III as k increases from 3 to the critical value $k_{crit}\cong 3.57$. Because it plots orbits as they approach their limit points, where they exist, it does not draw attention to repelling fixed and periodic points, but only their attracting counterparts. As we have seen, each bifurcation spawns two new attracting periodic points of period twice that of the attracting fixed or periodic point from which they sprang. The parent attracting limit point (thought of as changing continuously with k) does not disappear, but merely becomes repelling, and so ceases to be a limit point of orbits after that bifurcation. If repelling periodic and fixed points were plotted on the bifurcation diagram, they would continue on the trajectory they were on before each bifurcation splits the new periodic points off in opposite directions.

The structure of these successive splittings is an important geometrical feature of the logistic map. All bifurcations in the bifurcation diagram *look* similar because they really are: they turn out to be approximately self-similar in the same way that fractals are, with smaller instances being very nearly identical to scaled down versions of larger ones. In fact, the diagram itself is a fractal. For example, the bottom branch of the first bifurcation, when inverted, replicates almost exactly the structure of the entire diagram. (See Figure 24 and Figure 25 for illustrations of the fractal geometry of Regime IV.)

2.9 *Excursus: The Feigenbaum Constant*

The shape of bifurcations (disregarding their size) varies slightly, especially in the early stages of a bifurcation cascade. However, the shape changes less and less as these bifurcation splittings become smaller and smaller, approaching a limit bifurcation shape. The scaling constant that gives the ratio of sizes of adjacent bifurcations as they approach this limit shape was named after Mitchell Feigenbaum, a mathematician who has made many contributions to understanding the logistic map. He calculated it to be 4.6692016091029.... This means that each bifurcation of the cascade in Regime III of the logistic map is roughly 4.6692 times smaller than the previous one, and the approximation gets better the further through the cascade we go.

The same constant applies to all cascades in the logistic map (there are more of them in Regime IV), to the magnification ratios used to observe the fractal character of the bifurcation diagram as a whole, to the ratios of banding divisions (see below for more on this), and to other features of the bifurcation diagram. Thus, the Feigenbaum constant is an extremely efficient indicator of the behavior of the logistic map. Unsurprisingly, in view of these observations about scaling, the bifurcation region of the logistic and related iterated maps have important connections to the topic of renormalization in physics.

The following table indicates the way the ratios of adjacent bifurcation sizes approach the Feigenbaum constant.

bifurcation	begins	ends	width	ratio to next
period 2	3.0	3.449489...	.449489...	4.7514...
period 4	3.449489...	3.544090...	.094601...	4.6562...
period 8	3.544090...	3.564407...	.020317...	4.6683...
period 16	3.564407...	3.568759...	.0043521...	4.6687...
period 32	3.568759...	3.569692...	.00093219...	4.6692...
period 64	3.569692...	3.569891...	.00019964...	

It turns out that this same scaling constant applies to a whole family of maps on the unit interval: smooth maps with one hump graphs and non-zero second derivatives at the critical point (0.5, for the logistic map). Besides the logistic map, this family includes the sine wave, and even maps with non-symmetrical graphs. That makes Feigenbaum's number an intriguing mathematical constant with profound geometric significance, much like π and e.

The mystery was only to deepen, however, when it was discovered that the Feigenbaum constant also appears in nature. There have been numerous

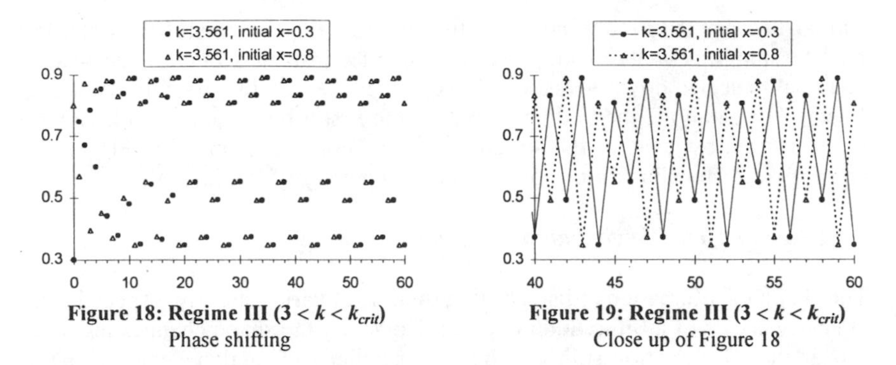

Figure 18: Regime III ($3 < k < k_{crit}$)
Phase shifting

Figure 19: Regime III ($3 < k < k_{crit}$)
Close up of Figure 18

experiments in the last two decades designed to calculate the same ratio for physical dynamical systems exhibiting period doubling cascades (see the bibliography for examples), and the results agree rather closely with Feigenbaum's number. It appears, therefore, that this number is more than an important mathematical constant. It also seems to be a kind of natural constant; the sense in which this is so is a pressing question for many scientists.

There are other scaling constants for the many classes of mathematical dynamical systems. In fact, above one dimension, there are as many families of maps with characteristic scaling constants as there are real numbers. Like the Feigenbaum constant, some of these may also be detectable in nature.

2.10 *Excursus: Dependence on Initial Conditions*

We have already seen one kind of dependence on initial conditions—trivial for $k \leq 3$—in that, for a given k, convergence is achieved in a finite number of steps when x_0 is in the finite convergence set of k, FC_k. That means that the value of k determines convergence of sequences not for *all* x_0, but only for *almost all* x_0. Dependence on initial conditions would occur in a non-trivial form if it were evident for almost all x_0, rather than the relatively few exceptional values in FC_k. This happens for the first time in Regime III, where initial conditions affect the order in which sequences bounce close to the attracting periodic points of its stable limit state. This effect is called 'phase shifting.'

Figure 18 shows phase shift effect for a limit cycle of 8 periodic points of period 8. Since plotting two orbits—$O^+(3.561,0.3)$ and $O^+(3.561,0.8)$—on the same diagram can be a little confusing, we have provided Figure 19, which is a blow-up of the data from Figure 18 beginning at the 40th iteration. In the blown up diagram, we can see that the sequence for x_0=0.3 cycles through 8 periodic points one step ahead of the iterated sequence beginning from x_0=0.8. That is, the 43rd iteration for the x_0=0.3 sequence registers a value of approximately 0.9, whereas this value occurs for the 42nd iteration of the x_0=0.8 sequence. *This shift is preserved indefinitely by successive iterations.*

The phase shift represents what can be called an *historical* element in these systems; the initial conditions continue to affect the system indefinitely, and the effect (the size of the phase shift—here one iteration) is independent of time. The size of the shift, however, differs for different starting values.

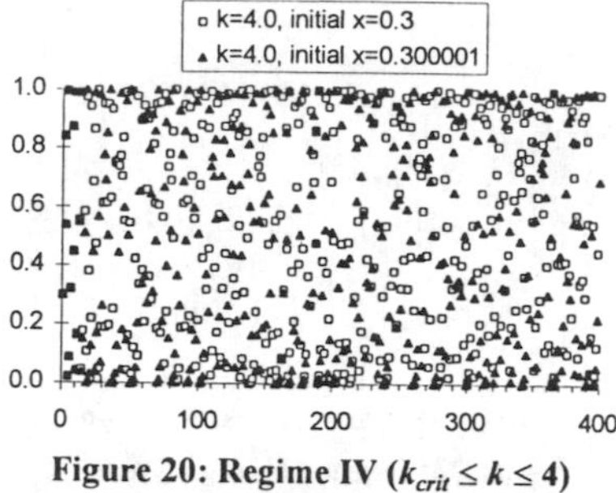

Figure 20: Regime IV ($k_{crit} \leq k \leq 4$)
Chaos over the entire unit interval (k=4)

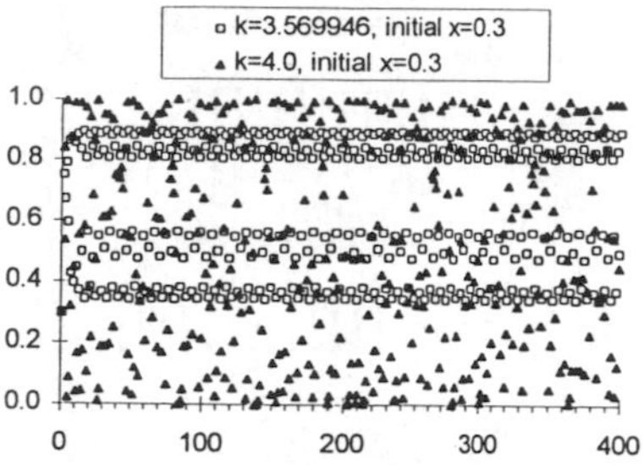

k=3.569946, initial x=0.3
k=4.0, initial x=0.3

Figure 21: Regime IV ($k_{crit} \leq k \leq 4$)
Chaos in bands (k≅3.57 and k=4)

2.11 *Regime IV: Chaos* ($3.57 \cong k_{crit} \leq k \leq 4$)

Chaotic orbits separated by densely packed bifurcation regions; maximal dependence on initial conditions.

In Regime III, bifurcations occur closer together as k increases until a critical value of k, k_{crit}=3.5699456..., is reached. At this point, the logistic map is chaotic, in a sense that will be made precise after we have presented a summary of the structures of Regime IV and Regime V. For now, we will be content with a rather limited but suggestive phenomenological definition: the iterated sequences for $k=k_{crit}$ and almost all x_0 are chaotic in the sense that, short of knowing that the logistic map produced them, no known mathematical test could distinguish these sequences from one that is random, such as the sequence of numbers obtained by listing, in order, the last digit of each telephone number in the New York telephone directory.

We want to illustrate the difference between dependence of Regime IV sequences on k, and their dependence on initial conditions, x_0. First we hold k fixed, and vary the initial conditions slightly, as shown in Figure 20 for k=4. Though the two plotted orbits are generated from starting values very close together (they differ by only .000001), the iterated sequences are evidently quite different from one another. This is suggestive of the exquisite sensitivity to initial conditions that characterizes chaos.

Next we hold the initial conditions fixed and vary k, as shown in Figure 21 for x_0=0.3, and $k=k_{crit}$ and k=4. Notice the way the orbit for $k=k_{crit}$ confines itself to bands (between what are called 'constraints'). These bands widen as k increases above k_{crit} through Regime IV until orbits for k=4 (and almost all x_0) cover the entire unit interval. The bifurcation diagram (Figure 17) gives an overview of banding in Regime IV, and we will pause presently to ask what these bands are. For most k in Regime IV, attractors for orbits of k are extraordinarily complex. Banding discloses one aspect of these attractors, however: they are constrained as illustrated in the bifurcation diagram, and—as we have seen for other convergence issues—the constraints depend only on k.

Regime IV, which begins when $k=k_{crit}$, is called the 'chaotic' Regime. The term 'chaotic' may be misleading, since it suggests that orbits for all k in Regime IV are chaotic. This is very far from being the case. In fact, while it-

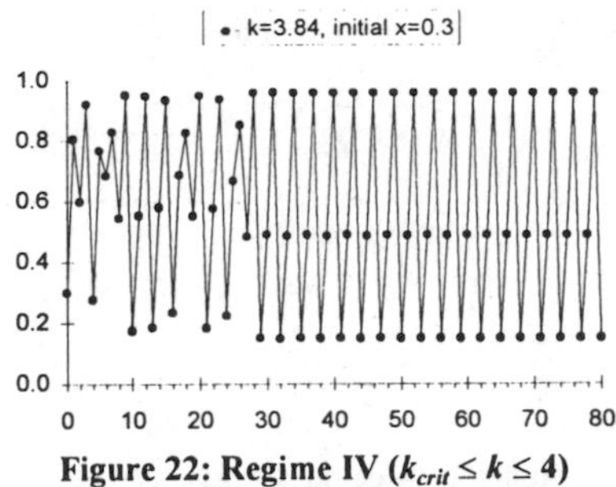

Figure 22: Regime IV ($k_{crit} \leq k \leq 4$)
Limit cycle, period 3

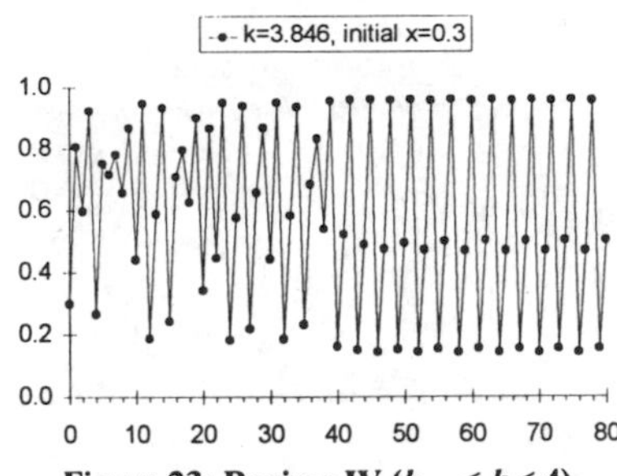

Figure 23: Regime IV ($k_{crit} \leq k \leq 4$)
Limit cycle, period 6

erated sequences *appear* to fluctuate entirely at random for most values of k in this Regime, and the limit cycles of attracting periodic points so obvious in the bifurcation cascade of Regime III *appear* to be absent, there are actually sub-regions in which the bifurcation structure of multiple attracting periodic points found in Regime III recurs.

For example, at $k \cong 3.82843$, stable oscillation among attracting periodic points of period 3 occurs for almost all x_0. As k increases slightly, the 3-fold structure bifurcates, with each periodic point giving birth to two attracting periodic points, and producing a 6-fold limit structure at $k \cong 3.8416$. This in turn splits into twelve-fold, twenty-four-fold, and forty-eight-fold structures with further slight increases of k. Feigenbaum's constant again describes the ratio of adjacent bifurcation sizes as they become smaller. The process continues indefinitely through a cascade of bifurcations, culminating at $k \cong 3.84945$ with chaotic orbits, much as the cascade of Regime III culminated at k_{crit}. Figure 22 and Figure 23 illustrate 3-fold and 6-fold limit cycles for values of k that lie between these bifurcations.

Other bifurcation cascades are visible. In Figure 24 it is possible to discern indications of smaller cascades beginning from limit cycles of period 18, 15, and 9. Each of these bifurcations is characterized by Feigenbaum's constant, and culminates in a value of k for which the logistic map is chaotic.

The main difference between the cascade of Regime III and those of Regime IV is that the logistic map has infinitely many repelling periodic points throughout the cascades of the chaotic regime, while only finitely many in Regime III. We can state the result in technical language using Sarkovskii's famous theorem.

Suppose that the natural numbers are ordered in the following way (using the symbol $\triangleright$ to mean 'precedes'):

$$3 \triangleright 5 \triangleright 7 \triangleright 9 \triangleright \ldots \triangleright 2 \cdot 3 \triangleright 2 \cdot 5 \triangleright \ldots \triangleright 2^2 \cdot 3 \triangleright 2^2 \cdot 5 \triangleright \ldots \triangleright 2^3 \cdot 3 \triangleright 2^3 \cdot 5 \triangleright \ldots \triangleright 2^3 \triangleright 2^2 \triangleright 2 \triangleright 1$$

Sarkovskii's theorem states that a continuous map with periodic points of period m will also have periodic points of period n for all n such that $m \triangleright n$. When the logistic map has periodic points of period 3, therefore—whether attracting or repelling makes no difference—it has periodic points of every period. The only way to have finitely many periodic points is for all periods to be powers of two, which is the situation in Regime III.

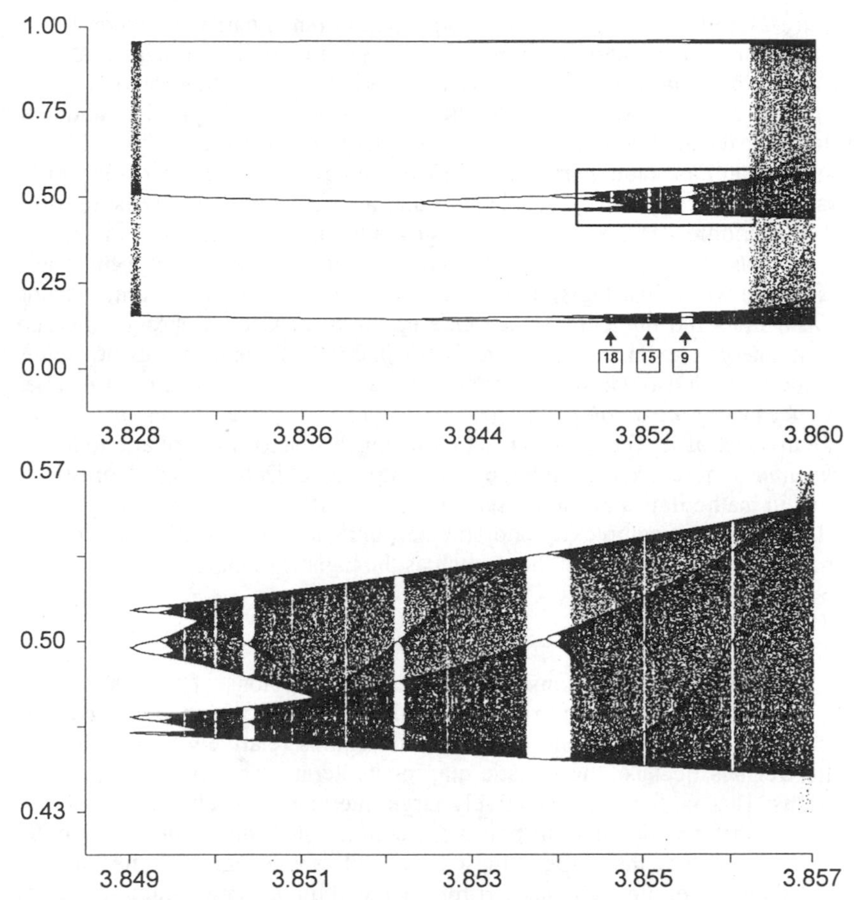

Figure 24: Bifurcation Diagram: Cascades in the Fractal Chaotic Regime
Plot of the limit points of orbits of the logistic map for k between 3.828 and 3.860, indicating the location of narrower 9-fold, 15-fold, and 18-fold bifurcation windows. The rectangle in the top diagram is blown up in the lower diagram. When inverted, this diagram is almost identical to the bifurcation diagram for Regime IV (see Figure 25).

The bifurcation cascades *visible* in Regime IV constitute only the tip of the structural iceberg. It is known that the set of k for which the logistic map has stable orbits with periodic attractors—we shall call this the set of *periodic k*—is open and dense in Regime IV. That is, this set is a (possibly infinite) union of open intervals, and its closure (the set plus its limit points) is the whole of Regime IV. For these values of k, the logistic map does not produce chaotic behavior even though it may have infinitely many repelling periodic points and its orbits may be so complex that there is no way visually to distinguish them from chaotic orbits. *Regime IV is made up of infinitely many bifurcation cascades.* Some bifurcation cascades develop in relatively wide windows of k, such as the one beginning with 3 periodic points illustrated in

Figure 24, but most begin and are completed in much narrower windows of k. In fact, for each n, there are approximately (and never more than) $2^n/2n$ windows in which the logistic map has stable periodic orbits of period n.

There is an extraordinarily difficult and technical question about the relative space in Regime IV taken up by periodic values of k (for which the logistic map has stable periodic orbits), and aperiodic values of k (for which it has no periodic behavior). It is known that the size (Lebesgue measure) of the set of aperiodic k is greater than zero—even though there can be no interval of such points, because any interval would contain at least part of a bifurcation cascade in which the logistic map exhibits periodic behavior. An apt analogy is with irrational numbers, which take up all the 'space' (Lebesgue measure) of an interval, even though there is no interval of them, since the rational numbers are dense. Unlike the irrational numbers, however, the set of aperiodic k cannot take up *all* of the space in Regime IV, since bifurcation cascades take up most of it. The question as to how much space the set of aperiodic k do consume is therefore a complex one, and it is the object of much sophisticated work in mathematics at the present time. The difficulty of this question is an indication of the complexity and amazing intricacy of chaotic dynamical systems—even the simplest of them, such as the logistic map.

2.12 *Excursus: Banding*

To understand banding, consider that the darker regions of vertical slices of the bifurcation diagram correspond to the values at which orbits have the greatest density. Orbits tend to visit these regions relatively more often than other regions because the logistic map or its iterates are *contracting* in these regions. That is, they map relatively large intervals into relatively small ones there, so that a wide range of points get concentrated into a small region. It is possible to deduce the shape of these bands with some degree of precision.

To venture into technical territory, we will do the small amount of calculus required for this deduction not merely because the calculations are simple, but also to show how a beautiful and complex aspect of the behavior of the quadratic map can be easy to describe.

The geometric indicator of a contracting region of a graph is a horizontal tangent (a zero derivative), which the logistic map has at its critical point, 0.5. Moreover, the flat regions of the graph of any iterate of the logistic map are directly caused by the flat region around the critical point. To see this, denote the logistic map by L and note that the flat regions of its nth iterate occur when $(L^{(n)})'(x)=0$. The chain rule implies that

$$(L^{(n)})'(x)=L'(L^{(n-1)}(x)) \cdot L'(L^{(n-2)}(x)) \cdot \ldots \cdot L'(L^{(2)}(x)) \cdot L'(L(x)) \cdot L'(x)$$

so we can deduce that those flat regions will occur when any of the factors in this product are zero, which in turn is when x is such that

$$L^{(n-1)}(x)=0.5,\ L^{(n-2)}(x)=0.5,\ \ldots,\ L^{(2)}(x)=0.5,\ L(x)=0.5, \text{ or } x=0.5.$$

These values of x are the first n-1 steps of the preimage of the logistic map's critical point, 0.5. The nth iterate has a crest or a dip that flattens out at

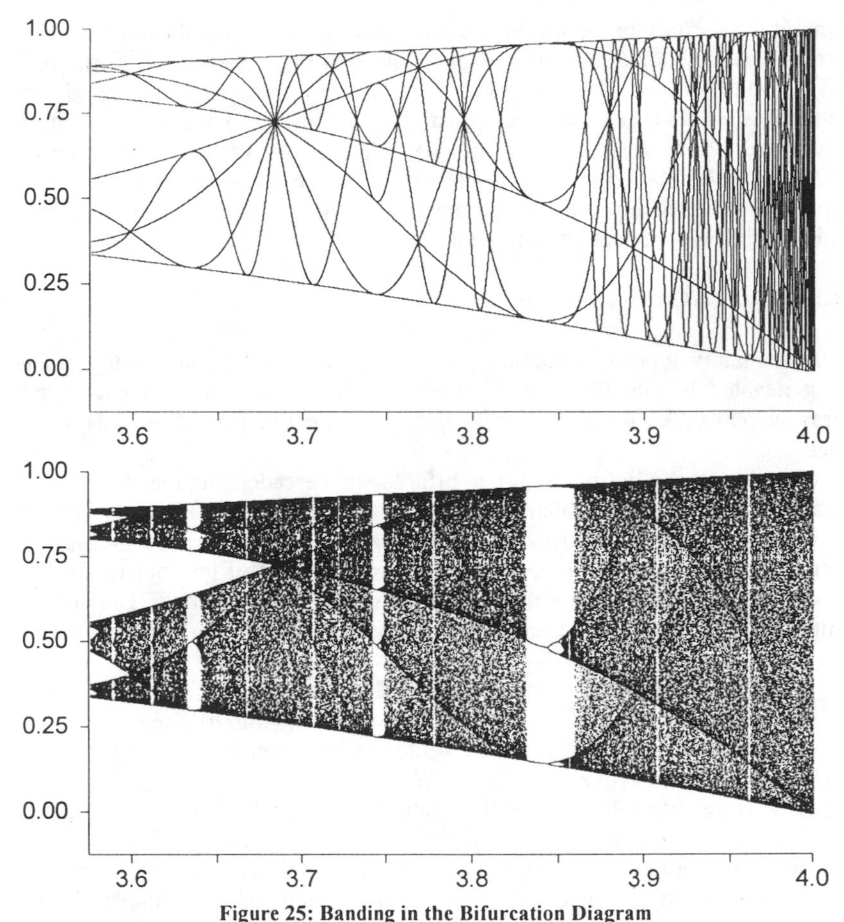

Figure 25: Banding in the Bifurcation Diagram
Plot of the first 10 iterates of the logistic map for k between k_{crit} and 4 above a plot of the bifurcation diagram for the same range of k.

each of these points, which may number as many as 2^n-1. The broadest crest or dip, and so the most powerful contraction, occurs when x=0.5, and at the value $L^{(n)}(0.5)$. This will be the major constraint contributed to orbits by the nth iterate of the logistic map. The second most powerful contracting region of the nth iterate occurs when $x=L^{(-1)}(0.5)$, at the value $L^{(n)}(L^{(-1)}(0.5))=L^{(n-1)}(0.5)$, but this is also the major constraint for the $(n-1)^{\text{th}}$ iterate.

From this it follows that the constraints evident in Regime IV of the bifurcation diagram trace (closely, but not precisely) the trajectories of iterates of the critical point through k-space. Figure 25 illustrates the first ten iterates of the critical point, and offers a comparison with the constraints in the bifurcation diagram. The flat region of the logistic map itself (the first iterate) is the strongest contractor, and will produce the highest orbit density. The resulting constraint (at the top) traces close to the graph of $k \rightarrow L_k(0.5)=k/4$. The flat regions of the second iterate produce lower orbit densities than those of the first iterate, though still higher than average. The resulting constraint is at the bottom of the bifurcation diagram, and traces close to the graph of $k \to L_k^{(2)}(0.5)=k\cdot(k/4)\cdot(1-k/4)$. The constraint running from top left to bottom right traces close to the graph of the third iterate of the critical point: $k \to L_k^{(3)}(0.5)=k\cdot[k\cdot(k/4)\cdot(1-k/4)][1-k\cdot(k/4)\cdot(1-k/4)]$. Successive iterates will be responsible for smaller increases in the average orbit density, and lighter constraints, all of which depend only on k.

2.13 *Excursus: Chaos in Regime IV*

There are many types of chaotic dynamics in Regime IV, and much work is being devoted to classifying and understanding the dynamics of each type. Here, we will make some inevitably rather technical comments on just a few types.

First, at the culmination of a bifurcation cascade, the logistic map is chaotic on a Cantor set, which is an uncountable closed set of isolated points of measure zero. The construction of a different Cantor set is described in the discussion of Regime V (see Figure 26). The dynamics of the logistic map at the culmination of bifurcation cascades is described as follows. Denote the Cantor set at k_{crit} by C. Then $C=P \cup Q \cup R$, where P, Q, and R are disjoint and:

P={repelling periodic points of period 2^n for some n}
Q={preimage of periodic points of period 2^n for some n}
R={closure of the forward orbit of the critical point, 0.5}

The set P is infinite and dense in the Cantor set, C, and its orbits are periodic. P and Q together constitute the finite convergence set for k_{crit}. The set Q is countably infinite in size, and orbits of points within Q are eventually periodic, for they are forced sooner or later into P, where they settle permanently into some unstable limit cycle. All other orbits—there are uncountably many of them—are pushed away from the dense repelling periodic points in P, and toward the attractor for the logistic map at k_{crit} (the set R). This is an uncountably infinite set, dense in C, called the Feigenbaum attractor, which can be constructed by closing (i.e. adding in the limit points of) the forward orbit of the critical point, 0.5. The logistic map is chaotic on C with this most peculiar attractor, the Feigenbaum attractor. This is a good example of the fact that attractors do not necessarily—and in the case of chaotic orbits do not—consist only of finitely many attracting fixed and periodic points. The dynamics of the logistic map are more or less the same for the other values of k at which a bifurcation cascade culminates—called Feigenbaum-like values.

The Feigenbaum-like values of k are countable, and so have measure zero in the set of aperiodic k, which has positive measure. Thus, there is a lot more chaos in Regime IV besides that occurring at the culmination of bifurcation cascades, even though the set of periodic k is dense in Regime IV. A second type of chaotic dynamics occurs when k is in a countable set of so-called Misiurewicz points, for which orbits of the critical point land on one of the repelling periodic points. In this case, the logistic map is chaotic not on a Cantor set, but on an interval. To be precise, it is chaotic for almost all x in the interval or union of intervals that is the closure of the set of periodic points. A Misiurewicz point occurs at k=4, for example, because the critical point is iterated to the fixed point at 0. This confirms what we know otherwise, namely, that chaos occurs over the entire unit interval for k=4.

The Misiurewicz points and the Feigenbaum-like points mark out extremes in relation to which the many other kinds of chaotic dynamics of the logistic map in Regime IV are something like middle cases. These middle cases make up all of the non-zero measure of the set of aperiodic k.

2.14 *Regime V: More Chaos* ($k \geq 4$)

Chaotic orbits on a Cantor subset of the unit interval.

Regime V is particularly complex, and can only be described in somewhat technical terms. We have seen that sequences generated by iterating the logistic map rapidly explode to infinity if x_0 is not within the unit interval. If the tuning constant, k, is increased above 4, then the feedback situation that has characterized the logistic map for $0 \leq k \leq 4$ breaks down for at least some x_0. For example, the logistic equation maps x_0=0.5 into k/4, which is greater than 1 for k>4. When k is greater than 4, therefore, the logistic equation will map an open interval around 0.5—and the preimage of this open interval—outside the unit interval, and the possibility of convergent, periodic, or chaotic orbits no longer exists. The size of this interval, I_k, will depend on k.

The first question about all this is, for a given k, whether the open interval I_k and its preimage exhaust the unit interval. It turns out that a special closed set of isolated points—a Cantor set, denoted Λ_k—remains after I_k and its preimage is removed from [0,1]. The set Λ_k changes shape as k increases above 4, because I_k grows larger as k does, but Λ_k is always a Cantor set.

To see the way Λ_k is formed from the preimage of I_k, suppose k is given and denote I_k as $A1$. Then denote by $A2$ the set of points mapped by the logistic equation into the open interval, $A1$. $A2$ is a pair of open intervals symmetric in that part of the unit interval left after $A1$ has been removed. Then denote by $A3$ the points mapped into $A2$. This is a set of four open intervals in the four parts of the unit interval left after the removal of $A1$ and $A2$. Continue this process, so that $A[n]$ is a set of 2^{n-1} open intervals lying within each of the 2^{n-1} parts of the unit interval left after $A1$,$A2$,$A3$,...,A[n-1] have been removed. In this way, all of the space (Lebesgue measure) of the unit interval is eventually removed, and Λ_k, an uncountably large set of measure zero, remains. Λ_k is a perfect set of isolated points, because it contains all of its limit points. It is also a fractal, since any part of it is approximately a scaled-down version of the whole. Figure 26 illustrates its construction.

The second question is about the behavior of the logistic map on the set Λ_k. The answer is that the logistic map is chaotic on Λ_k for all k>4. There are no bifurcation cascades in this Regime, as in Regime III and Regime IV. This is because the flat part of the graph of the logistic equation is the geometric feature necessary for any stable behavior, including stable limit cycles of peri-

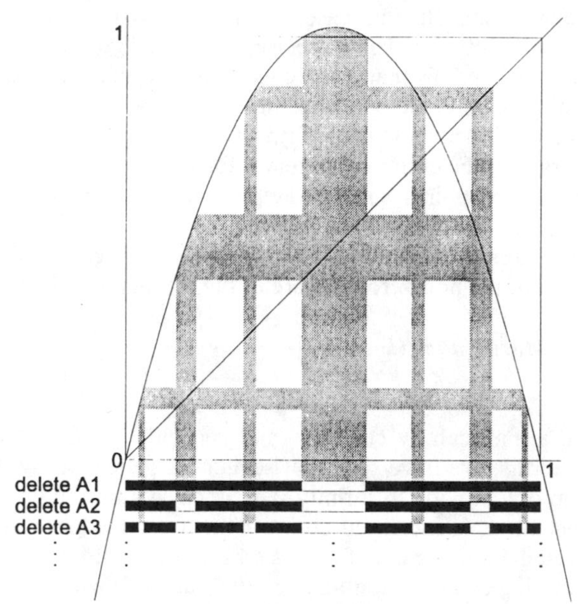

Figure 26: Cantor Set for Regime V ($k > 4$)
Formation of the set Λ_k for k=4.1, using the cobweb method.

odic points, but all points heading into the flat region of the logistic equation are eventually mapped outside of [0,1] when k>4.

Chaos in Regime V cannot be illustrated with a computer, because computing devices are not precise enough either for specification of an initial value in Λ_k, or for avoiding rounding errors in calculation. Either limitation causes iterations to be pushed out of the unit interval, and sends the iterated sequence accelerating toward infinity.

There is an interesting comparison to be made between chaos on Λ_k for k>4, and chaos for Feigenbaum-like values of k, such as k_{crit}, both of which occur on a Cantor set. Orbits of the logistic map for k>4 have no attractor smaller than all of Λ_k; all orbits in Λ_k range throughout Λ_k. Orbits of the logistic map for k=k_{crit}, by contrast, do have an attractor; at k_{crit} it is the Feigenbaum attractor, a strict subset of the Cantor set. Moreover, the logistic map is never chaotic (whether on an interval, a Cantor set, or something else) for an *interval* of values of k in Regime IV, whereas in Regime V chaos occurs for all k>4, even though only on a Cantor set.

Early in the investigation of chaos, it was discovered that the constant breaking up of chaotic dynamics by other sorts of dynamics is a quirk of the one-dimensional case. In higher dimensions (even in the complex plane, in fact) chaos frequently occurs in entire regions and for intervals of 'tuning' constants. The virtue of chaos in higher dimensions is that it is more conducive to research using mathematical modeling and computers, because perturbations of the chaotic map tend to have the same chaotic dynamics. Attractors could never be found for chaos in natural systems modeled with one-dimensional maps, because perturbations of these maps in computer simulations or in the specification of initial conditions totally alter the dynamics of the model. In higher dimensional systems, however, chaotic dynamics frequently vary continuously, and perturbations of these sorts do not upset dynamics.

Chaotic orbits are no more predictable because of this, of course, but modeling and the identification of attractors underlying chaotic orbits are made possible. For example, an attractor could not have been found for the dripping faucet if time-series data had been analyzed only one-dimensionally. The attractor for a higher dimensional model, however, agreed tolerably well with that suggested by apparently chaotic orbits of the natural system. The stability of chaos in higher dimensional systems is the key to this type of analysis.

2.15 *Excursus: Chaos*

What, precisely, is chaos? We began with a phenomenological definition that likened chaos to randomness by means of the unpredictability of chaotic orbits. This is problematic, however, for reasons that we shall discuss in the next section. It also fails to come to grips with some of the most important characteristics of chaos. Geometrically speaking, the key to chaos in mathematics is the stretch-and-fold character of certain maps. They expand rather than contract most of their domain (stretch), and then map the stretched parts back into their domain (fold), ready for more stretching and folding. But this is very far from being a precise definition of chaos, so the question remains: how should chaos be characterized?

There are many definitions of chaos, some stronger than others. The most commonly used characterization refers to the rate at which a typical set of nearby orbits of a map diverge from one another: for maps with chaotic dynamics, this divergence is exponential—that is, similar to e^{an} as $n \to \infty$ for some $a>0$. In the case of the logistic map, because the difference between chaotic orbits never exceeds 1, this exponential condition must be interpreted in terms of the rate at which the difference between two orbits becomes statistically indistinguishable from the two orbits themselves (see the discussion of Figure 27, below).

Exponential deviation of orbits is not, however, a sufficient condition for chaos. The exponential map itself, for instance, amplifies orbit errors by a factor of e, and so has this characteristic, even though it is not chaotic. Other properties must also be important, and are essential in designing a definition of chaos that includes of all chaotic dynamical systems of interest. To appreciate what these properties might be, recall that the attractors of chaotic orbits have constraints and that they occur (roughly speaking) where repelling periodic points do not. Thus, there is something *organized* about chaotic orbits. But the apparent randomness of chaotic orbits also suggests a kind of *disorderly* character, too. And the way nearby orbits diverge suggests a kind of steady, systematic breakdown of all topological features of a set under iteration of a chaotic map, in much the same way as increasing *entropy* in closed thermodynamic processes indicates their increasing disorder.

There is as yet no generally accepted definition covering all instances of what mathematicians would like to call chaos. However, for the discrete case of iterated maps, the three characteristics just mentioned—orderliness, disorderliness, and topological entropy—correspond to three properties that are jointly sufficient to characterize chaos. These properties—mixing, density of periodic points, and sensitivity—are defined as follows. A function, f, is chaotic on a set, S, if it exhibits:

1. *Mixing*, a property characterizing the *disorderliness* of the dynamical system. We say that the function f *mixes* S by analogy with the way a pinch of spices will be spread throughout a lump of dough if the stretch-and-fold operation of kneading is executed properly.

   Technically, if U and V are open subsets of S, then f eventually iterates some points from U into V. For maps on compact intervals, such as the logistic map on [0,1], this is equivalent to there being an orbit of f that is dense in S.

2. *Density of periodic points*, a property characterizing the *orderliness* of the dynamical system. Think of the way sour cream curdles in hot coffee: the cream moves in all directions away from points throughout the coffee cup, which are like densely distributed repelling periodic points, in order to clump at certain other points, which are akin to points in a chaotic attractor.

   Technically, each arbitrarily small open set in S contains periodic points for the function f on S. Of course, all of these periodic points are repelling, for otherwise the attractor of the orbits of f on S would be a limit cycle of periodic points and not chaotic.

3. *Sensitive dependence on initial conditions*, a property characterizing the *topological entropy* of the dynamical system. Think of the butterfly effect, which describes the way an intricately connected system allows tiny influences to have large effects.

   Technically, given any point in S, there is an arbitrarily close point in S such that f iterates the two points arbitrarily far apart. This is a weaker condition than exponential divergence of orbits, since it only demands separation by an arbitrary distance. The linear map $f(x)=ax$ for $a>1$ satisfies this condition, for example, whereas this map does not meet the exponential divergence condition. In the case of iterated maps on compact sets, however, the exponential condition is stronger than needed, and sensitive dependence does the job.

This definition of chaos, with its three simple properties, captures the examples of interest in the vast family of discrete (iterated) chaotic maps. However, in the case of one dimensional, discrete maps on compact sets, such as the logistic map on [0,1], these conditions are *not* independent. *In this special case, mixing implies density of periodic points, which in turn implies sensitivity to initial conditions.* In higher dimensional discrete systems, however, these implications do not necessarily hold, and all three conditions are needed.

From all of this we learn that sensitivity, and even exponential divergence, do not by themselves indicate chaos. In the discrete case, it is only when the properties of mixing and dense periodic points give rise to sensitivity that the kind of unpredictability associated with chaos appears. More vaguely and generally, chaos requires the simultaneous presence of a measure of organization, disorganization, and topological entropy. This is achieved by many maps that stretch and fold their domains in a feedback mechanism.

2.16 *Excursus: Unpredictability*

The behavior of chaotic systems is in some sense both predictable and unpredictable. With regard to predictability, for example, pick k=4, so that the logistic map is chaotic. Then pick any number, x_0, so that the resulting iterated sequence is chaotic—that is, x_0 is in the infinite convergence set for k=4 (IC_4). What are the succeeding numbers in the sequence, the numbers to which the logistic map bounces after x_0? Well, that's easy; just calculate: $x_1=4x_0(1-x_0)$. Then calculate x_2 and x_3 and all the rest in the same way. *In this sense, the behavior of chaotic sequences is completely predictable.*

But how should the unpredictability of chaotic dynamical systems be characterized? Consider that neither k nor x_0 can be specified precisely in almost all instances. Indeed, in general, a number can be expressed precisely only if it has a finite decimal representation, and such numbers take up no 'space' (Lebesgue measure). Even if x_0 and k could be specified precisely, so that x_n would have a finite decimal representation for every n, the number of decimal places in x_n would increase exponentially with iteration. This information cannot be stored or calculated with in practice without some kind of rounding errors being introduced.

Such errors are not important in Regimes I-III of the logistic map, and in most bifurcation windows of Regime IV. To see this, consider two iterated sequences, beginning from x_0 and y_0. If y_0 is a *close* approximation to x_0, the sequence iterated from y_0 will eventually bounce between the same 'limits' that the sequence beginning from x_0 does, albeit perhaps in a different order—and if y_0 is *sufficiently close* to x_0, the iterated sequence will hit a limit cycle *in phase* with the sequence generated from x_0.

At the chaotic values of k such as k=4, however, details are everything. Every last decimal place of information is important as the iterations continue. If y_0 approximates x_0 to 6 decimal places, we might achieve tolerable accuracy for a certain number of iterations, but the sequence generated from y_0 will bear no relation to the sequence generated from x_0 beyond that point. Figure 27 illustrates this by plotting the first 50 values of the sequences generated from x_0=0.3 and y_0=0.300001, together with the difference between the two sequences. After 18 iterations, the sequence representing the difference between the orbits is qualitatively indistinguishable from either of the sequences themselves, showing that all information about initial states has been lost.

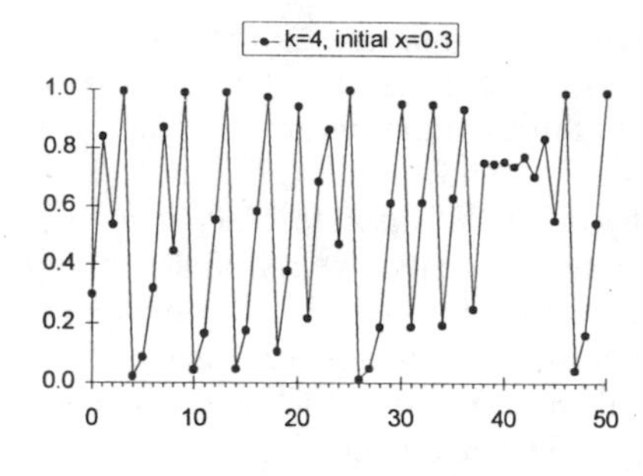

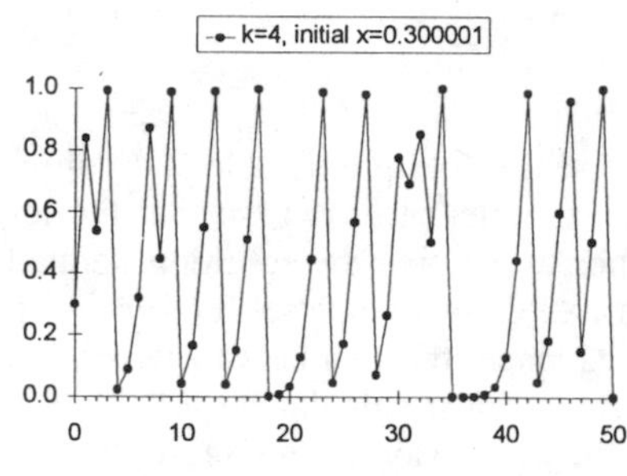

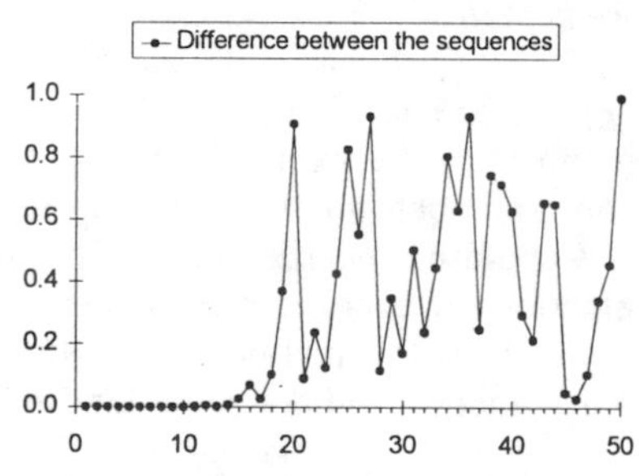

Figure 27: Eventual Unpredictability
Two iterated sequences and the difference between them (k=4)

In sum, there is no way to predict the value of successive iterations for a chaotic value of k except actually to calculate them, and this depends on our having complete knowledge of the initial conditions. Thus, the chaotic behavior of the logistic map is unpredictable in the sense that the map's sensitivity at chaotic values of k, and the impossibility of infinitely precise specification of most starting points, blocks indefinite prediction of future iterations. We shall call this *eventual unpredictability*.

It is vital to know that 'eventual' unpredictability entails 'temporary' predictability. In fact, because the chaotic orbits of the logistic map (for k=4) are generated by an equation, it is even possible to deduce the number of iterations before predictions will fail to match the orbit, as a function of the precision of initial conditions. To see this, note that it can be shown that the amplification of small errors proceeds at the rate of roughly a factor of two for each iteration of the logistic map. Now, in Figure 27 the difference between the two orbits is 0.000001. If the error doubles with every iteration, a simple calculation shows that it should surpass 0.5 after 19 iterations. From our computer simulation, we can see that there is in fact an error of about 0.4 after 19 iterations, and about 0.9 after 20 iterations, confirming the error estimate. This error estimate implies that, if the prediction were to be accurate (within a small positive number, ε) after n iterations, then the error in specification of initial conditions would have to be less than $\varepsilon 2^{-n}$—and the computing machine would have to be capable of the corresponding order of precision (roughly $n+|log_2\varepsilon|$ bits) to avoid increasing the error.

It is the *predictability in principle* afforded by the equation for the logistic map that makes possible *temporary predictability*, the concommitant of *eventual unpredictability*. These correlated features of chaotic maps allow us to say that chaotic randomness, even with its eventual unpredictability, is nevertheless predictable in principle, and temporarily (though not indefinitely) predictable in practice. This form of predictability in practice is weak, to be sure, but it is important to know that eventual unpredictability entails it.

3 *Reflections on Randomness, Determinism, and Chaos in Nature*

There are many questions affected by chaos theory, including those concerning complex systems, evolutionary novelty, and reductionism in scientific method. We will refrain from extending our reflections into these areas, however. Our aim in this section is the limited one of connecting the mathematical discussion of chaos to both the phenomenological characteristics of chaos (the way chaos *seems* to us), and to philosophical questions surrounding metaphysical determinism. To that end we will make some more abstract remarks on the themes of randomness, determinism, and chaos in nature.

By way of indicating the direction of the argument, we shall first give chaotic randomness a precise definition in terms of eventual unpredictability, drawing on the discussion in the previous section. We shall then show that chaotic randomness so understood is a *tertium quid*; it is mathematically distinguishable from both strict randomness and the complete absence of randomness. Finally, by means of a discussion of chaos in nature, we shall argue that chaotic randomness strengthens the case for metaphysical determinism while simultaneously introducing a fundamental limitation on how complete this case can be made. The concluding section of this paper draws out the theological implications of this result.

3.1 *Randomness and Determinism*

It is often said that chaotic orbits look *random*, but the meaning of randomness is murky given chaos. In common usage, to declare that a sequence of numbers is random means either that it has no discernible pattern, or that it is unpredictable. As we have seen, however, chaotic orbits are often quite structured, with banding constraints and a close relationship to repelling periodic points. Calling chaotic orbits random in the sense of having no discernible pattern thus seems inappropriate. If there is to be a connection with intuitive ideas of randomness, therefore, it will be through the idea of unpredictability.

At this level, we are justified in saying that chaotic orbits are random in the limited sense of *eventual unpredictability*. However, randomness usually means not eventual unpredictability but *total unpredictability*—the absence of any equation for a sequence that consistently permits, given one number, even an approximate calculation of the next. We call this kind of randomness *strict* randomness. If chaotic randomness means 'eventual unpredictability,' then strict randomness means 'total unpredictability.' Chaotic orbits of the logistic map are *not* random in the strict sense since, obviously, they are produced by an equation (namely, the logistic equation). On the other hand, the sequence recording the count of each kind of item on the shelves of an indoor market at a particular moment, arranged alphabetically, *is* random in the strict sense.

These examples of randomness would make sense mathematically whether or not Laplace was correct that the natural world was a mechanistic, unfree, unbroken web of causal connections, and so entirely predictable to the great divine mathematician whose calculations are not limited by practical considerations. Some other uses of 'randomness,' however, seem to presuppose that Laplacian determinism is false. For example, consider radioactive decay. The average number of alpha particles emitted by a uranium sample is predictable; it decreases exponentially with time. The exact number of alpha decays each second, however, is random. Randomness here always means at least that no known equation determines this exact number, but it sometimes carries additional suggestions of metaphysical indeterminism. The connection to metaphysics is hazardous, however, because it depends upon choosing between conflicting interpretations of quantum physics—and both deterministic interpretations (such as non-local hidden variables) and indeterministic interpretations are presently defended.

Other examples of the word 'randomness' being used so as to presuppose metaphysical indeterminism follow the same pattern: *speaking of randomness in mathematics does not by itself presuppose metaphysical indeterminism*; instead it must be linked to metaphysics by an additional layer of interpretation. Thus, we must find a definition of randomness in terms of unpredictability alone, considering it neutral toward metaphysical questions about determinism.

With this, then, we are justified in speaking of two metaphysically neutral kinds of randomness (chaotic and strict) defined in terms of two varieties of unpredictability (eventual and total), and correlated with several types of predictability. ***Chaotic randomness is neither absence of randomness nor srict randomness but a tertium quid.*** This is one way in which the phenomenon of chaos in mathematics is remarkable. *Whereas it might once have been supposed that predictability and unpredictability were directly opposed, chaos theory opens up a nether-world in which this supposedly sharp distinction is blurred to the extent that a particular kind of unpredictability (eventual) occurs in the context of predictability-in-principle and even what we have called temporary predictability-in-practice. We are thus justified in speaking of a more or less unforeseen albeit weak type of randomness, namely chaotic randomness.*

These types of randomness exist in mathematics and apparently also in nature, and they have been defined in metaphysically neutral fashion with respect to the question of determinism, so we may consider this central metaphysical question separately. Thus, we are led to the following table, noting that 'determinism' means Laplacian determinism, and many opposed views are collected under 'indeterminism.'

types of randomness	types of unpredictability	types of predictability		metaphysical scenarios
		in-principle	in-practice	
none	none	yes	yes	*Is determinism or indeterminism the best hypothesis?*
chaotic	eventual	yes	temporary	
strict	total	no	no	

Types of mathematical randomness can be defined precisely in terms of types of unpredictability and correlated with types of predictability. However, there are no *direct* logical connections between the these types of randomness and the thesis of metaphysical determinism.

Now we may pose the basic metaphysical question: which metaphysical scenario—determinism or indeterminism—offers the best explanation of the fact that non-randomness and two kinds of randomness appear in mathematics and apparently also in nature? Even with the distinctions introduced up until now, this question is poorly framed, because it begs the question of the relation between chaos in mathematics and chaos in nature. This question must be answered before it is possible to raise the deeper question of the relation between chaos and an ultimate account of nature such as metaphysical determinism.

For example, it is superficially possible to say—indeed, some writers appear to have said—that, since the mathematical equations for chaotic systems are deterministic, nature *must* be metaphysically deterministic if chaos actually occurs in nature. This, however, is to mistake a metaphorical use of 'deterministic' in mathematics for a literal one; the precision of a mathematical equation does not imply that the use of an equation in a modeling physical systems only makes sense if nature is metaphysically determined. Thus, we must consider questions surrounding chaos in nature in order to press the metaphysical question any further.

Idealizing natural systems when using mathematics to model them has proved to be an extremely useful strategy for centuries, sponsoring profound understandings of nature. When the aim is to match natural and mathematical dynamical systems in detail, however, the complexity of nature and the exquisite sensitivity of chaos profoundly complicate modeling, and an entirely new range of questions appears. One group of these questions concerns the *meaning* of such modeling: What is the status of mathematical models as explanations of nature's dynamics? Does chaos actually occur in nature at either the macro- or quantum levels? Another group of questions surrounds the *difficulties* of using mathematical dynamical systems to model natural dynamical systems, such as specifying initial conditions for natural systems with great precision, and keeping exquisitely sensitive mathematical dynamical systems in step with enormously complex natural systems over time. We will consider these two groups of questions in what follows, and then reopen the issue of chaos and metaphysics, hopefully better prepared to address it.

3.2 *Chaos in Nature: The Meaning of Modeling*

Modeling nature with mathematics always involves making simplifying assumptions. In a good model, these simplifying assumptions render the model mathematically manageable, while still useful for predicting the behavior of the natural system in question. When a model proves serviceable in this way, the hope is that it illumines not just the future behavior of the system but also the reasons why the natural system works the way it does. To understand the complexities surrounding the status of chaotic models as explanations, a simpler example will be useful.

When we use Newtonian mechanics to model the causal relationship between gravity and falling objects the mathematics of the model is simplest when we ignore all atmospheric effects and the gravitational effects of everything except the earth. The model yields accurate predictions in spite of these approximations because we stay with billiard and cannon balls rather than feathers and polystyrene packaging, thus minimizing the impact of the atmosphere. The model also produces a powerful explanation. It supports an hypothesis about what is in nature (forces and massive objects), says something about how these constituents operate (with an equation for gravitational force), and links the theory to observations with fair precision. When the problem is modeling complex natural dynamical systems, however, the form of explanation is quite different.

Consider the moth colony population case, for example. No attempt is made to model this system with forces and objects; the model would be absurdly complex and utterly unenlightening. A description in the form of a statistical catalogue of moth biographies seems to be the only option: so many moths were born this generation, a percentage of these died from disease, a percentage starved, a percentage reproduced with an average of such and such offspring, and so on.

While such a description is interesting, and enables computer simulations, it leaves the apparent patterns of population in successive generations (decline to extinction, growth to equilibrium, stable oscillation, apparently random fluctuation) something of a mystery. When it is recalled that such patterns are characteristics of the system as a whole and not of individual moths, however, it makes sense to use as a model a mathematical dynamical system that ignores details of moth life and only attends to system-wide properties such as the total population, and generationally invariant environmental conditions. We saw that the logistic map suggests itself as a serviceable model at this overarching system level, and that the model exhibits patterns of population change (extinction, equilibrium, oscillation, random fluctuation) similar to those of the moth system. But in what sense does the mathematical dynamical system *explain* the natural dynamical system?

The form of explanation appears to be similar to that of statistical mechanical models for well-behaved gases. System properties such as pressure, volume, and temperature follow statistically precise laws that can be deduced by averaging the behavior of the numerous components of the system. Pressure, for instance, is the same throughout the system: higher pressure in one place is quickly diffused on average, rather than being intensified, so it can be spoken of as a reliable system characteristic. Thus, statistical mechanics supports the amazingly simple and eminently testable hypotheses that pressure is proportional to temperature and inversely proportional to volume ($P \propto T/V$).

The use of the logistic map to model moth colonies is more complex than statistical models of gases, both because the interaction of system components is far more complex in the moth case than in the gas case, and because system-wide features of moth colonies are more difficult to quantify than pressure, temperature and volume. Nevertheless, the same form of explanation seems to be at work: the moths with their different biographies are assumed to have behavior that averages out in large collectives so as to justify speaking about reliable system variables—in this case, population. Then changes in population can sensibly be thought to be modeled by the logistic equation, even as $P \propto T/V$

for most gases. The same form of explanation applies to other instances in which mathematical dynamical systems are used to model natural dynamical systems; the bibliography lists several examples.

There is another respect in which applying the logistic map to a moth colony is more complex than applying statistical mechanics to a well-behaved gas, and this consideration affects the validity of explanations advanced on the basis of the model. In the chaotic regime of the logistic map, the premise for the statistical form of explanation—namely, that behavior of system components averages out to make reliable system variables—breaks down because of the exquisite sensitivity of the mathematical model. Quite apart from the practical problems surrounding whether the natural system can be modeled, therefore, there is a very serious reason to think that chaos in an otherwise useful mathematical model does not *explain* very much about the natural system it models. It may well explain after the fashion of statistical mechanics when the model accurately predicts fairly straightforward behavior, such as equilibrium and simple periodicity, but the logic of this kind of explanation does not extend without further ado to chaotic or highly complicated periodic behavior.

We are driven by this argument directly into asking an important question: does chaos occur in nature, or only in mathematics? Scientists are led to wonder if a mathematical model in the form of a chaotic dynamical system might apply to a natural system at the macro level when they notice 'regimes' with characteristic behavior in the natural system, and especially bifurcation cascades. On many occasions such clues have led to effective modeling and prediction using chaotic dynamical systems. In what sense is this kind of successful modeling evidence for chaos in nature?

On the one hand, testing such models in detail is impossible unless the system exhibits equilibriated or simple periodic behavior. Add to this the breakdown of the logic of the statistical form of explanation when the model is displaying exquisite sensitivity, and we have strong arguments that models based on chaotic dynamical systems have limited usefulness and questionable explanatory power at precisely those times when we would most like to rely on them, namely, when we are trying to model apparent randomness in natural processes using mathematical chaos.

On the other hand, the fact that a model tracks a natural system through several regimes and types of behavior suggests that the model might also apply to the apparent randomness of the same natural system. Moreover, while predictive testing is not feasible in any chaotic dynamical system, indirect evidence of chaos in nature can still be secured. As mentioned earlier, chaos appears stably in some higher dimensional mathematical models, in the sense that the attractor for one orbit is nearly identical to the attractor for any slightly perturbed orbit. Eventual unpredictability still applies, so testing predictions of the model is not made possible by this form of stability. However, such a model does sometimes make possible the identification of attractors for apparently chaotic orbits of the natural system. Thus, geometric or statistical confirmation of mathematical models for chaotic dynamical systems is still a possibility when direct confirmation of orbit predictions is not. This kind of confirmation has been used in a number of experiments, including the chaotic regime of the dripping faucet, but it remains somewhat indirect because it has no recourse to testing orbit predictions; the model could be quite wrong in detail, therefore, and we might never know.

In spite of the problems with the status of explanation in models based on chaotic dynamical systems, and in spite of the impossibility of testing their orbit predictions in detail, there is strong evidence—not unequivocal, to be sure—for chaos in nature. Of course, the most likely state of affairs is that the global features ('tuning') of natural dynamical systems are in continual subtle migration, so that an ordinary chaotic orbit that could be sensibly compared with a model's chaotic orbit would never be encountered. *Furthermore, three kinds of behavior—chaos, sufficiently complicated periodicity, and what we have called strict randomness—can rarely be phenomenologically distinguished from each other, even though they are mathematically quite distinct alternatives.* Consequently, a natural system could move among all three with subtle shifts of its tuning, or under the influence of other factors unknown to us, and there would be no way for us to tell in most cases which kind of behavior was occurring. Indeed, this constant subtle migration, and the washing out of chaotic orbits that it entails, may be one of the factors accounting for the omnipresence of equilibrium and other forms of stability in nature, despite the ubiquity of chaos in mathematics.

The question of the occurrence of chaos in nature is more difficult at the quantum level. The Schrödinger equation, for example, is linear, and so cannot be a chaotic map, which is one of several reasons to think that quantum chaos is impossible. This is a superficial conclusion, however, because quantum systems are only governed *indirectly* by this equation. The Schrödinger equation describes the time-development of a mathematical function, the wave function, whose absolute value indicates the relative probability of finding the physical system in a specific state. The extent, if any, to which the wave function refers to the physical system as such is one of several contentious issues in the interpretation of quantum physics. To date, no single such interpretation is universally agreed upon. Moreover, the experimental search for evidence of quantum chaos is only in its infancy. The basis at the quantum level for chaos in nature at the macro level remains a research challenge.

3.3 *Chaos in Nature: The Difficulty of Testing Models*

We now turn to the intriguing factors complicating the modeling of natural dynamical systems with chaotic dynamical systems from mathematics. Suppose that there is a chaotic dynamical system in nature, and that this system has high-level features whose varying moves the system from one regime to another, changing its characteristic behavior. Suppose further that we are trying to assess the virtues of a mathematical model that has already proved itself serviceable for predicting equilibrium and simple periodic behavior of the system, by virtue of one or more tuning variables that correspond to the tuning factors in the natural system. There are several problems facing any attempt to determine the predictive adequacy of this mathematical model when it is tuned to produce complicated periodic or chaotic behavior.

Firstly, modeling the tuning of the system with the tuning constant is difficult, even when the tuning of the system can be quantified with reasonable precision, as it can in some well-controlled laboratory experiments. This is important because testing predictions of the model only makes sense if we are sure that the tuning of system and model match. But the model's orbits vary dramatically with even a miniscule variation of its tuning constant in chaotic regimes. Of course, outside of the discrete, one dimensional case, chaos can occur stably, but this only applies to the fundamental dynamics of the system; the eventual unpredictability of chaotic orbits still interferes with the testing of the model's predictions, and the tuning of the model plays into this.

Secondly, supposing (contrary to fact) that it were possible to be sure that the model and the natural system were identically tuned, the matching of initial conditions in system and model is similarly complicated. Practical limitations include heat noise and the openness of all systems to minuscule, ineradicable effects, even influences as extraordinarily minute as the gravitational attraction of an electron in a neighboring galaxy. The ultimate limitation is obviously quantum uncertainty.

Thirdly, supposing (again, contrary to fact) that it were possible to match tuning and initial conditions in both natural system and model, we would still be faced with the problems surrounding calculation that apply in the mathematics of chaos, and have already been discussed. Moreover, because exponentially greater precision is needed to express successive iterations, exponentially greater time would be required to perform the calculations. For example, in the logistic equation, with errors increasing by a factor of 2 each iteration, calculation will eventually be slower than the evolution of the system, since twice as much precision in initial conditions would be required for the same precision in each successive iteration.

The literature on chaos theory has tended to discuss these impediments to testing chaotic models against natural systems under different terms, speaking of *algorithmic complexity*, *computational complexity*, the *continuum problem*, and the *butterfly effect*. Algorithmic complexity means that there is typically no analytic solution by which the development of a mathematical chaotic system can be expressed directly in terms of initial conditions, so that the only way to track its development is painstakingly to calculate each step; there are no shortcuts. Computational complexity refers to the exponentially increasing time needed to perform calculations of chaotic orbits, so that a natural system eventually develops faster than the predictions of the model can be calculated. The continuum problem refers to the impossibility of representing most numbers (and so most initial conditions) with the finite decimal representations needed to express them precisely in computing machines. The butterfly effect refers to the practical problems flowing from sensitive dependence of chaotic orbits on initial conditions, including the openness of natural chaotic systems to ineradicable, miniscule effects of many kinds.

We have presented these customary considerations slightly differently so as to reflect more adequately the fact that some factors interfering with testing of chaotic models are purely mathematical, while others derive from nature. However they are classified, these reflections establish that there are insurmountable problems facing any attempt to evaluate specific predictions derived from the behavior of orbits in a chaotic dynamical model. When these problems are added to those connected with the meaning of modeling discussed earlier, we can see that modeling apparent randomness in a natural system with the aid of a chaotic dynamical system from mathematics is extraordinarily difficult. *The epistemic and explanatory limitations at the root of this difficulty are important considerations in weighing evidence for and against metaphysical determinism, because they place a fundamental, new limit on the extent to which it can be known that a chaotic dynamical system in mathematics, with its 'deterministic' equations, provides the right explanation for 'randomness' in a natural system.*

3.4 *Chaos and Metaphysical Determinism*

Now we are in a more appropriate position to pose the question suggested by the table defining types of randomness in terms of types of predictability: Which scenario—metaphysical determinism or its opposite—offers the best explanation of chaos? The intervening discussion has made clear the difficulties attending the movement from chaos in nature to a metaphysically deterministic explanation of apparent randomness in nature, and this has an intriguing influence on the question.

We can say without hesitation that chaos in nature gives no evidence of any metaphysical openness in nature. The fact that a natural dynamical system is open to its environment, which is sometimes described in terms of a whole/part causal relationship, does not entail metaphysical openness, for the entire environment may be causally determined. Neither does the butterfly effect imply metaphysical openness, to attack a linkage dear to the popular reception of chaos theory. In fact, sensitive dependence—a feature of chaotic dynamical systems in mathematics—is attributed to natural systems on the basis of the power of mathematical dynamical systems to model them. To the extent that this modeling works—and we have seen that it is problematic in chaotic regimes—the natural presupposition is that the (metaphorical) 'determinism' of mathematical chaotic dynamical systems corresponds to the metaphysical determinism of nature. Put bluntly, the butterfly effect testifies to the high degree of causal connectedness in certain natural systems, and so is most naturally exploited in support of the thesis of metaphysical determinism.

More generally, the hypothesis of metaphysical determinism is strengthened by chaos theory because it enables the apparent randomness of even simple natural dynamical systems potentially to be brought under the umbrella of determinism. Instead of the theoretical instability of planetary orbits being a bone in the throat of apologists for determinism, for example, it is construed as an ordinary consequence of a simple deterministic system. Laplace would have been delighted with chaos theory.

While chaos theory offers not one shred of evidence against metaphysical determinism, it does introduce an imposing limit on how well the deterministic hypothesis can be supported. This is the significance of the epistemic and explanatory problems surrounding the question of the occurrence of chaos in nature that have just been discussed. To see this, consider two opposed scenarios from among the many conceivable metaphysical possibilities. On the one hand, it is possible that what we have called chaotic randomness never occurs in nature, that all randomness is of the strict variety, as we defined it earlier, and that nature is metaphysically open to acts by free causal agents.

On the other hand, it may be that nature is in fact metaphysically determined, and chaos omnipresent within it. If so, then only the free agency of God—and not human free agency, which seems ruled out by assumption—could act to change the course of nature. God might do so and have significant macroscopic effects by taking advantage of the phenomenon of sensitive dependence, as only an unlimited super-calculator being could. But this still begs the question of the presumed natural determinism governing the setting of initial conditions by *their* prior natural conditions, *ad infinitum*. Ultimately, then, even a super-calculator being would have to intervene in nature so determined, either by breaking or suspending its natural causal sequences. The intervention of law-suspending miracles would differ only in that sensitive dependence permits *these* acts of interference or suspension to be at an extremely small scale, rendering them undetectable. Why one form of interference would be preferable to the other from God's point of view—assuming there is some divine choice in the matter, an assumption process metaphysicians might be likely to reject—would then be a question for theologians.

Chaos theory insists with its introduction of an epistemological limitation that neither of these metaphysically opposed scenarios can be ruled out. *It must be said, therefore, both that the hypothesis of metaphysical determinism has never been as well attested as it is after chaos theory, and that it cannot be made stronger than chaos theory makes it.*

Of course, there are other domains in which one might venture to test the hypothesis of metaphysical determinism. It comes off rather badly in those regions of inquiry that must presuppose that humans are free causal agents, such as ethics and the various studies of the phenomenology of human experience. But there are competing deterministic visions of human consciousness as epiphenomenal and of human societies as inculcating a mistaken habit of interpreting humans as free, responsible agents. Metaphysical determinism also seems to come off rather badly in the quantum world, where one widely held interpretation assumes that genuine metaphysical openness grounds both quantum uncertainty and the probabilistic nature of the wave function. But here, too, there are alternatives, for some interpretations of quantum mechanics suppose that the uncertainty principle shields a deeper deterministic process.

Metaphysical determinism and its opposite, it seems, are locked in battle. *There can be no question that chaos theory adds its considerable weight to the side of determinism. However, just as there is the epistemic limitation of self-reference in the study of human existence, and the epistemic limit of the Heisenberg uncertainty principle in the quantum world, chaos theory highlights an epistemic limit in the macro-world of dynamical systems, tethering the deterministic hypothesis even as it advances it.*

4 *Conclusion: The Relevance of Chaos Theory for Theology*

The development of chaos theory has been welcomed by some theologians as powerful evidence that the universe is metaphysically open (i.e., not completely deterministic) at the macro-level. Metaphysical indeterminacy at the quantum level does not even need to be assumed, on this view, for chaos theory makes room for human freedom and divine acts in history that work wholly within nature's metaphysical openness and do not violate natural laws. At a number of places in this essay, this interpretation of chaos theory is shown to be without justification. It makes little sense to appeal to chaos theory as positive evidence for metaphysical indeterminism when chaos theory is itself so useful for strengthening the hypothesis of metaphysical determinism: it provides a powerful way for determinists to argue that many kinds of apparent randomness in nature should be subsumed under deterministic covering laws.

On the other hand, we have seen that chaos theory also imposes a fundamental limit on how well the deterministic hypothesis can be supported, thus preserving the relevance of chaos theory for forming a philosophy of nature, and assessing the possibility of free divine and human action in accordance with natural laws. Setting aside the mistaken enthusiasm just mentioned, the relevance of chaos theory for theology can be assessed rather differently depending on one's starting point. To appreciate this difference concretely, consider the reaction to chaos theory of two hypothetical (and, sexes aside, probably archetypal) thinkers, one a natural scientist and the other a theologian from a theistic tradition, such as Christianity, with a commitment to divine action.

Laplace thought that metaphysical determinism was the best way to explain the apparently well-founded Newtonian assumption that nature is an unbroken causal web. Laplace's determinism has been one of the assumptions guiding scientific inquiry into nature at least since Newton's time, and our hypothetical scientist is inclined, notwithstanding quantum mechanics, to see the world in deterministic terms as a first approximation to her actual experience. Our scientist's assumption that determinism is the most adequate account of things will be strengthened by chaos theory, since it allows her to give explicit deterministic accounts of many seemingly random processes. Of course, she always *assumed* that some enormously complex deterministic account of each apparently random dynamic process would apply, but chaos theory allows her to describe the deterministic behavior of some of those systems with relatively simple dynamic models, thus furnishing her with evidence that many random processes can actually be given deterministic explanations. Though we have seen that this form of explanation is far from straightforward, it strengthens the case for metaphysical determinism by bringing apparent randomness (of some kinds at least) into its explanatory reach.

On the other hand, our theologian is accustomed to assuming that nature is metaphysically open, since he is committed to interpreting the free actions of God and people within history, and must assume that metaphysical determinism is a flawed hypothesis if his task is to make sense. Even though he does not know how to speak of any specific mechanism by which God influences nature, his project calls for the assumption that God acts in 'natural,' law-like ways. For our theologian, the strengthening of the deterministic hypothesis by the development of chaos theory will be bad news, so long as he thinks that what we will call (purely for convenience) the *see-saw principle* is true, namely, that a stronger case for the deterministic hypothesis necessarily implies a weaker case for the possibility of free divine and human acts in history.

Not all theologians accept the see-saw principle, of course. Immanuel Kant's dual argument that we are forced by our mental apparatus to understand nature as a determined, closed causal web, and that things in themselves have to be assumed to be undetermined to make sense of our moral life, has long been the standby for many theologians, radical and traditional, liberal and conservative alike. If our theologian accepts this central tenet of Kant's system—and it can be affirmed quite independently of some of the more problematic aspects of Kant's philosophy—then the see-saw principle will be rejected and chaos theory will be irrelevant: even if the case for determinism were completely conclusive, divine and human freedom would not be impugned, for we cannot help seeing things in terms of a closed causal system, no matter what they are like in themselves. Moreover, Kant's is not the only way to avoid assuming the see-saw principle. Coarsely speaking, the philosophy of organism ('process' metaphysics) and some forms of panentheism and pantheism (including some versions of 'the world is God's body') are alternative philosophies of nature that, when appropriately tweaked, provide less clear cut, but still reasoned, ways of avoiding it.

So far, then, we have the scientist who is inclined to see chaos theory as simply specific confirmation of what she believed all along, and so not helpful for theological purposes. And we have a theologian who sees chaos theory either as irrelevant because of a supervening solution to the problem of divine action, or as bad news, for the same reason that it would have been good news for Laplace: it strengthens the case for the hypothesis of metaphysical determinism. This goes part of the way to account for the negative reactions to the attempt to bring chaos theory and the problem of divine action into conversation: both scientists and theologians have legitimate reasons to think of it as irrelevant or as bad news for the theological attempt to justify the traditional assertion that God acts freely in accordance with natural laws.

As we have seen, however, there is another level at which chaos theory is good news for this theological project and perhaps bad news for polemical determinists. We will conclude this essay by summarizing what has been said about how this can be so.

Consider a natural system that exhibits apparently random behavior under certain conditions. Could this system be described with relatively simple dynamics, so that its apparent randomness would be only the expected behavior of that simple dynamical system in its chaotic regimes? Trying to answer this question is fraught with difficulties, as we have shown, but suppose that we came up with a decent hypothesis, and set about testing it. If the model were any good at all, (1) it would be fairly easy to show that it predicted system behavior in non-chaotic regimes; and (2) if we were lucky, we could even find an attractor for some chaotic regimes of the natural system, as has been done for the dripping faucet. But (1) it is impossible to show that the model's predictions are accurate when the system is in a chaotic state; and (2) the basis of the statistical form of explanation needed to interpret the model breaks down when the dynamics are chaotic. Though the model might predict system behavior perfectly in situations of equilibrium or relatively simple periodicity, therefore, it is impossible to be sure that another, unknown model of the dynamical system (say, one that involves God as a causal factor) does not finally provide a better description of the system's behavior. Thus, there is a limit to how well the hypothesis of metaphysical determinism can be supported.

Again, it is important not to misinterpret this result, as some enthusiasts have done. We are not here speaking of chaos theory warranting the assertion of any degree of metaphysical openness in nature. Nor is the obvious result that chaos theory strengthens the case for determinism being called into question. The point is just that chaos theory discloses a limitation in how compelling the case for determinism can be made, a limitation unnoticed before chaos theory drew attention to it. For some theologians (those who assume the see-saw principle), this is a kind of limited good news. Chaos theory might have made the case for metaphysical determinism stronger than it has ever been before, but chaos theory also guarantees that the case cannot get any stronger. The small degree of ambiguity in the analysis of dynamical systems is, therefore, irreducible. Perhaps that is not much to get excited about, especially since so much theological work seems to rely on a usually anonymous Kantian compatibilism (the strategic denial of what we have been calling the see-saw principle), making any result of chaos theory irrelevant to the problem of divine action. But it does open a window of hope for speaking intelligibly about special, natural-law-conforming divine acts, and it is a window that seems to be impossible in principle to close.

5 *A Bibliographical Note*

Most of the mathematics in this discussion of the logistic map appears in many other places. Rather than arbitrarily citing sources, therefore, a list of selected works organized according to key themes seems more appropriate. We would like to acknowledge the help we received in conversations with Bob Devaney and Karl Young, who saved us from several errors. We, of course, and not they, are responsible for any that remain.

4.1 *Text Books*

A number of books give detailed coverage to chaos and iterated maps on the unit interval. Major examples, from quite different perspectives, are:

[1] Ralph H. Abraham; Christopher D. Shaw. *Dynamics: The Geometry of Behavior*. 2nd ed. Redwood City, California: Addison-Wesley, 1992. A unusual and fascinating visually oriented introduction to dynamical

systems and chaos.

[2] Pierre Collet; Jean-Pierre Eckmann. *Iterated Maps on the Interval as Dynamical Systems*. Boston, Basel, Stuttgart: Birkhäuser, 1980. A detailed discussion of most of the interesting characteristics of iterated maps on the interval, focusing on kneading theory, renormalization, and the classification of one dimensional maps. Includes a lengthy discussion in Part I of the motivation for and interpretation of the mathematical details of the difficult remainder of the book.

[3] Robert L. Devaney. *An Introduction to Chaotic Dynamical Systems*. 2nd ed. Redwood City, California: Addison-Wesley, 1989. A well organized introduction to the mathematics of chaos, accessible to college graduates in mathematics. Includes proofs of all results, and an emphasis on the logistic map in the examination of one-dimensional dynamics in Part I.

[4] John Guckenheimer; Philip Holmes. *Non-linear Oscillations, Dynamic Systems, and Bifurcations of Vector Fields*. New York: Springer-Verlag, 1983. A standard reference for chaos in continuous systems.

[5] Stephen H. Kellert. *In the Wake of Chaos: Unpredictable Order in Dynamical Systems*. Science and Its Conceptual Problems Series, David L. Hull, Gen. Ed. Chicago: University of Chicago, 1993.

[6] F. C. Moon. *Chaotic Vibrations: An Introduction for Applied Scientists and Engineers*. New York: John Wiley & Sons, 1987. A introductory textbook for the audience named in the title.

[7] Heinz-Otto Peitgen; Hartmut Jürgens; Dietmar Saupe. *Chaos and Fractals: New Frontiers of Science*. New York, Berlin, London: Springer-Verlag, 1992. A well presented textbook of almost 1,000 pages aimed at college level mathematics and science students. Includes clear explanations of key mathematical ideas, imaginative diagrams illustrating all aspects of chaos, several sections devoted to the logistic map, and a detailed bibliography.

4.2 *Key Technical Articles*

A few of the key articles containing breakthrough work on the logistic map and other iterated maps are:

[8] J. Banks; J. Brooks; G. Cairns; G. Davis; P. Stacey. "On Devaney's Definition of Chaos." *American Mathematical Monthly* 99/4 (1992), pp. 332-334. Reporting the result that topological transitivity and density of periodic points implies sensitive dependence on initial conditions, as Devaney defines these terms in [3].

[9] M. J. Feigenbaum. "Quantitative Universality for a Class of Nonlinear Transformations." *Journal of Statistical Physics* 19 (1978), pp. 25-52. Reporting Feigenbaum's discovery that the bifurcation cascades of a very large class of iterated maps are characterized by the same scaling (renormalization) constant, together with an estimate of that constant.

[10] M. J. Feigenbaum. "Universal Behavior in Non-Linear Systems." *Physica* 7D (1983), pp. 16-39. A useful statement of the major results concerning universality.

[11] M. Jakobsen. *Communications in Mathematical Physics* 81 (1981), pp. 39-88. Reporting the result that the set of $k<4$ for which the logistic map has no periodic orbits has positive Lebesgue measure.

[12] Tien-Yien Li; James A. Yorke. "Period Three Implies Chaos." *The American Mathematical Monthly* (November, 1975), pp. 985-992. Reporting the vital result that, if a particular type of map has a periodic point of period three, then it has periodic points of all periods and chaotic behavior on some set.

[13] E. N. Lorenz. "Deterministic Non-Periodic Flow." *Journal of Atmospheric Science* 20 (1963), pp. 130-41. A breakthrough paper later recognized as initiating contemporary research in chaos theory.

[14] M. Metropolis; M. L. Stein; P. R. Stein. *Journal of Combinatorial Theory*, 15A (1973), pp. 25-44. Presenting some key results in the symbolic dynamics of maps on the unit interval.

[15] S. Smale. "Differentiable Dynamic Systems." *Bulletin of the American Mathematical Society* 73 (1967), pp. 747-817. The first report of the vital result that chaos appears stably (perturbations do not alter dynamics) on a whole space in some higher dimensional systems.

4.3 *Experimental Applications of Iterated Maps*

There have been many experiments relating iterated maps to physical dynamical systems. Here is a selection covering a wide range of topics; the titles are usually self-explanatory.

[16] F. T. Arecchi; R. Meucci; G. Puccioni; J. Tredicce. "Experimental Evidence of Subharmonic Bifurcations, Multistability, and Turbulence in a *Q*-switched Gas Laser." *Physical Review Letters* 49/17 (1982), pp. 1217-1220.

[17] J. P. Gollub; Harry L. Swinney. "Onset of Turbulence in a Rotating Fluid." *Physical Review Letters* 35/14 (1975), pp. 927-930.

[18] Michael R. Guevara; Leon Glass; Alvin Shrier. "Phase Locking, Period-Doubling Bifurcations, and Irregular Dynamics in Periodically Stimulated Cardiac Cells." *Science* 214 (1981), pp. 1350-1353.

[19] A. Libchaber; C. Laroche; S. Fauve. "Period Doubling Cascade in Mercury, a Quantitative Measurement." *Le Journal de Physique-Lettres* 43 (1982), pp. L211-L216.

[20] Robert M. May; George F. Oster. "Bifurcations and Dynamic Complexity in Simple Ecological Models." *The American Naturalist* (July-August, 1976), pp. 573-599.

[21] Edgar E. Peters. *Chaos and Order in the Capital Markets: A New View of Cycles, Prices and Market Volatility*. New York: Wiley & Sons, 1991.

[22] Alvin M. Saperstein. "Chaos—A Model for the Outbreak of War." *Nature* 309, pp. 303-305.

[23] Reuben H. Simoyi; Alan Wolf; Harry L. Swinney. "One-Dimensional Dynamics in a Multicomponent Chemical Reaction." *Physical Review Letters* 49/4 (1982), pp. 245-248.

[24] James Testa; José Pérez; Carson Jeffries. "Evidence for Universal Chaotic Behavior of a Driven Nonlinear Oscillator." *Physical Review Letters* 48/11 (1982), pp. 714-717.

4.4 *Useful Introductions and Surveys*

There are a large number of introductory and survey articles and even some books covering the topics of chaotic dynamical systems, chaos, and the logistic map. Here are some of the most useful:

[25] James P. Crutchfield; J. Doyne Farmer; Norman H. Packard; Robert S. Shaw. "Chaos." *Scientific American* 255 (December, 1986), pp. 46-57. This article is included in the present volume. It furnishes a non-technical account of the identification of a strange attractor for the dripping faucet.

[26] Joseph Ford. "What is Chaos, that We should be Mindful of it?" Paul Davies, ed., *The New Physics*, pp. 348-372. New York and Cambridge: Cambridge University Press, 1989.

[27] Joseph Ford. "How random is a coin toss?" *Physics Today* (April, 1983), pp. 40-47.

[28] James Gleick. *Chaos: Making a New Science*. New York: Penguin Books, 1987.

[29] John Holte, ed. *Chaos: The New Science*. Nobel Conference XXVI. Lanham, Maryland: University Press of America, 1993.

[30] John Horgan. "In the Beginning...." *Scientific American* (February, 1991), pp. 117-125.

[31] John T. Houghton. *Does God Play Dice?* Leicester: IVP, 1988.

[32] Leo P. Kadanoff. "Roads to Chaos." *Physics Today* (December, 1983), pp. 46-53.

[33] Robert M. May. "Simple Mathematical Models with Very Complicated Dynamics." *Nature* 261 (June 10, 1976), pp. 459-467. This survey article focuses particularly on the logistic map.

[34] Ilya Prigogine; Isabelle Stengers. *Order out of Chaos*. New York, Bantam Books, 1984. An introduction to the theory of complex systems, which is intimately connected to chaos theory.

[35] Ian Stewart. "Recipe for Chaos." *Does God Play Dice?* pp. 348-372 Cambridge: Blackwell, 1989.

4.5 *Key Works on Chaos Theory and Theology*

During the 1980s, theologians and philosophers interested in such questions as providence and divine action began discussing the connections with chaos theory and, more often, the connected theory of complex systems. Some of the important articles and books including extended discussions of chaos theory and theology are:

[36] Steven Dale Crain. *Divine Action and Indeterminism: On Models of Divine Action that Exploit the New Physics*. University of Notre Dame, Dissertation, 1993.

[37] Niels Henrik Gregersen. “Providence in an Indeterministic World.” *CTNS Bulletin* 14/1 (Winter, 1994).

[38] Philip J. Hefner. “God and Chaos: The Demiurge Versus the Ungrund.” *Zygon* 19 (1984), pp. 469-485.

[39] J. T. Houghton. “New Ideas of Chaos in Physics.” *Science and Christian Belief* 1/1 (1989), pp. 41-51.

[40] Bernd-Olaf Küppers. “On a Fundamental Paradigm Shift in the Natural Sciences.” W. Krohn, et. al., eds., *Selforganization: Portrait of a Scientific Revolution*, pp. 51-63. The Hague: Kluwer Academic Publishers, 1990.

[41] D. M. Mackay. *Science, Chance and Providence*. Oxford: Oxford University Press, 1978.

[42] John Polkinghorne. “The Nature of Physical Reality.” *Reason and Reality: The Relationship Between Science and Theology*, chapter 3, pp. 34-48. London: SPCK, 1991.

[43] John Polkinghorne. “A Note on Chaotic Dynamics.” *Science and Christian Belief* 1/2 (1989), pp. 126ff.

[44] John Polkinghorne. *Science and Providence: God's Interaction with the World*. London: SPCK, 1991.

[45] Keith Ward. “The Constraints of Creation.” *Divine Action*, chapter 7, pp. 119-133. London: Collins, 1990.

[46] Walter J. Wilkins. “Chaos and the New Creation.” A paper delivered in the Religion and Science Section of the American Academy of Religion Annual Meeting, November 21, 1988.

II

CHAOS, COMPLEXITY, AND THE PHILOSOPHY OF LIFE

UNDERSTANDING COMPLEXITY

Bernd-Olaf Küppers

1 *Introduction*

One of the most impressive features of contemporary science is the enormous progress that has been made in the understanding of complex phenomena. An outstanding example of this is the development of modern biology, which at present is as rapid as the development of physics at the beginning of our century.

Above all, the results of molecular biology have become the focal point of interest, because they have led to deep insights into the physical basis of living organisms, including the physical principles of inheritance. Furthermore, the results of molecular biology have induced the development of a physical theory of evolution and thus the development of a coherent concept of the physical origin of life. Biologists are going to elucidate and understand the molecular principles of differentiation and morphogenesis. Neurobiologists are learning more and more about the physical structure and function of the central nervous system. Some day, biologists will be able to simulate the intelligent achievements of the brain by means of artificial neural networks. Last but not least, the plan to determine the complete sequence of the human genome seems to promise an all-embracing knowledge of the physical organization of the human being.

What has been paraphrased here is the present success and the future goals of the so-called reductionistic research program. This program is based on the central working hypothesis that all biological phenomena can be explained totally within the framework of physics and chemistry. In other words: the reductionistic research program starts from the epistemological premises that there is no principal difference between non-living and living matter, and that the transition from the non-living to the living must be considered as a quasi-continuous one, in which no other principles are involved than the general principles of physics and chemistry. If there are certain limitations with regard to our physical understanding of living matter, then these are supposed to have a temporary but not a fundamental character.

Within the framework of the reductionistic research program, such limitations are attributed exclusively to the material complexity of the phenomena under consideration, in the same way as certain restrictions of the physical foundation of chemistry are attributed to the restricted computability of molecular structures. To give just one example of this kind of thinking: Although the Schrödinger equation cannot be solved for complex molecules, a solid reductionist will have no serious doubts about his working hypothesis that chemistry can be reduced in principle completely to the laws of quantum physics.

According to its own conception of itself, the reductionistic research program claims that it is purely and simply the increasing material complexity in the transition from atoms to molecules and from molecules to living systems that gives chemistry and biology the status of semi-autonomous sciences, each with its own vocabulary and law-like statements. Nevertheless, the reductionistic research program adheres to its central doctrine that all phenomena of chemistry and biology can *in principle* be reduced to physics.

Despite the undeniable and overwhelming success of the reductionistic research program in the past, a number of objections to its central doctrine have been put forward. And it is more than only a historical note that those objections have been made mainly with respect to the relationship between biology and physics, although the relationship between chemistry and physics seems also to be a target of those objections. This already indicates that, to some extent, the objections to the reductionistic research program have psychological roots. Indeed, one can understand very well the discomfort that an anti-reductionist must feel. For, if the reductionistic view of the world is correct and all biological phenomena can be—at least in principle—explained by the laws of physics, one seems to arrive at the ontological conclusion that life is nothing but a complex interplay of a large number of atoms and molecules. According to this picture, even the phenomenon of consciousness must be considered as a direct consequence of material complexity that is thought to be governed exclusively by physical laws.

Now, in order to immunize themselves against the desperate picture of the world drawn by reductionism, some biologists and associated philosophers of science have invented the concept of "emergence" and the closely related concept of "downward causation." The main idea behind these concepts can be summarized by the general claim that, if a material system reaches a certain level of complexity, manifested in a high degree of relatedness of its components, then genuinely novel properties and processes may emerge, which cannot be explained by the material properties of the components themselves. Consequently, the emergentists claim that physics is not a sufficient tool for explaining complex phenomena, even though it is a necessary one.

In short, one can express the quintessence of the concepts of emergence and downward causation by two theses:

(1) The whole is more than the sum of its parts.

(2) The whole determines the behavior of its parts.

From the first thesis follows the concept of emergence. From the second thesis follows that of downward causation. Both concepts have been presented more precisely in the literature and have been discussed there at a sophisticated level. For a philosopher of science who is interested in the mind-brain problem, the concepts of emergence and downward causation are quite appealing. But from the perspective of a scientist actually working at the borderline of physics and biology, both concepts remain rather vague and mysterious.

It will be the main goal of this article to draw attention to a general problem of reduction in biology that has recently become apparent, and which may shed new light on the concepts of emergence and downward causation and their relevance for the understanding of complex phenomena. However, the reference point of the following discussion will not be the mind-brain problem, but the more general problem of the transition from the non-living to the living. The last one is almost paradigmatic for biology's challenge to physics, and its analysis during recent years has led to completely new insights into the possibilities and limitations of understanding biological complexity.[1]

In this article, the problem of understanding complexity will be approached in the following way: first it is shown what kind of complexity one encounters in non-living systems. Then the nature of complexity encountered in living systems will be described. And finally the capacity of the reductionistic research program in closing the explanatory gap between simple and complex systems will be investigated.

2 *Emergence and Downward Causation in Non-Living Systems*

If we ask the general question of how physics differs from biology, we usually get the answer that physics deals with simple systems, whereas biology deals with complex systems. At first glance, this answer seems to be convincing, in that it is consistent with our intuitive ideas about the subject matter. But, on closer inspection, this view is hardly tenable.

Let us consider a so-called "simple" system, typical of physics, for example, one mole of an ideal gas. Under standard conditions, this amount of gas contains about 10^{24} particles—the exact number is given by the well-known Avogadro number. If we wish to describe the system in its microphysical details, that is, starting from the fundamentals of classical mechanics, then we must specify at a given time the position and velocity (or momentum) of each of the 10^{24} particles. Or, expressed technically: we have to specify the so-called phasc space. The dynamical behavior of the system is then characterized by the movement of a cloud of points in this phase space. However, the specification of all points in phase space is a task that can only be solved in a "*Gedankenexperiment*," and cannot be solved in reality. So, even in this allegedly simple case, we are already dealing with an extraordinarily complex system.

However, there is a way out of this complexity. Instead of analyzing the microstate of the system, which is given by the positions and velocities of all particles in the phase space, we can restrict the analysis to certain macrostates of the system. The macrostates of the system are given by defined mean values of the particle distribution. For example, instead of determining the velocity of each individual particle, one can consider the mean velocity of all particles of the distribution. Such considerations lead to certain macroscopic parameters like pressure and temperature, that is, to parameters that can be observed and measured directly. Since each microstate has a uniquely defined macrostate, this form of abstraction is a legitimate and useful procedure to reduce the complexity of the system by reducing the complexity of its description.

This procedure, which is typical for equilibrium physics, allows us to speak in this case of a "simple" system. Of course, a price must be paid for such an

---

[1]Bernd-Olaf Küppers, *Information and the Origin of Life* (Cambridge, MA: MIT Press, 1990).

abstraction. If we change the level of description, in that we now consider the macrostate of the system, we lose, as is unavoidable for any statistical consideration, a certain amount of information. In the present case, we lose the information about the microstates. And there is even a measure for this loss of information: the entropy.[2]

The relationship between macroscopic thermodynamics and statistical mechanics, which I have sketched here, has been the subject of many investigations concerning the problem of intratheoretical reduction in physics. If one adopts the reductionistic point of view, one is compelled to claim that the macroscopic observables like pressure and temperature can be reduced by means of statistical mechanics to the microscopic behavior of the system's particles. Ernst Nagel, in his paradigmatic investigations of the problem, follows essentially this reductionistic argument, although he sees himself compelled to make use of so-called bridge laws.[3] These are additional premises that link the vocabularies of the two theories that are to be reduced.

The emergentists, however, would use this thermodynamic example in order to demonstrate the opposite, namely, the occurrence of emergent properties like temperature and pressure in physical systems on the one hand and the inherent weakness of reductionism in explaining these properties on the other hand. The bridge laws are the principal object of their attack. Most emergentists refuse to accept the bridge laws as a sufficient tool for carrying out the reduction. Emergentists would at least deny the availability of such bridge laws for the psychophysical case.

It is not useful to go into the details of the arguments. However, we must conclude that for a true emergentist physics itself cannot be reduced to physics, that is, physics itself must be considered as being irreducible. This may be the curious meaning of the notion that the concept of emergence is equivalent to nonreductive materialism, or, as Popper has expressed it, that "materialism transcends itself."[4]

The same line of argument holds for the concept of downward causation. One can demonstrate this very easily by means of the law of mass action. Consider two reactants in chemical equilibrium. None of the molecules in such a system "knows" that it is in a chemical equilibrium and has to behave dynamically in a certain way, so that the equilibrium is preserved. Nevertheless, the equilibrium state is the stable

---

[2]This can easily be seen. It is true that each microstate has a uniquely defined macrostate. But, on the other hand, it is also true that a given macrostate can be realized by more than one microstate. The entropy of a system is just the measure of the number of microstates contained in its macrostate. This means that the entropy, and thus our ignorance about the system, is the larger the more microstates are contained in their corresponding macrostates. For the singular case in which a macrostate contains one and only one microstate, the entropy is zero and our knowledge about the system is at a maximum.

[3]Ernest Nagel, *The Structure of Science: Problems in the Logic of Scientific Explanation* (New York: Harcourt, Brace, and World, 1961).

[4]Karl R. Popper and John C. Eccles, *The Self and its Brain* (Berlin: Springer, 1977), chap. 1.

state of the system. In fact, it is some kind of downward causation or macro-determination that causes the regular behavior of the molecules. The law of mass action ensures that any fluctuation that disturbs the equilibrium becomes dampened. The larger the deviations from the equilibrium, the larger is the force restoring the original state. In other words: with respect to its fluctuations, any chemical equilibrium is self-regulating. But the self-regulating power of an equilibrium system is a property of the system as a whole, and it cannot be deduced from the material properties of its components.

Thus, it seems to be a banal truth that emergence and downward causation are intrinsic phenomena of our material world, and that these phenomena by no means are restricted to the sphere of biological complexity. If this view is correct, one can draw some more interesting conclusions concerning the phenomena of emergence and downward causation. In order to do so, let us forget for a moment the mind-brain problem and consider only the problem of life itself. The present theory of the origin of life yields no indication that there is a non-physical principle involved in the transition from the non-living to the living. Nowadays, the origin of life can be completely understood on physical grounds as a quasi-continuous process of molecular self-organization in the course of which its material complexity steadily increases.[5] Consequently one has to give up the idea that the phenomena of emergence and downward causation at once emerge when matter reaches a certain level of complexity. Instead, both phenomena must be thought of as epiphenomena of self-organizing matter that continuously emerge when matter unfolds its complexity by organizing itself.[6] It will be the task of the last part of this article to express this idea in precise physical terms. First, however, some light must be shed on the nature of the complexity that one encounters in biology.

3 *Complexity in Living Systems*

For this purpose, let us consider a system which is of intrinsic complexity and which cannot be simplified by changing the level of description or abstraction. The example to be discussed here is taken from molecular biology and is known by molecular biologists as sequence space.[7] In order to present this concept we must make a short digression into the mechanism of genetic information storage. As is well known, the information required for the construction of a living organism resides universally in a particular class of the cell's molecules: the nucleic acids, usually DNA. DNA molecules are macromolecules: that is, they are made by

---

[5]Küppers, *Molecular Theory of Evolution: Outline of a Physico-Chemical Theory of the Origin of Life*, 2d.. ed., trans. Paul Woolley (New York: Springer-Verlag, 1985).

[6]Thus, the extended version of emergentism, according to which the history of nature is a process of emergent evolution, has to be brought into agreement with the modern concepts of physical evolution as a quasi-continuous process of molecular self-organization.

[7]M. Eigen, "Macromolecular Evolution: Dynamical Ordering in Sequence Space," *Beratungen der Bunsengesellschaft für physikalische Chemie*, 89 (1985): 658-67.

joining smaller molecules, the so-called nucleotides, into a long chain. In such an informational macromolecule, the monomers have the same function as the letters in a language. In particular, the detailed sequence of the monomers of a DNA molecule determines completely its genetic information.

If, by spontaneous or site-directed mutagenesis, a letter in such a genetic blueprint becomes altered, this may lead—as a typographical error in a book—to a distorted text and thus to the death and decay of the organism. One can speak here of a distorting mistake, insofar as, in general, the disturbed substructure of the organism can no longer fulfill its specific function for the whole system. From this it becomes clear that the genetic information possesses a certain semantic aspect that encodes the structural and functional properties of the organism and that represents a measure of its material complexity.

Many insights into the problem of biological complexity arise from the concept of sequence space. Sequence space is a high-dimensional hypercube for the representation of all possible variants of a DNA sequence. By analogy with phase space, we can assign to the sequence space a certain number of microstates, namely the number of sequences that can be built up form a finite class of monomers in a macromolecule of finite length. Even for extremely short macromolecules, the sequence space has a considerable complexity. However, as soon as we reach biologically significant orders of magnitude, we venture into inconceivable dimensions of complexity. Even the informational sequence of the bacterial genome has about $10^{2,400,000}$ microstates in its sequence space, whereby only a vanishingly small fraction of all these microstates can be assumed to encode biologically meaningful information.

Thus, in contrast to physical systems of the type discussed before, in biological systems the detailed knowledge of the microstate of their informational macromolecules is of essential importance for the understanding of those systems. It is the specific sequence of its DNA that gives a biological system the properties of living matter. Obviously, there is no possibility here of referring to any mean values in order to reduce the complexity of the sequence space.[8] There is no possibility of finding a macrostate that yields sufficient information about the system. Moreover, if one wishes to understand how a living system could have arisen when there existed only non-living matter, one has to analyze the whole sequence space for the system and to investigate its evolutionary pathway through all its possible microstates.

Thanks to the tremendous complexity of sequence space, biology forces physicists to a new intellectual approach. Certain kinds of abstraction and idealization, as are usual within the framework of equilibrium physics, are no longer meaningful for biology. This statement is the more important as the phenomenon of intrinsic complexity is found in the living world on all levels of

---

[8]This can be illustrated by the following example. A sonnet by Shakespeare is represented by a defined sequence of letters. All possible sequence alternatives of this poem build up a certain sequence space. But the mean value of all those sequences will hardly represent a Shakespeare sonnet as well.

organization, whether we consider the sequence space of informational macromolecules or the state space of the neurons in a neural network.

Can one explain and understand physically the immense complexity of living matter or do there exist definite limits of the physical method? Does the reductionistic research program in this case fail, or is physics undergoing a new paradigm shift, thereby transforming the explanatory capacity of the whole program? And finally, how can one correlate the phenomenon of complexity with those of emergence and downward causation? These questions now lead to the central part of this article.

4 *Boundary Conditions and Downward Causation*

For a long time, the investigation of complex phenomena by physicists was characterized by the expectation that new, fundamental laws of physics might be found in this area. However, these expectations have not been realized, and the prospect of the discovery of such laws is dwindling from day to day. If one therefore concludes the converse, that is, that even complex phenomena must be explicable on the sole basis of the known laws of physics, then one has to look for new aspects revealing the physical roots of complexity.

Such a new aspect becomes clear in outline when one directs one's attention to the logical structure of physical explanatory models. As is well known, physical theories possess a dual structure: they contain both general statements in the form of laws, which possess at least a local validity, and so-called initial or boundary conditions. The latter conditions are not fixed by the theory itself but are specified by the given facts within the scope of application of the theory. Mathematically speaking, the initial or boundary conditions are certain selection conditions of possible solutions to the differential equations by which the physical laws are described.

We shall now attempt to justify the central assertion of this article, which is that the complexity of a system or a phenomenon lies in the complexity of its conditional complex, that is, in the complexity of its boundary conditions.

The first to emphasize the significance of boundary conditions for an understanding of complex phenomena was the distinguished physicist Michael Polanyi.[9] Although his final conclusions with regard to biology ultimately proved to be erroneous, it will serve our purpose well to follow his analysis for some of the way.

First, let us consider a complex system made by human hands, such as a sophisticated machine. We can use various criteria to characterize such a system. Above all, we can state the purpose of the machine and explain how it functions. The purpose that the machine is intended to fulfill is dictated by human goals. Its mechanism, however, is dictated by the obedience of matter to natural laws. We cannot construct a machine that functions in a manner contrary to natural laws.

---

[9]M. Polanyi, "Life's Irreducible Structure," *Science* 160 (1968): 1308-12.

There exists, for example, no "*perpetuum mobile*" that can suspend the law of conservation of energy or the second law of thermodynamics.

The function of a machine is laid down by the manner of its construction. This includes both the driving force, that is, the source from which the machine draws its energy, and the detailed arrangement of the components of the machine. Let us ignore the energy source for a moment. The structure of a machine is then given unambiguously by the internal relatedness of its components, which in turn is determined by the special shape of the components and by the way in which they are put together. The relationship between the material components of the machine thus represents a boundary condition under which the natural laws become operational in the machine. It is the joint action of all processes induced and constrained by these boundary conditions that finally realizes the functional purpose that the machine is to serve.

Every machine works under such a principle of dual control: the higher control principle is that of the machine's design, and this harnesses the lower one, which consists in the physical processes on which the machine relies. The design of the machine finds its expression in the internal boundary conditions. The boundary conditions, which give the machine its special properties, are designed by the constructor of the machine. It is obvious that these cannot be deduced from the general laws of physics and chemistry.

Thus, the structure of a machine and the working of its structure show all the features of an emergent system: the system represents a whole, which is neither additive nor subtractive. Moreover, the properties of the system are not reducible to the material properties of its components. And, last but not least, all processes going on in the system are induced and constrained by the system's boundary conditions, indicating the presence of some kind of downward causation.

Thus, the example of the machine focuses attention upon the fundamental importance of the boundary conditions for an understanding of complex systems and their emergent properties. This is of particular value here because the example of the machine can immediately be transferred to the archetype of complex systems, the living organism. In fact, the living organism is comparable in many ways to a complex machine. And it is not a matter of pure chance that an entire field of research, namely cybernetics, has its origin in the machine theory of the organism. As in the case of the machine, the organism is subservient to the manner in which it is constructed, whereby the functional order of all its parts is preordained. Its principle of construction represents a boundary condition under which the laws of physics and chemistry become operational in such a way that the organism is reproductively self-sustaining. Again, as in the case of the machine, the phenomenon of emergence as well as that of downward causation can be observed in the living organism and can be coupled to the existence of specific boundary conditions posed in the living matter.

Clearly, it is the special structure of the boundary conditions that determines the overall construction of the system, both for the machine and for the living organism. We must therefore ask by what means the boundary conditions are exactly determined. In the case of the machine the answer to this question is obvious. The boundary conditions are defined in the blueprints drawn up by the constructor of the machine. But even for the living organism such a blueprint does exist. It resides in the genome of the organism and is—in contrast to the machine—an inherent part of the material organization of the system. If the genetic material of an organism is placed in a suitable physico-chemical environment, such as an egg cell, then expression on the phenotypic level results; that is, the construction plan of the organism encoded in its genome is realized in matter, step by step. Thus, the structures of all the boundary conditions that appear in the complete, differentiated organism are implicitly laid down in the genome of the organism. The genome therefore represents a primary boundary condition, which determines all other boundary conditions appearing in the living organism.

5 *Boundary Conditions in Physical Systems*

The above arguments underline the importance of the concept of boundary conditions for the understanding of complex phenomena. However, up to now they have been based exclusively upon plausibility considerations. We must now crystallize the term 'boundary condition' so as to bring it into line with its corresponding concept in physics.

For this purpose, let us enlarge a little on the boundary conditions that lie behind the order of living matter. As already noted, all boundary conditions of a living organism reside in particular physical structures, the DNA molecules. It is the microstate, that is, the detailed sequence of the monomers in the DNA, that encodes all structural and functional properties of the living organism. The monomer sequence of a DNA molecule is a genuine boundary condition in the physical sense. It can be specified by stating the coordinates in space of all the atoms in the structure of the molecule, along with other physico-chemical environmental conditions such as the temperature, the ionic strength and so on. The structure represents a precise set of physical parameters, which—according to the doctrine of genetic determinism—governs all the physical and chemical processes that take place in the system.

It should be mentioned, however, that there is an important difference between the boundary conditions of complex and simple systems. Complex systems can be defined as systems that depend very sensitively upon their boundary conditions. Thus, in living systems, even the exchange of a single nucleotide in a genome can lead to instability and collapse of the system. Simple physical systems behave differently in this respect. Simple systems may be defined as systems in which the nature of the physical processes induced by the boundary conditions will not change fundamentally if the boundary conditions are altered. For example, a body in free fall will always show a qualitatively similar behavior, irrespective of the initial conditions from which the process starts. The boundary conditions of simple physical systems, such as the position and momentum of a particle at a particular time, are contingent quantities. 'Contingent' means in this context that the boundary conditions can be chosen arbitrarily. They are marginal quantities that may be chosen freely in an experiment and whose values are restricted only by the range of validity of the natural laws in question.

The boundary conditions of complex systems, on the other hand, are non-contingent quantities, since any significant alteration in these boundary conditions would also change significantly the dynamics of the system and thereby its properties. In addition, the boundary conditions are usually unique, in that they represent a defined selection from a gigantic number of physically equivalent alternatives. The reader will recall that the number of alternative sequences or microstates available to the genome of a bacterial cell is no less than $10^{2,400,000}$.

So far, we have classified simple and complex systems according to the nature of their boundary conditions. We have associated complex systems with non-contingent boundary conditions and simple systems with contingent boundary conditions. However, it is important to notice that the transition from one of these classes to the other is a continuous one, in that the concept of contingency itself allows this. The concept of contingency is the logical antithesis of that of law, and therefore leaves open the question to what degree a phenomenon is restrained or unrestrained by natural law. Incidently, this view fits very well into the modern theory of molecular self-organization, according to which the transition from non-living to living systems must be regarded as a quasi-continuous one.

6 Boundary Conditions and the Reductionistic Program

After discussing the concept of boundary conditions, we shall now use it to obtain deeper insight into the explanatory capacity of the reductionistic research program.[10] It will turn out that the concept of boundary conditions yields a key to the comprehension of the paradigm shift that is induced in physics in connection with the investigation of complex phenomena.

Let us start with some remarks concerning paradigm shifts in general. The conquest of new phenomena by a branch of science very often results in changes in its theoretical frame. This is because a novel class of phenomena will usually require a new set of terms and concepts for its description. However, terms and concepts only gain a precise meaning within the framework of a theory. Thus, the decisive step in a paradigm shift is always the expansion or the rebuilding of its theoretical frame.

The rearrangement of the theoretical frame of a paradigm can take place at two different levels. It can, for example, be a change consisting in the modification and/or expansion of the theoretical frame. It can alternatively be the case that the theoretical frame requires no modification or expansion, but merely an internal shift of emphasis. The former will always be the case when recurrent anomalies of explanation bring a paradigm into an internal crisis, one that can be resolved only by a radical rebuilding of the theoretical frame of the paradigm. It was such instances that Kuhn had in mind when he spoke of scientific revolutions occurring

---

[10] Küppers, "On a Fundamental Paradigm Shift in the Natural Sciences," in *Self-Organisation—Portrait of a Scientific Revolution*, Wolfgang Krohn, Gunter Kuppes, and Helga Nowotny, eds., vol. 14, *Sociology of the Sciences* (Dordrecht: Kluwer, 1990), 51-63.

from time to time.[11] The other case, that of internal shifts of emphasis, appears at first sight to be of less philosophical interest. However, we shall see that precisely this seemingly uninteresting case forms the basis for the paradigm shift under discussion here, and that its consequences will be in no way less revolutionary than those of Kuhnian scientific revolutions.

In order to see this, we must first express more precisely the idea of an internal shift of emphasis. This shift of emphasis need not necessarily refer to a change in the fundamental laws of physics. But it clearly means more than the mere extension of interest from simple to complex phenomena. We must rather regard the internal shift of emphasis as a change in the inherent explanatory structure of the paradigm itself.

But what is a scientific explanation? In physics it is customary to say that there is a scientific explanation for a class of natural events if a general law can be formulated that allows the appearance of such an event to be deduced from its boundary conditions. For example, in a physical experiment nature is subjected to certain restrictions in order to observe its behavior under these restrictions. By variation or by narrowing or widening the experimental boundary conditions, the law-like behavior of matter can be detailed.

This explanatory scheme has been formalized by Hempel and Oppenheim.[12] Their model of scientific explanation is based fundamentally upon two classes of statements: (1) a class of statements that describe the so-called antecedent conditions that precede the *explanandum*, that is, the event to be explained and (2) a class of statements that describe the general laws lying behind the *explanandum*.

According to the Hempel-Oppenheim scheme, the actual process of explanation consists in the logical deduction of the explanandum from the antecedent conditions and the general laws, which together make up the so-called *explanans*. The antecedent conditions are nothing other than the boundary conditions constituting the conditional complex for a natural phenomenon. In the traditional models of physical explanation, the boundary conditions are irreducible premisses for formulated natural laws. They represent the physical quantities that express the contingency of the natural event. Now, within the framework of the reductionistic research program, the boundary conditions of complex phenomena are no longer regarded as contingent quantities but rather as non-contingent ones, which must themselves be explained so that they move into the center of the explanation.

This has brought us to the decisive characteristic feature of the shift of emphasis within the reductionistic research program, a shift that has led to the formulation of the so-called paradigm of self-organization and thereby to a fundamental shift in the explanatory capacity of the reductionistic research

---

[11]Thomas S. Kuhn, *The Structure of Scientific Revolutions*, 2d ed., enlarged (Chicago: University of Chicago Press, 1970).

[12]C.G. Hempel and P. Oppenheim, "Studies in the Logic of Explanation," *Philosophy of Science* 15 (1948): 135-75; reprinted in C.G. Hempel, *Aspects of Scientific Explanation: and Other Essays in the Philosophy of Science* (New York: Free Press, 1965), 245-90.

program itself. The contingency of natural processes, previously firmly anchored in the contingency of the boundary conditions of a natural phenomenon, now itself becomes the object of scientific explanation.[13]

However, this does not mean that the contingent quantities can in this way be eliminated completely from our model of scientific explanation. For, when the boundary conditions of a system or process become the explanandum, this in turn presupposes the existence of other boundary conditions on a higher level of explanation. The resulting iteration either continues in an indefinite recursion, or else breaks down when a level is reached on which the boundary conditions are no longer explicable and thus adopt the status of classical contingent initial conditions. In either case, this recursive explanatory procedure will reduce considerably the number of contingent phenomena. This is exemplified well by the theory of the self-organization of living systems, which represents an attempts to explain the origin of non-contingent boundary conditions, namely information-carrying DNA molecules, by appeal to contingent initial conditions, namely random molecular structures.

The decisive step leading to the new paradigm of self-organization consists in the recursive shift of the explanandum to the boundary conditions. The paradigm of self-organization, which has become a central part of the reductionistic research program, is still based upon the basic explanatory model as formalized by Hempel and Oppenheim, but now a whole series of explanatory steps are linked consecutively, so that one now can speak of the nested or hierarchically-organized structures and processes of living matter as directly reflected in the hierarchical explanatory structure of the overall model of self-organization. The paradigm shift, discussed here, puts physics into a position to realize in a novel way its inner goal of demarcating the region of contingent phenomena and of tracing it back, by local inference, to the law-like behavior of nature.

7 *Explanation of Boundary Conditions*

Let us finally ask to what extent the existence of the specific boundary conditions characteristic of living matter can actually be explained within the framework of physics. To answer this question we have to refer once more to the concept of sequence space. This is because the boundary conditions of living organisms are materialized at the molecular level in the DNA sequence of its genome and thus represent defined microstates in sequence space.

As we have seen, even for the simplest organisms the number of dimensions of the sequence space is tremendously large. We must therefore ask whether the existence of a specific microstate can be explained by the laws of physics. Up to now we have found no support for the hypothesis that the molecular boundary

---

[13]Similar conclusions have been reached by P. Hoyningen-Huene, "Zu Problemen des Reducktionismus der Biologie," *Philosophia Naturalis* 22 (1985): 271-86.

conditions of a living organism are a direct consequence of physical laws. Instead, we have a lot of experimental evidence for the assumption that all microstates in sequence space have the same probability of becoming realized under prevailing physical laws. Thus, since the total number of all possible microstates is extremely large, the probability of the realization of a certain predefined microstate becomes vanishingly small. This in turn means that the selection and stabilization of a predefined microstate, for example a microstate that carries biological information, cannot be explained within the framework of equilibrium physics.

Thus, a selective self-organization of the microstates in sequence space seems only to be possible under the conditions of nonequilibrium processes, in the course of which an a priori indeterminable number of microstates is narrowed down to a few biologically relevant ones.[14] The present results of the new paradigm of self-organization show unambiguously that the process of molecular self-organization is essentially subject to certain principles of selection and optimization, and these can be reduced completely to the known laws of physics. However, the results also indicated an inherent limitation of the reductionistic research program. Thus, although the existence of specific boundary conditions can be completely explained as a general phenomenon within the framework of physics, it is not possible to deduce their detailed physical structure. This has been revealed by a thorough analysis of evolutionary dynamics in sequence space. The fine structure of biological boundary conditions reflects the historical uniqueness of the underlying evolutionary processes, which, by definition, cannot be described by natural laws.

We conclude that the concept of boundary conditions is indispensable for the understanding of complex phenomena. Many questions of philosophical relevance still waiting for an answer fit closely to this concept. These concern above all the relationship between the concept of boundary conditions and that of information, as well as the possible connection between the notions of contingency and emergence. Furthermore the question of the relevance of boundary conditions for the understanding of neural networks has yet to be analyzed—a question which may yield a novel approach to the old problem of the relationship between mind and brain.[15]

---

[14]Küppers, "On the Prior Probability of the Existence of Life," in *The Probabilistic Revolution*, vol. 2, ed. L. Krüger, G. Gigerenzer, and M.S. Morgan (Cambridge, MA: MIT Press, 1987), 355-69.

[15]I wish to thank Paul Woolley for stimulating discussions on the subject of this article.

CHAOS, PROBABILITY, AND THE COMPREHENSIBILITY OF THE WORLD

Michael Heller

1 *Introduction: Compressibility and Comprehensibility*

There are quotations which mark important steps in the history of human thought. One of them is certainly the passage from Einstein:

> The very fact that the totality of our sense experiences is such that by means of thinking (operations with concepts, and the creation and use of definite functional relations between them, and the coordination of sense experiences to these concepts) it can be put in order, this fact is one which . . . *we shall never understand.* One may say *"the eternal mystery of the world is its comprehensibility"* (italics mine).[1]

This mystery is seminally present in our prescientific cognition, but it reveals itself in full light only when one contemplates, as it has been expressed by E. Wigner, "the unreasonable effectiveness of mathematics in the natural sciences."[2]

What is meant here by the *effectiveness* of mathematics in the natural sciences is rather obvious (at least for those of us who are accustomed to the methods of modern physics): we model the world in terms of mathematical structures, and there exists an admirable resonance between these structures and the structure of the world; by means of experimental results the world responds to questions formulated in the language of mathematics. But why is this strategy *unreasonable*? In constructing mathematical theories of the world we invest into them information we have gained with the help of the joint effort of former experiments and theories. However, our theoretical structures give us back more information than has been put into them. It looks as if our mathematical theories were not only information processing machines, but also information creating devices.

Let us consider an outstanding example. In 1915, after a long period of struggle and defeat, Einstein finally wrote down his gravitational field equations. He succeeded in deducing from them three, seemingly insignificant, effects by which his general theory of relativity differed from the commonly accepted Newtonian theory of gravity. These effects were so small that the majority of physicists at that time could see no reason to accept a theory that required such a huge mathematical structure and yet explained so little. However, "the equations

---

[1] Albert Einstein, "Physics and Reality," chap. in *Ideas and Opinions* (New York: Dell, 1978), 283-315.

[2] E. Wigner, "The Unreasonable Effectiveness of Mathematics in the Natural Sciences," *Communications in Pure and Applied Mathematics* 13 (1960): 1-4.

are wiser than those who invented them."[3] This is certainly true as far as Einstein's equations are concerned. In about half a century physicists and mathematicians found a host of new solutions to these equations. Some represent neutron stars, gravitational waves, cosmic strings, stationary and rotating black holes, and so on. Fifty years ago nobody would even have suspected the existence of such objects. Now some of them have been discovered in the Universe,[4] and our confidence in Einstein's equations has grown so much that we are sure that the existence of at least some others will soon be experimentally verified. New information seems not only to be created by the mathematical equations, but surprisingly often it corresponds well to what we observe if we focus our instruments on domains suggested by the equations themselves. It looks as if the structure of Einstein's equations somehow reflects the structure of the world: information about various strata of the world's structure seems to be encoded in the equations. By finding their correct solution, and correlating it, through suitable initial or boundary conditions, with the given stratum of the structure of the world we are able to decipher this information, and it often happens that this information was unavailable before we solved the equations.

We often read in philosophy of science textbooks that the mathematical description of the world is possible owing to *idealizations* made in the process of constructing our theories (we neglect the air resistance, the medium viscosity, or we invent non-existing motions along straight lines under the influence of no forces, and so on). This is a typical half-truth. At least in many instances, it seems that the idealization strategy does not consist in putting some information aside, but instead it is one of the most powerful mechanisms for the creation of information. For instance, the law of inertia (uniform motion under the influence of no forces!) has led us into the heart of classical mechanics. We should also notice that there were no experimental results that suggested which "influences" should be neglected, but it was the form of the equations of motion that selected those aspects of the world upon which the experiments should focus. The quantum world would remain closed to us forever if not for our mathematical models and idealizations on which they are based. Here we had no possibility at all to choose what should be taken into account and what should be left aside. We are totally at the mercy of mathematical structures. Almost all of the more important concepts of our every-day experience—such as localization, motion, causality, trajectory in space and time, individuality—drastically change their meanings when we move from the macroscopic world to the quantum world of elementary interactions. The only way to visualize what happens in this world is to force our imagination to follow mathematical structures and surrender to their explicative power.

Mathematics, as employed to reconstruct physical situations, enjoys another "unreasonable" property—it has enormous unifying power: in an almost miraculous way it unifies facts, concepts, models, and theories far distant from

---

[3]This saying is ascribed to Heinrich Hertz.

[4]'*U*niverse' will be used to refer to the Universe in the maximal possible sense. Later we shall see that the Universe may contain many *u*niverses.

each other. The huge field of phenomena investigated by contemporary physics has been divided into a few subdomains, with each subdomain governed by a single equation (or system of equations). The equations of Einstein, Schrödinger, and Dirac are the best known representatives of this aristocratic family of equations. One printed page would be enough to write down the entirety of physics in a compressed form. We prefer fat volumes because we want to explore the architecture of these mathematical structures: we gain understanding by analyzing, step by step, the system of inferences, and by interpreting the formal symbols, triggering this subtle resonance between the logical structure and the results of measurement. We feel entitled to believe that these subdomains of physics, which until now were separated from each other, are but different aspects of the same mathematical structure. Although still remaining to be discovered, it is often credited with the name "Theory of Everything."

To express a law of physics in the form of a differential equation means to collect a potentially infinite set of events into a single scheme, in the framework of which every event, by being related to all other events, acquires a significance and is explained. This is an example of what is called *algorithmic compressibility*. However, what is really important is that it is always possible to disentangle what has been compressed: if needed, each event can be extracted from the entirety (but already in its reprocessed, significant form) by finding a suitable solution and choosing the corresponding boundary conditions. Today Einstein's question "Why is the world so comprehensible?" is often formulated: "Why is the world algorithmically compressible?"[5] Indeed:

> [W]ithout the development of algorithmic compressions of data all science would be replaced by mindless stamp collecting—the indiscriminate accumulation of every available fact. Science is predicated upon the belief that the Universe is algorithmically compressible and the modern search for a Theory of Everything is the ultimate expression of that belief, a belief that there is an abbreviated representation of the logic behind the Universe's properties that can be written down in finite form by human beings.[6]

All these properties of mathematics, when applied to physical theories, often evoke in scientists the feeling of encountering something extremely beautiful. One could ask: Is mathematics beautiful because it is effective? This would be a utilitarian theory of beauty. Or more in the Platonic vein, do only beautiful mathematical structures prove to be effective in physics? Probably these questions have no straightforward answer, but the fact that they are so often asked points

---

[5]I shall not discuss here the question of whether the statements: "The world is comprehensible," and "The world is algorithmically compressible" are equivalent, or whether the latter is but a part of the former.

[6]John D. Barrow, *Theories of Everything: The Quest for Ultimate Explanation* (Oxford: Clarendon Press, 1991), 11. See also Joseph Ford's essay, "What is chaos, that we should be mindful of it?" in *The New Physics*, ed. Paul Davies (New York: Cambridge University Press, 1989), 348-72, for clarification regarding science's belief that the Universe is algorithmically compressible.

toward the significant (albeit not yet sufficiently acknowledged) role of esthetics in the philosophy of science.

Was Einstein right when he expressed his belief that the comprehensibility of the world will remain its "eternal mystery"? There is an attempt to neutralize Einstein's puzzlement over the question of why the world is so comprehensible by reducing all regularities present in the Universe to the blind game of chance and probability.

> [I]t is just possible that complete anarchy may be the only real law of nature. People have even debated that the presence of symmetry in Nature is an illusion, that the rules, governing which symmetries nature displays, may have a purely random origin. Some preliminary investigations suggest that even if the choice is random among all the allowable ways nature could behave, orderly physics can still result with all the appearances of symmetry.[7]

Two essentially different implementations of this philosophy have been envisaged. The first and less radical attempt seeks to explain all regularities observed in the present Universe by reducing them to the chaotic (i.e., "most probable") initial conditions. The second, maximalistic one, claims that the only fundamental law is the "game of probabilities," and all the so-called natural laws are but averages which won in this game. Although only partial results have been obtained so far in both of these approaches, the philosophical ideas lying behind them seem to be an interesting counterproposal with respect to Einstein's philosophy, and certainly are worthwhile to discuss. This is the goal of the present paper.

The problem is of key importance for the topic of this publication. In its most fundamental sense, God's action in the world consists in giving to the world its existence, and giving it in such a way that everything that participates in existence also participates in its rationality, that is, is subject to mathematically expressible laws of nature. If Einstein's "mystery of comprehensibility" is indeed neutralized by the "pure game of chance and probability," then the central meaning of God's action in the world seems to be in jeopardy: anarchy takes over, the world at its foundations is not rational. Since rationality and existence are very close to each other, the existence of the world, in turn, no longer seems to be the most profound *locus* of God's action but a random outcome of a degraded mystery.

Hence, I shall show that such an attempt to neutralize Einstein's fascination with the comprehensibility of the world leads us even deeper into the mystery. Probability calculus is as good as any other mathematical theory, and even if chance and probability lie at the core of everything, the important philosophical and theological problem remains of why the world is *probabilistically comprehensible*. Why has God chosen probability as God's main strategy? In fact, the theory of probability permeates all aspects of our present understanding of the world. In particular, deterministic chaos theory and the theories of complexity and self-

---

[7]John Barrow and Joseph Silk, *The Left Hand of Creation: The Origin and Evolution of the Expanding Universe* (London: Unwin, 1983), 213.

organization—the main subject of this volume on divine action—work because the world enjoys certain probabilistic properties.

We begin by presenting in more detail those approaches which attempt to explain the present world's regularities probabilistically. Hence, in section 2, we turn to the question of whether the chaotic initial conditions for the Universe are able to explain its actual structure. In section 3, we discuss the program of reducing all physical laws and symmetries to pure chance and randomness. To deal responsibly with the problems posed in the two preceding sections we must undertake a thorough discussion of the foundations of probability calculus. This is the aim of section 4. The conclusions are drawn and their theological implications are discussed in section 5.

2 *The Sharpest Needle Standing Upright on its Point*

Although the need to justify certain large-scale properties of the observed Universe (such as its spatial homogeneity and isotropy) was noticed rather early by many authors, the paradigm of the "anthropic principle" stressed the fact that the initial conditions for the Universe had to be extremely "fine-tuned" to produce a world which could be the subject of exploration by a living observer. There is no need to repeat here all arguments which have been quoted on behalf of this thesis.[8] All these arguments point to the fact that the present state of the Universe is as hard to produce from random initial conditions as it is hard to make "the sharpest Needle stand Upright on its Point upon a Looking-Glass" (this is Newton's expression introduced in, essentially, the same context).[9]

The additional difficulty is that each mechanism proposed to explain the large-scale properties of the Universe must first be able to overcome the barriers created by the limiting velocity of the propagation of physical interactions (the so-called "horizon problem"). Within the standard world model, to answer the question why a certain property of the Universe (for instance, the temperature of the microwave background radiation) is the same in regions which were never able to communicate with each other, one essentially needs to postulate the fine-tuning of the initial conditions responsible for this fact.

An early proposal to overcome these difficulties goes back to C.W. Misner's classical works[10] on the so-called Mixmaster program in cosmology, nowadays more often known under the name of the chaotic cosmology. The idea is that it was the "mixing" character of physical processes in the very young Universe that led

---

[8]See Barrow and Frank J. Tipler, *The Anthropic Cosmological Principle* (Oxford: Clarendon Press, 1968); and John Leslie, *Universes* (New York: Routledge, 1989).

[9]Sir Isaac Newton, "Letter 2," in *Isaac Newton's Papers and Letters on Natural Philosophy and Related Documents*, ed. I.B. Cohen and R.E. Schofield, (Cambridge, MA: Harvard University Press, 1958), 292.

[10] C.W. Misner, "The Isotropy of the Universe," *Astrophysical Journal* 151 (1968): 431-57; and *idem*, "Mixmaster Universe," *Physical Review Letters* 22 (1969): 1071-74.

to its large-scale properties in their present shape, independently of any special initial conditions. To put it briefly, the "mixing" processes end up always producing identical universes regardless of their initial state. Various processes were tried as mixing candidates: hadron collisions, particle creation, neutrino viscosity, but all these mechanisms are strongly constrained by the "horizon problem." This means that they can work efficiently only in those cosmological models which enjoy a very special geometric property: The expansion rate of the Universe must be related to the velocity of propagation of the mixing process such that the mixing should be able to reach distant regions of the Universe before they are too distant to be affected (the goalpost cannot recede faster than the runner can run). This means that there must exist a large-scale property of the Universe which controls the mixing process by synchronizing it with the global expansion rate and does this without exchanging physical signals between distant parts of the Universe. But this is exactly what we wanted to avoid.[11] Let us note that this problem is strictly connected with the phenomenon of deterministic chaos, which is one of the main topics in this volume. In fact, the Misner program does not work since—as deterministic chaos theory predicts—the relaxation time in the Mixmaster model is reached after an infinite lapse of time (the so-called Omega time appearing in Misner's equations).

A newer attempt to solve these difficulties, proposed by A. Guth[12] and A. Linde,[13] is known as the "inflationary scenario." In its standard version, at the epoch when the Universe was 10^{-35} seconds old, the splitting of the strong nuclear force from the electroweak force made the factor driving the world's evolution[14] negative and caused a rapid (exponential) expansion of the Universe to be superimposed on its ordinary expansion. In a fraction of a second the radius of the Universe increased from about 10^{-23} cm to 10 cm (22 orders of magnitude!), that is, from something which was 10 billion times smaller than the size of a proton to something about the size of an orange. After this dramatic inflation phase the Universe came back to its standard, much slower expansion. Such a rapid inflation erased from the Universe all vestiges of its pre-inflationary state; in this way, the initial conditions are unimportant. On the other hand, regions of the Universe, now

[11]See, Z.A. Golda, M. Szydlowski, and M. Heller, "Generic and Nongeneric World Models," *General Relativity and Gravitation* 19 (1987): 707-18; and A. Woszczyna and M. Heller, "Is a Horizon-free Cosmology Possible?," *General Relativity and Gravitation* 22 (1990): 1367-86.

[12]A. Guth, "Inflationary Universe: A Possible Solution to the Horizon and Flatness Problems," *Physical Review* D23 (1981): 347-56.

[13]A.D. Linde, "A New Inflationary Scenario: A Possible Solution of the Horizon, Flatness, Homogeneity, Isotropy and Primordial Monopole Problems," *Physical Letters* 108B (1982): 389-93.

[14]Strictly speaking, the factor is $\rho+3p$, where ρ is the energy density and p the pressure of the fluid filling the Universe.

very distant from each other, remember information from the epoch when they were in mutual contact. In this way the horizon problem can be overcome.[15]

However, one should notice that the inflationary strategy is able to explain probabilistically the present large-scale properties of the Universe only if the set of initial conditions leading to the inflationary phase is "large enough" in the space of all initial conditions. There are strong suspicions that this is not the case.[16] If this is true we again face the problem of fine-tuning to explain the inflation itself. And so, difficult questions return through the back door.

3 *Laws from No-Laws*

The standard view today, underlying all efforts to achieve the final unification of physics, is that at extremely high energies, somewhere beyond the Planck threshold, everything had been maximally symmetric, and that the subsequent breaking of this primordial symmetry led the Universe to its present diversified richness of forms. There is, however, another possibility: there could be no symmetry at high energies at all or, equivalently, all possible symmetries could coexist on an equal footing, with order and law emerging only later from the primordial chaos. Barrow and Tipler, considering this possibility, ask the following questions: "'*Are there any laws of Nature at all?*' Perhaps complete microscopic anarchy is the only law of Nature?"[17] What we now call the laws of nature would be the result of purely statistical effects, a sort of asymptotic state after a long period of averaging and selecting processes. "It is possible that the rules we now perceive governing the behavior of matter and radiation have a purely random origin, and even gauge invariance may be an 'illusion': a selection effect of the low-energy world we necessarily inhabit."[18]

There are several attempts to implement this philosophy in working physical models. I shall mention two of them.

(1) Within the so-called *chaotic gauge* program only preliminary results have been obtained so far. The idea is to show that physical laws and symmetries should arise, by some averaging processes, from a fundamental, essentially lawless and non-symmetric, level. In this approach, at low energies (i.e., on our macroscopic scale) one sees maximum symmetry, but this gradually disappears if we penetrate into more fundamental levels of high energies.[19] In particular, this approach should refer to gauge symmetries, which seem to play an ever increasing role in

---

[15]See L.F. Abbot and So-Young Pi, eds. *Inflationary Cosmology* (Singapore: World Scientific, 1986).

[16]See George F.R. Ellis and William Stoeger, "Horizons in Inflationary Universes," *Classical and Quantum Gravity* 5 (1988): 207-20.

[17]Barrow and Tipler, *Anthropic Cosmological Principle*, 256.

[18]Ibid.

[19]See J. Iliopoulos, D.V. Nanopoulos, and T.N. Tomaras, "Infrared Stability or Anti-Grandunification," *Physics Letters* 94B (1980): 141-44.

contemporary physics. The proponents of this program write: "It would be nice to show that gauge invariance has a high chance of arising spontaneously even if nature is not gauge invariant at the fundamental scale."[20] Or more technically, it would be nice to show that if the Lagrangian, from which physical laws are to be derived, is chosen at random, then at low energies local gauge invariance will emerge, and it will be a stable property in the space of all possible Lagrangian-based theories. However, the same authors were only able to show that a gauge theory arises at low energies from a theory which at high energies differs from the exact gauge invariant theory by no more than a specified amount of non-invariant interactions, that is, that it is necessary to assume an approximately gauge-invariant theory at high energies in order to obtain the usual gauge theory on our scale. Advocates of this program express their hope that, by using this strategy, it will be possible to estimate the order of magnitude of at least some fundamental constants and to demonstrate their quasi-statistical origin.

(2) Another possibility is Linde's *chaotic inflationary cosmology*.[21] The dynamics of the Linde universe is dominated by a non-equilibrium initial distribution of a non-interacting scalar field ϕ with the mass m, $m \ll M_P$, where $M_P \sim 10^{19}$ GeV is the Planck mass. If the Universe contains at least one domain of the size $l \geq H^{-1}(\phi)$ with $\phi \geq M_P(M_P/m)^*$, where $H = R'/R$, R being the scale factor of the locally Friedman universe, it endlessly reproduces itself in the form of inflationary mini-universes. In fact, this reproduction process leads to an exponentially growing number of causally non-interacting universes. Since the birth of each new mini-universe is independent of the history of the mother Universe, "the whole process can be considered as an infinite chain reaction of creation and self-reproduction which has no end and which may have no beginning."[22] When, during such a birth process, the Universe splits into many causally disconnected mini-universes of exponentially growing sizes, "all possible types of compactification and all possible vacuum states are realized."[23] This leads to various kinds of physics in various daughter-universes. Linde writes:

> Whereas several years ago the dimensionality of spacetime, the vacuum energy density, the value of electric charge, the Yukawa couplings, etc., were regarded as true constants, it now becomes clear that these "constants" actually depend on the type of compactification and on the mechanism of

---

[20]D. Foerster, H.B. Nielsen, and M. Ninomiya, "Dynamical Stability of Local Gauge Symmetry," *Physics Letters* 94B (1980): 135-40.

[21]See, A.D. Linde, *Fizika Elementarnykh Chastits I Infliatsionnaia Kosmologiia* [Physics of Elementary Particles and Inflationary Cosmology] (Moscow: Nauka, 1990); and *idem*, "Inflation and Quantum Cosmology," in *300 Years of Gravitation*, S.W. Hawking and W. Israel, eds. (Cambridge: Cambridge University Press, 1987), 604-30.

[22]Linde, "Inflation and Quantum Cosmology," 618.

[23]Ibid., 627. Roughly speaking, by compactification Linde understands the process by which the number of space-time dimensions is established inside the newly-born mini-universe; this number may be different from that of the mother Universe.

symmetry breaking, which may be different in different domains of our universe.[24]

In this way, a "chaos"[25] is realized not within the one Universe but within the ensemble of many universes, and some sort of the anthropic principle is necessary if our "local universe" is to have the physical laws we now discover and the structure we now observe.[26]

4 *Probabilistic Compressibility of the World*

The strategy presented in the two preceding sections was aimed at understanding the Universe by reducing its laws and structure to a pure game of probabilities. Our first reaction to such strategies is that if one of them succeeds (especially one of their stronger versions presented in section 4), then the "eternal mystery of the world's comprehensibility" that Einstein stressed would disappear: comprehensibility would give place to probability and mystery would change into an averaging mechanism. However, to go beyond "first reactions" and to critically assess such an approach to the "rationality of the world," we must turn to the foundations of the probability calculus. This is the aim of the present section.

Many branches of modern mathematics have their origin in an interplay of theory and application. This is also true as far as the probability calculus is concerned. Moreover, one would be inclined to say that in this case more depends on application than on theory. This is not only because the probability calculus originated from experience but mostly because it is very difficult to separate the very notion of probability from its empirical connotations. This fact gave rise to many philosophical discussions concerning the foundations of probability. In what follows I shall try to avoid entering into these discussions; instead I shall trace the meaning of some fundamental concepts by placing them within the mathematical structure of the probability theory in its standard (Kolmogorov) formulation.[27]

In the contemporary standard approach, probability theory is a special instance of measurement theory. *Measure*, in the mathematical sense, is a function defined on subsets of a certain space called the *measure space*. These subsets, called *measurable subsets*, can be thought of as objects to be measured. The function defined on these objects ascribes to each of them the result of a measurement (i.e., its measure, a number). For instance, the objects in question could be subsets of the Euclidean space and the measure a function ascribing to

---

[24]Ibid.

[25]In this case, as in the case of the chaotic gauge program, the term 'chaos' is not used in the technical sense of deterministic chaos, although one could expect that in both cases deterministically chaotic phenomena (in the technical sense) are involved.

[26]See Barrow, *The World within the World* (Oxford: Clarendon Press, 1988), 281-89.

[27]Regarding different views and philosophies of probability, see D. Home and M.A.B. Whitaker, "Ensemble Interpretations of Quantum Mechanics: A Modern Perspective," *Physics Reports* 210 (1992): 223- 317.

each subset its volume. From the mathematical point of view, the essential circumstance is that outside the measure space the concept of measuring is meaningless.

Some cases are known in which not every subset of a given space is measurable. In such a space there are "things" (subsets) which cannot be measured, that is, no measurable result can be meaningfully ascribed to them. This runs counter to the common view that "what cannot be measured does not exist." Such subsets might indeed seem rather unusual, but one can find them even in the open interval (0,1) of real numbers.[28]

Probability is just a measure satisfying one additional condition: the measure of the entire space should be equal to one. Consequently, the measure of any of its subsets is either zero or a fraction between zero and one. If this axiom is satisfied the measure space with its measurable subsets is called *probability space*, and the measure defined on it the *probability distribution*.

Let us notice that so far there is nothing in our theory which would suggest the uncertainty or indeterminacy we intuitively connect with the idea of probability. All consequences follow from the axioms in a strictly apodictic manner, exactly the same as in other mathematical theories. Intuitions we connect with the concept of probability enter our theory *via* its reference to reality, that is, *via* its interpretation. The standard method of referring mathematics to reality is by the intermediary of physics: some mathematical structures are used as building blocks of a physical theory, and the task of this theory is to investigate the world. The mathematical theory of probability, however, seems to relax this rule. It often makes references to reality with no direct help of a physical theory. For instance, when making probabilistic predictions of the outcome of throwing dice or of a price increase in an approaching fiscal year, a certain physical-like interpretation of a mathematical structure must intervene, but it is so natural and so closely linked to the mathematical structure itself that we prefer not to call it a physical theory but rather a *probabilistic model* of a given situation (with no reference to physics).

To be more precise, physical intuition enters the probabilistic model through the definition of the probability distribution. For instance, if we want to model playing with ideal dice mathematically, we define the probability distribution as a function which ascribes to each *elementary event*, that is, to each of six possible outcomes, the value of the probability measure equal to 1/6. This particular value is taken from experience, namely from a long series of throwing dice, but once put into the definition of the probability distribution, it becomes a structural part of the mathematical theory itself.

The feeling of a "probabilistic uncertainty" is connected with the *frequency interpretation* of the distribution function defined in the above way. The value 1/6

---

[28]For example, let a and b be real numbers in the open interval (0,1). If a-b is a rational number we write $a\#b$. This is clearly an equivalence relation. We define A to be a subset of real numbers consisting of exactly one number of each equivalence class. It can be shown that A is not measurable. See Robert Geroch, *Mathematical Physics* (Chicago: University of Chicago Press, 1985), 254-55.

of the distribution function at a given (elementary) event, for instance at the event "outcome three," is interpreted as giving the relative frequency of the "outcome three" (i.e., the ratio of the number of fortuitous events, in our case "outcome three," to all possible events) in a long series of throwing dice. Indeed, experiments show that, in this circumstance, relative frequencies are approximately equal to 1/6. The longer the series of throws, the closer the relative frequency approximates this value. This property of the world is known as its *frequency stability*. It is a property of the world, and not of the mathematical theory, since it is taken from experience and has no justification in the theory itself.

The frequency stability of the world is of fundamental importance for our analysis. In both everyday life and in physics we often meet random events or random experimental results. The result of an experiment is said to be random if it is not uniquely determined by the conditions under which the experiment is carried out and which remain under the control of the experimenter. Subsequent results of such an experiment are unpredictable. If in a series of n such experiments, n_A experiments give the result A, and n-n_A give some other results, the number $f(A) = n_A/n$ is called the *frequency* of A. It turns out that as n is larger and larger, $f(A)$ approaches a certain number more and more closely. This tendency to certain numerical results reflects the world's frequency stability.

This is indeed an astonishing property. One cannot see any *a priori* reason why the world should be stable in this respect. But the world is frequency stable, and it is clear that without this property the probability calculus could not be applied to analyze the occurrence of events in the world. We can say that owing to its frequency stability the world is *probabilistically compressible. A priori* we could expect that truly chaotic or random phenomena would evade any mathematical description, but in fact the description of phenomena we call random or chaotic is not only possible but can be compressed into the formulae of the probability theory. The probabilistic compressibility of the world turns out to be a special instance of its algorithmic compressibility, and one would dare to say that it is the most astonishing (or the most unreasonable) instance of it.

This is even more the case if we remember that the applicability of probabilistic ideas to the real world underlies much of the foundations of statistical physics, and also the derivation of the classical limit of quantum theory, as well as the analysis of observations. All these aspects of probability applications are closely related to the problem of the arrow of time. Since the laws of fundamental physics are time reversible they must be involved in a subtle game of probabilities in order to produce irreversible phenomena on a macroscopic scale. There are strong reasons to suspect that the answer to the question of why the cosmic process evolves in time, rather than being reduced to an instant, is but another aspect of the probabilistic compressibility of the world.

And we should not forget that probability theory is as good as any other mathematical theory. The distribution function is defined by idealizing some experimental results, but the probabilistic model, once constructed, produces uniquely determined results. The frequency interpretation of the probabilistic axioms does not influence formal inferences or the manipulation of formulae; it only allows us to look at the Universe in a special way—in a way in which events are not just given but seem to have a certain potentiality to happen, and the cosmic process does not just unfold but seems to have the possibility of choosing various branches in its unfolding.

The above considerations have shown that even if we were able to reduce the comprehensibility of the world to its probabilistic compressibility (as it was presupposed by the strategies and philosophies presented in sections 2 and 3), the questions would remain: Why does probability theory apply to our world? Why has our world the property of being frequency stable?

When we ask the question "Why is the world mathematical?" we should also wonder why it is subject to the "game of probabilities." Clearly the riddle of probability does not eliminate the mystery of comprehensibility.

5 *God of Probabilities*

The ideas presented in sections 2 and 3 have their origin in an interesting property of the human mind, for which the high probability of an event is a kind of sufficient reason for its occurrence, but low probabilities always call for some special justification. One could guess that this property of our mind has evolved through an intricate agglomeration of selection effects in the world, the structure of which is predominantly shaped by frequency-stable processes.

In classical natural theology, the justification of low probability events was often sought in the direct action of God. The low probability itself was considered to be a gap in the natural course of events, a gap that had to be filled in by the "hypothesis of God." In this way, high probability becomes a rival of God. We hear the echo of such views in metaphors contemporary scientists sometimes evoke to impress the reader with how finely the initial conditions should be tuned to produce the Universe in which the reader-like being could be born and evolve. For instance, in a famous book by Roger Penrose, the caption under the picture of God pointing with the pin to the initial conditions (or equivalently to the point in phase space) from which God intends to create the world, reads:

> In order to produce a universe resembling the one in which we live, the Creator would have to aim for an absurdly tiny volume of the phase space of possible universes—about $1/10^{10^{123}}$ of the entire volume.[29]

On the contrary, the attempt to reduce phenomena to the random events hiding behind them (e.g., to random initial conditions) is often thought of as supporting an atheistic explanation. For instance, the main argument of Leslie's book on the anthropic principles[30] is that the principal competitor of the God hypothesis is the idea of multiple worlds in which all possibilities are realized, along with some observational selection effects which would justify our existence as observers of our world. The God hypothesis relies on the argument from design,

---

[29]Roger Penrose, *The Emperor's New Mind: Concerning Computers, Minds, and the Laws of Physics* (New York: Oxford University Press, 1989), 343.

[30]Leslie, *Universes*. See also Barrow and Tipler, *Anthropic Cosmological Principle*.

which is "based on the fact that our universe looks much as if it were designed." However, there might be immensely many universes.

> And their properties are thought of as very varied. Sooner or later, somewhere, one or more of them will have life-permitting properties. Our universe can indeed look as if designed. In reality, though, it may be merely the sort of thing to be expected sooner or later. Given sufficiently many years with a typewriter even a monkey would produce a sonnet.[31]

A different view of probability came with the advent of quantum mechanics. The Hilbert space, an arena (in fact, the phase space) in which quantum processes occur, is a very beautiful and very solid mathematical structure, but when interpreted in a standard probabilistic way it reveals the unexpected image of the micro-world: Wave functions, containing all information about a quantum object, are essentially non-local entities. They are defined everywhere; for instance, from the wave function you can compute the probability of finding an electron at any place in the Universe. Wave functions evolve in time in a strictly deterministic way, but when a measurement is performed deterministic evolution breaks down, all available information reduces to the unique measurement result, an infinite number of possibilities collapses to the single eigenvalue of the measurement operator.[32] Less informed philosophers speak about the free will of electrons; better informed ones see that the time-honored antinomy between lawfulness and probability should be reconsidered *ab initio*.

Recent developments in deterministic chaos theory have shown that this is also true as far as the macroscopic world is concerned. An instability of the initial conditions leads to unpredictable behavior at later times, and there are strong reasons to believe that a certain amount of such a randomness is indispensable for the emergence and evolution of organized structures.

The shift we have sketched in our views on the significance of probability has had its impact on modern natural theology. Randomness is no longer perceived as a competitor of God, but rather as a powerful tool in God's strategy of creating the world. For instance:

> God is responsible for ordering the world, not through direct action, but by providing various potentialities which the physical universe is then free to actualize. In this way, God does not compromise the essential openness and indeterminism of the universe, but is nevertheless in a position to encourage a trend toward good. Traces of this subtle and indirect influence may be discerned in the progressive nature of biological evolution, for example,

---

[31]Leslie, *Universes*, 1. However, Leslie clearly expresses his own opinion: "While the Multiple Worlds (or World Ensemble) hypothesis is impressively strong, the God hypothesis is a viable alternative" (p. 1).

[32]In quantum theory, any measurement is represented by an operator acting on the corresponding wave function. Eigenvalues of this operator represent possible results of the measurement.

> and the tendency for the universe to self-organize into a richer variety of ever more complex forms.[33]

Or:

> On this view God acts to create the world *through* what we call "chance" operating within the created order, each stage of which constitutes the launching pad for the next. However, the actual course of this unfolding of the hidden potentialities of the world is not a once-for-all pre-determined path, for there are unpredictabilities in the actual systems and processes of the world (micro-events at the "Heisenberg" level and possibly non-linear dynamical complex systems). There is an open-endedness in the course of the world's "natural" history. We now have to conceive of God as involved in explorations of the many kinds of unfulfilled potentialities of the universe(s) he has created.[34]

Still, either a God with a sharply pointed pin in hand choosing the improbable initial conditions for the Universe, or a God exploring the field of possibilities by playing with chance and randomness, seem to be but a Demiurge constrained by both a chaotic primordial stuff and the mathematical laws of probability (just as Plato's Demiurge was bound by the preexisting matter and the unchanging world of ideas). Of course, we could simply identify the laws of probability with God (or with the ideas present in God's mind), but this would bring us back to all traditional disputes surrounding the Platonic interpretation of God and mathematics.

Instead of immersing ourselves in risky disputes, I believe we should once more ask Einstein's question: Why is the world so comprehensible? As we have seen, there is no escape from this question *via* the "game of probabilities," since if we reduce comprehensibility to probability, new questions will emerge: Why should the theory of probability be privileged among all other mathematical theories?[35] Why is the world probabilistically compressible? And if the answer to the last question is: The world is probabilistically compressible since it enjoys the property of being frequency stable, we will then ask: Why is it frequency stable?

Any natural theology is sentenced to the "God of the gaps" strategy. But if there are no gaps in the natural order of things, if the world is a self-enclosed entity, then there is no way from the world to its maker. The essential point is to distinguish between spurious gaps and genuine ones. Spurious gaps are temporary holes in our knowledge usually referring to an incomplete scientific theory or hypothesis and to a restricted domain of phenomena. Genuine gaps are truly disastrous; they overwhelm everything. I think that all gaps are spurious except for the following two or three:

---

[33]Paul Davies, *The Mind of God: The Scientific Basis for a Rational World* (New York: Simon & Schuster, 1992), 183. Davies refers here to Whitehead's philosophy of God.

[34]Arthur R. Peacock, "God as the Creator of the World of Science," in *Interpreting the Universe as Creation: A Dialogue of Science and Religion*, ed. Vincent Brümmer (Kampen, The Netherlands: Kok Pharos, 1991), 110-11.

[35]To see that this question is not trivial, see section 4.

First is the *ontological gap*. Its meaning is encapsuled in the question: Why is there something rather than nothing? The problem at stake is sheer existence. Even if we had a unique theory of everything (and some physicists promise us we shall have it in the not too distant future), the question would remain who or what "has breathed fire into the equations" to change what is merely a formally consistent theory into one modeling the real universe.

Second is the *epistemological gap*: Why is the world comprehensible? I have dealt with this question in the present paper. It is truly a gap. Science presupposes the intelligibility of the world, but does not explain it. Philosophy of science can at most demonstrate the non-trivial character of this question, but remains helpless if one further asks "Why?"

From the theological perspective both gaps, the ontological gap and the epistemological one, coincide: everything that exists is rational, and only the rational is open to existence. The source of existence is the same as the source of rationality.

I strongly suspect that there is the third genuine gap; I would call it the *axiological gap*—it is connected with the meaning and value of everything that exists. If the Universe is somehow permeated with meaning and value, they are invisible to the scientific method, and in this sense they constitute the real gap as far as science and its philosophy are concerned. Here again, by adopting the theological perspective, I would guess that the axiological gap does not differ from the remaining two: the sources of existence, rationality, and value are the same.

Modern developments in science have discovered two kinds of elements (in the Greek sense of this word) shaping the structure of the Universe—the *cosmic elements* (integrability, analycity, calculability, and predictability), and the *chaotic elements* (probability, randomness, unpredictability, and various stochastic properties). I think I have convincingly argued in the present paper for the thesis that the chaotic elements are in fact as "mathematical" as the cosmic ones, and if the cosmic elements provoke the question of why the world is mathematical, the same is true as far as the chaotic elements are concerned. On this view, *cosmos* and *chaos* are not antagonistic forces but rather two components of the same *Logos* immanent in the structure of the Universe.[36] Einstein's question: "Why is the world so comprehensible?" is a deeply, and still not fully understood, theological question.

---

[36]For examples of such a cooperation between "cosmic" and "chaotic" elements see my paper "The Non-Linear Universe: Creative Processes in the Universe" (especially section 5), to be published in the *Proceedings of the Plenary Session of the Pontifical Academy of Sciences* (1992).

CHANCE AND LAW IN IRREVERSIBLE THERMODYNAMICS, THEORETICAL BIOLOGY, AND THEOLOGY[1]

Arthur R. Peacocke

There is an element of "necessity" in the universe, the giveness, from our point of view, of certain of its basic features: the fundamental constants, the nature of the fundamental particles (and so of atoms, and so of molecules, and so of complex organizations of molecules), the physical laws of the interrelation of matter, energy, space, and time. We are in the position, as it were, of the audience before the pianist begins his extemporizations—there is the instrument, there is the range of available notes, but what tune is to be played and on what principle and in what forms is it to be developed?

1 *Chance*

Given the limiting features which constitute our necessity, how are the potentialities of the universe going to be made manifest? Jacques Monod's answer was that it is by "chance": indeed humankind's emergence in the "unfeeling immensity of the universe" is said to be "*only* by chance."[2] So the question to which we turn is that of the roles of chance and necessity, or "law," in the evolutionary process, in particular in the origin and development of living forms, and of the implications of this balance-and-interplay for discourse about belief in God as creator. (By 'law' here is meant the law-like framework constituted by the constraints of the laws of physics, chemistry, etc., and of the fixed values of the fundamental constants.) It will transpire that, by and large, I agree that chance, appropriately defined, is the means whereby the potentialities of the universe are actualized but that from this I shall draw conclusions different from those of Monod.

Chance has often been apotheosized into a metaphysical principle threatening the very possibility of finding meaning in human life, as recognized in the bitter

---

[1]Paper given at the conference on "Our Knowledge of God and Nature: Physics, Philosophy, and Theology," organized by the Vatican Observatory and held at Castel Gandolfo, September 21-26, 1987. The proceedings of that workshop, which were published as *Physics, Philosophy and Theology: A Common Quest for Understanding*, ed. Robert John Russell, William R. Stoeger, and George V. Coyne (Vatican City State: Vatican Observatory, 1988), did not contain this paper, which was published separately in *Philosophy in Science*, 4 (1990): 145-80.

[2]Jacques Monod, *Chance and Necessity: An Essay on the Natural Philosophy of Modern Biology*, trans. Austryn Wainhouse (London: Collins, 1972), 267.

comment of the author of Ecclesiastes: "Time and Chance govern all."[3] In the ancient Greek myths Chance reigned in Chaos, that state of affairs which preceded the Cosmos we now inhabit. Chaos was the mythical state of affairs which preceded the emergence of the world order, of Cosmos, which was thought to manifest itself in the totality of natural phenomena.

Even before Darwin, many were disturbed by this ancient fear of chaos ruled by chance, and the publication of his ideas gave an impetus to the anguish of those already despairing of finding meaning or purpose in the universe. To attribute the processes of the universe to "chance" can trigger off in the sensitive a profound sense of despair at the meaninglessness of all life, and of human life in particular. Such an emotive word warrants closer analysis, for there are more precise meanings which may be given to it in the context of the sciences.

1.1 *Two Meanings of 'Chance'*

For our present purposes we can distinguish usefully two meanings of 'chance.'

(1) When we toss a coin we say that the chances of it coming down heads or tails are even. We mean that in any long run of coin tossings 50% will come down "heads" and 50% "tails" to a proportional accuracy which increases with the number of throws we make. But we also know that, had we sufficient knowledge of the exact values of the relevant parameters, the laws of mechanics would enable us in fact to say in any particular toss which way the coin would fall. In practice we cannot have all the information needed to analyze these multiple causes, and all we can know is that their net effect is equally likely to produce "heads" as "tails" after any individual tossing. So to apply 'chance' in this context is simply to recognize our ignorance of the multiple parameters involved. It is a confession of our partial ignorance, partial because we do know enough from the symmetry of the problem to say that in any long run of such tossings there will be an equal number of heads and tails uppermost at the end of the process. The use of the word 'chance' in this context does not imply a denial of causality in the sequence of events.

(2) A second use of the word 'chance' is that of the intersection of two otherwise unrelated causal chains. Suppose that when you leave the building in which you are reading these pages, as you step onto the pavement you are struck on the head by a hammer dropped by a man repairing the roof. From this accidental collision many consequences might follow for your mental life and for the welfare of your families. In ordinary parlance we would say it was due to "pure chance." The two trains of events—your leaving the building at the time you did and the dropping of the hammer—are each within themselves explicable as causal chains. Yet there is no connection between these two causal chains except their point of intersection, and *when* the hammer hits you on the head could not have been predicted from within the terms of reference of either chain taken by itself. In this case causality is again not denied, but because there is no cross-connection between the two causal chains we could not, unlike the previous case of the tossing

---

[3]Eccl. 9:11.

of a coin, make any accurate prediction of the chance of it happening. (The second instance is sometimes more properly called 'accident,' and some authors distinguish between chance and accident in this sense.)

Much more needs to be said (and indeed is said in the vast literature on the mathematical theory of probability); at least this initial simple analysis serves to show that when, in ordinary parlance, some event is said to be 'due to chance' this phrase is really not giving an explanation of the event in question or saying what its cause is but is simply acting as a stop card. It is saying in effect "the event in question has many multiple causes or seems to have been the result of the intersection of unrelated causal chains, so that we cannot attribute any *particular* cause to it." It is therefore a phrase to be avoided in our discussion. No doubt the phrase 'due to chance' has acquired currency because many of the laws in natural science are statistical in character. They do not take the form of statements to the effect that event or situation *A* will be followed by event or situation *B* but rather of the form that *A* will be followed by B', B'', or B''' with different respective probabilities. Whether this incomplete knowledge of the consequence of *A* arises from a fundamental absence of causality in the old sense or is the consequence of the incompleteness of our knowledge of the operative multiple causes (as in the coin-tossing example) will depend on the particular situation. The first of these two alternatives, a fundamental absence of causality, is sometimes called 'pure chance.' Any event whose cause has not yet been discovered may be viewed either as a pure chance event, which possesses no cause, or as a complex event of cause as yet unknown. However, the very notion of pure chance, of uncaused events in the sense of absolutely unqualified disorder, is self-contradictory as well as running counter to a basic assumption of scientists in their, not unsuccessful, work. As is well known, the Heisenberg uncertainty principle raises important questions concerning causality at the sub-atomic level, but even such micro-events follow statistical probabilistic laws.

2 *The Life Game*

Until the recent past, chance and law have often been regarded as alternatives for interpreting the natural world. But the interplay between these principles is more subtle and complex than the simple dichotomies of the past would allow. For any particular state of a system we have to weigh carefully what the evidence is about their respective roles in determining its present behavior and for interpreting its past. The origin and development of living organisms is no exception, and we now must consider some interpretations of the life game which have emerged from scientific work of the last three decades—I refer to the ideas of Monod, in molecular biology, and Ilya Prigogine in irreversible thermodynamics and theoretical biology.

2.1 *Mutations and Evolution*

The contemporary debate was sparked off in the early 1970s by Monod, the French molecular biologist and Nobel Laureate, in his book *Chance and Necessity*, which enjoyed a *succès de scandale*. He contrasted the "chance" processes which bring about mutations in the genetic material of an organism and the "necessity" of their consequences in the well-ordered, replicative, interlocking mechanisms which constitute that organism's continuity as a living form. He pointed out that mutations in the genetic material, DNA, are the results of chemical or physical events, and their locations in the molecular apparatus carrying the genetic information are entirely random with respect to the biological needs of the organism. Thus, one causal chain is a chain of events, which may be the chemical modification of one of the nucleotide bases in DNA or its disintegration through absorption of a quantum of ultraviolet or cosmic radiation. These changes in the nucleotide bases, and so in the information which the DNA is carrying, are incorporated into the genetic apparatus of the organisms only if they are not lethal and if, on interacting with its environment, they produce an organism which has a higher rate of reproduction than before. This sequence represents a second causal chain—the interplay between the genetic constitution (and behavior) of a living organism and the pressures to which it is subjected by the environment. These two causal chains are entirely independent, and it is in the second sense of chance that Monod is correct in saying that evolution depends on chance. It also qualifies for this description, in the other sense of chance, since in most cases we are not now in a position to specify all the factors which led to the mutated organisms being selected and, even less, the mechanism by which mutation was induced in the first place. (Indeed this latter is at a level at which quantum considerations begin to operate and is fundamentally precluded from any exact prediction.)

The molecular biology of recent years has thus been able to give a much more detailed picture of the process of the interplay between mutation and environment. This is the basis on which Monod stressed the role of chance:

> Pure chance, absolutely free but blind, at the very root of the stupendous edifice of evolution: this central concept of modern biology is no longer one among other possible or even conceivable hypotheses. It is today the *sole* conceivable hypothesis, the only one compatible with observation and tested fact. And nothing warrants the supposition (or the hope) that conceptions about this should, or ever could, be revised.[4]

Monod went on to draw the conclusion that *Homo sapiens*, and so all the works of human mind and culture, are the products of pure chance and therefore without any cosmic significance. The universe must be seen not as a directionally ordered whole (a cosmos) but as a giant Monte Carlo saloon in which the dice have happened to fall out in a way which produced humanity. There is no general purpose in the universe and in the existence of life and so none in the universe as a whole. It need not, it might not, have existed—nor might we!

---

[4]Monod, *Chance and Necessity*, 110.

However, *pace* Monod, I see no reason why this randomness of molecular event in relation to biological consequence, which Monod rightly emphasized, has to be raised to the level of a metaphysical principle interpreting the universe, for, as we have already seen, in the behavior of matter on a larger scale many regularities, which have been raised to the level of being describable as "laws," arise from the combined effect of random microscopic events which constitute the macroscopic. So the involvement of chance at the level of mutations does not, of itself, preclude these events manifesting a lawlike behavior at the level of populations of organisms and indeed of population of biosystems that may be presumed to exist on the many planets throughout the universe which may support life. Instead of being daunted by the role of chance in genetic mutations as being the manifestation of irrationality in the universe, it would be more consistent with the observations to assert that the full gamut of the potentialities of living matter could be explored only through the agency of the rapid and frequent randomization which is possible at the molecular level of the DNA. In other words, the designation 'chance' in this context refers to the multiple effects whereby the (very large) number of mutations is elicited that constitute the "noise," which, via an independent causal chain, the environment then selects for viability. This role of chance is what one would expect if the universe were so constituted as to be able to explore all the potential forms of organizations of matter (both living and non-living) which it contains. Moreover, even if the present biological world were only one out of an already large number of possibilities, it must be the case that the potentiality of forming such a world is present in the fundamental constitution of matter as it exists in our universe. The original primeval cloud of fundamental particles must have had the potentiality of being able to develop into the complex molecular forms we call modern biological life. It is this that I find significant about the emergence of life in the universe, and the role of chance, in both its forms, seems to me neither repulsive nor attractive but simply what is required if all the potentialities of the universe, especially for life, were going to be elicited effectively.

Since Monod wrote *Chance and Necessity* there have been developments in theoretical biology which cast new light on the interrelation of chance and law in the origin and evolution of life. In these developments it is possible to see more clearly than Monod was able to analyze, in his consideration of the mutation of the genetic material and their consequences for natural selection, the way in which chance processes can operate in a law-regulated system to produce new forms of organized and information-carrying systems of the kind which life requires. To these more recent ideas we must now turn.

3 *Irreversible Thermodynamics and Theoretical Biology*

"All kinds of private metaphysics and theology have grown like weeds in the garden of thermodynamics."[5] This pungent judgment of a distinguished historian of science serves to remind us that, although thermodynamics has an austere and lofty intellectual and architectonic framework, it has nevertheless frequently generated a plethora of gloomy emotions in those who have attempted to apply it on a cosmic scale.

What, in fact, is this aspect of modern physics and physical chemistry to which is attributed such dire consequences? R.A. Butler wrote a book on politics a few years ago which he called *The Art of the Possible*.[6] Thermodynamics is the *science* of the possible—it stands in relation to science as a whole very much as logic does to philosophy. It does not invent; it proscribes. Paradoxically, classical thermodynamics has been most successful in deriving the relationships that characterize physical systems at equilibrium, that is, just those systems in which all processes are reversible, unlike most actual systems which usually involve natural irreversible processes. The state of equilibrium is that to which all processes tend to lead any actual system. So classical thermodynamics deals accurately and powerfully with a kind of limiting world—that to which the actual tends. It is not, on that account, to be underestimated. The thermodynamic account of equilibrium and its interpretation of natural processes may be regarded as one of the best established pillars of modern science.

It was one of the great achievements of nineteenth-century science which led to the formulation of a property, the entropy (S), that not only measured the extent of irreversibility in a natural process in a particular system but was also a property of the state of that same system. The second law could then be stated in the form: "In any real process, there is always an increase in entropy." The greater the value of the increase in entropy (ΔS) in a natural process occurring in a particular system, the greater the extent of its irreversibility—that is, the more "degraded" and less available for performing mechanical work had the energy become in the process, and the more it had become chaotic, *thermal* energy. Thus, changes in entropy measured something to do with the "character," and not simply the quantity, of energy in a system.

What this "character" was, that was related to availability to perform directed mechanical work, only became clear with the later development of *statistical* thermodynamics, particularly at the hands of L. Boltzmann. This development is most easily understood in the context of the realization that the energy states of any system are all discrete and not continuous (even for translational energy). The "character" of the energy that is related to the entropy was proposed by Boltzmann to be the "spread" or distribution of matter over the possible available energy

---

[5]Erwin H. Hiebert, "The Uses and Abuses of Thermodynamics in Religion," *Daedalus* 95 (1966): 1075.

[6]Richard Austen Butler, *The Art of the Possible: The Memiors of Lord Butler* (London: Hamilton, 1971).

states—the degree of randomness or "disorderliness," as it were, of the matter-energy distribution. To be more precise, the entropy (*S*) is related to a quantity, denoted as *W*, which is the "number of complexions" of the system, that is, the number of possible dispositions of matter over the (equiprobable) available energy states (or, more precisely, the number of *micro*-states corresponding to a given *macro*-state identified through its macroscopic properties). The Boltzmann relation was simply:

$$S = k_B \ln W$$

where k_B, the Boltzmann constant, is equal to the gas constant (R) divide by Avogadro's number (N_o); and ln W is the natural logarithm of W (i.e., base *e*). Note that *W*=1 represents a maximum state of "order" with minimum entropy, namely zero, since ln1=0.

"Disorderliness" is, in this context, the zero value of a variable, "orderliness," which is the extent to which "any actual specimen of an entity approximates to the 'ideal' or pattern against which it is compared."[7] Such "orderliness" reaches its maximum extent in the "state of order," which is "an ideal reference state, laid down and specified, according to certain rules or conventions, which is taken as having 100% orderliness."[8] This "state of order" might well be that of the geometrical order of a crystal lattice at absolute zero. Orderliness would decrease (i.e., disorderliness would increase) according to the extent to which the atoms or molecules of the crystal were displaced from the lattice points and/or the extent of the spread of its quantized energy states relative to the ground state. For populations rather than single entities, 100% order with respect to any parameter or property would be characterized by all the members of the population exhibiting the same value of that parameter, or the same property. So when, loosely, entropy is said to be a measure of "disorder" or "randomness" it is this kind of orderliness that is being referred to. It is, to use an example of Denbigh,[9] the kind of order exemplified by a perfect wall-paper pattern rather than that of an original painting. Such "order" is therefore scarcely adequate as a measure of the complexity and *organization* of biological systems. Nevertheless it could be affirmed that the kind of *dis*order measured by entropy is incompatible with biological complexity and organization and, indeed, the state of maximum entropy, of maximum disorder in the sense defined, is equivalent to biological death. Thus, we may say that a state of low entropy is a necessary but not sufficient condition for biological complexity and organization to occur.

---

[7]K. Denbigh, "A Non-Conserved Function for Organized Systems," in *Entropy and Information in Science and Philosophy*, ed. Libor Kubat and Jiri Zeman (Oxford: Elsevier, 1975).

[8]Ibid.

[9]Ibid.

3.1 *Evolution and Thermodynamics—The Problem?*

One of the implications of the classical thermodynamic account of natural processes, just outlined, is that time has, in relation to natural events, a unidirectional character so that entropy has often been called "time's arrow," for the increase in entropy displayed by natural processes apparently specifies (or, better, is closely linked conceptually with) the direction of the flow of time. All of this has a curious and apparently problematic relation with that other great scientific development of the nineteenth century—namely the "discovery of time" in biology, more precisely, the idea of biological evolution. Time had already been "discovered" in the eighteenth century in the sense that the development of geology as a science had vastly extended the time-scale of the history of the earth as a planet and of the living organisms, including human beings, upon it. Then, with Darwin and Wallace, and the activity their proposal engendered, an understanding of both the inter-connectedness of all living forms and of their progressive development from single cells (themselves not *so* simple) to more and more complex forms was surprisingly rapidly established, in spite of the opposition we all know about. Yet, according to one version of the laws of thermodynamics (Clausius): "The energy of the universe is constant; the entropy of the universe is increasing to a maximum." As we saw, entropy, and so "disorder" in the sense we have defined, increases in all natural processes. So how is it that living organisms can come into existence and survive, swimming, as it were, against the entropic stream carrying all to thermal equilibrium and "heat-death"?

In living organisms, we see natural objects in which, while they are alive, complex organization is being maintained, and even enhanced, against the universal tendency of all processes to occur with an overall increase in "disorder." Are living organisms actually, in some way, breaking the second law of thermodynamics by maintaining systems in a high state of organization and so a low state of disorder, of low entropy? The brief answer to this question is "No," when one recalls that, for a natural process or change, one must take into account everything that changes. Now, living organisms are *open* systems exchanging energy and matter with their surroundings, and the changes in entropy in both the organism and the surroundings have to be assessed. So it is perfectly possible for there to be a *decrease* in entropy associated with the processes of metabolism, etc., occurring in a living organism while at the same time this decrease is more than offset by an *increase* in the entropy of the surroundings of the organism on account of the heat that passes to these surroundings from the organism.[10] So living organisms, once in existence, do not in any sense "break" the second law of thermodynamics—any more than these laws are broken in certain purely physical processes when a more ordered form of a system is generated with the evolution of heat, for example, the freezing of super-cooled water to form ice, or the generation of a density gradient in the molecular concentration along a tube of gas which is in contact with a source

---

[10] Recall the basic classical definition of entropy increase as heat absorbed divided by temperature, for a reversible process.

of heat at one end of the tube and a sink for heat at the other end. In such systems, a steady state is eventually reached with respect to the through-put of heat and the distribution of the molecules and in this state there is both energy flow and a steady rate of production of entropy in the flow processes. In these physical samples it is, of course, an "ordering" that is occurring, at the expense of a "disordering" (= increase in entropy) of the surroundings—and not strictly an increase in organization of the kind required to maintain living systems. But at least, an increase in order, thus physically defined à la Boltzmann, is a necessary pre-requisite, if not a sufficient one, for the maintenance of biological organization.

But it also has to be recognized that there are some natural processes, in which there is an overall increase in entropy, which nevertheless manifest an increase in molecular "complexity" of a kind not entirely captured by the simple concept of "order" already defined, though related to it. Molecular complexity *can* increase in chemical reactions that involve association to more complex molecular forms, in full accordance with the second law. For any decrease in entropy that results from the decrease in the number of complexions (W, above) when atoms or molecules combine, that is consequent upon the loss of translational modes of molecular motion, with their closely packed energy levels, to (by and large) vibrational modes, with their more widely spaced levels, is offset: (i) by an increase in the entropy of the surroundings resulting upon the heat generated by a decrease during the chemical combination in the potential (electronic) energy of chemical bonds; and (ii) by an increase in entropy due to the increase in configurational possibilities that occur, in spite of reduction in the number of molecules, when there is an increase in molecular heterogeneity with the formation of new chemical species (and this contribution is greater the greater the number of possible new chemical structures).

Wicken denotes (i) and (ii), respectively, as "energy-randomization" and "matter-randomization" to emphasize that it is the randomizing tendencies which the second law of thermodynamics formalizes that drive forward the formation of more complex structures.[11] The earth's biosphere is in a steady-state of free energy flux, from the sun to the earth (with its biosphere) to outer space, with the rhythm of the earth's diurnal rotation. Within the biosphere itself there is a continuous steady flow of energy through the various trophic levels of the ecology, with concomitant transfer of heat to the non-living environment and so to outer space. The formation of chemical bonds by process (i) then, entirely in accord with the second law and in conjunction with (ii), provides opportunity for the increase in molecular, and so organizational, complexity upon which natural selection then operates. So the second law of thermodynamics, far from prohibiting any increase in complexity, necessitates its increase at the molecular level. But does the recognition of this provide any thermodynamic basis for the actual coming into

---

[11]J. Wicken, "Information Transformations in Molecular Evolution," *Journal of Theoretical Biology* 72 (1978): 191-204; "The Generation of Complexity in Evolution: A Thermodynamic and Information-Theoretical Discussion," 77 (1979): 349-365; and "A Thermodynamic Theory of Evolution," 87 (1980): 9-23.

existence, in the first place, of living, organized systems of matter, as distinct from providing an interpretation of their continued existence, growth and death?

In the first half of the twentieth century, classical thermodynamics had been extended to natural irreversible processes at the hands of L. Onsager, J. Meixner, A. Katchalsky, S.R. de Groot, T. de Donder, and I. Prigogine and his colleagues of the Brussels school. Although this extension reaped many rewards in other fields, it did not assist much in the interpretation of biological processes. For the thermodynamics of irreversible processes was developed for the situation in which the flows and rates of the processes were linear functions of the "forces" (temperature, concentration, chemical potential gradients) that impelled them. Such linear, non-equilibrium processes can, as we just saw, lead to the formation of configurations of lower entropy and higher order—so that non-equilibrium can be a source of order. But the "order" so created is not really structural and far from the organized intricacies of biology. Moreover, and more pertinently, biological processes depend ultimately on biochemical ones and these, like all chemical reactions not at equilibrium, are intrinsically non-linear (in the relation between reaction rates and driving forces, the "affinity," or free-energy difference) and fall outside the scope of irreversible thermodynamics at this stage of its development.

To show that the Boltzmann equilibrium ordering principle—coupled with $\exp(-E/k_BT)$ being proportional to the probability of occupation of a state of energy E at temperature T—is inadequate to explain the origin of biological structures, it suffices to take an example of Eigen.[12] Consider a protein chain of one hundred amino acids, of which there are twenty kinds—and biological systems are orders of magnitude more complex than this. The number of permutations of the possible order of amino acids in such a protein, and on which its biological activity and function depend, is 20^{100} ($\simeq 10^{130}$), assuming all sequences are equally probable. This is the number of permutations necessary to obtain a given arrangement starting from an arbitrary initial distribution. If a change of structure occurred at the (impossibly) high rate of one every 10^{-8} seconds, then 100^{122} seconds would be needed altogether to produce a given sequence—yet the age of the earth is 10^{17} seconds. So the chance of "spontaneous" formation of the protein, à la Boltzmann, through processes at equilibrium, is indeed negligible. Equilibrium cannot give rise to biological order. There is no chance of an increase in "order" of the Boltzmann kind by equilibrium consideration.

However, account now has to be taken of the earth not being a system in equilibrium but being, and always to have been, an *open* system through which there is a major flux of energy inwards from the sun and outwards into space (and perhaps some minor flux of matter, too, though this seems in practice to be negligible). Thus, at present, the earth receives energy from the sun during the day and absorbs this heat both physically and chemically through the green chlorophyll of plants; at night-time much of this energy is radiated out again but some is stored in the carbohydrates of plants and so finds its way into other living organisms

---

[12]Manfred Eigen, "Selforganization of Matter and the Evolution of Biological Macromolecules," *Naturwissenschaften*, 58 (1971): 465-523.

(including ourselves) as the intermediate source of their energy—before again being given up as heat to the atmosphere, and so to space again. Hence biological evolution has to be considered thermodynamically in relation to the openness of this whole system, and in particular, of the biosphere located near to and on the earth's surface. So it will be a thermodynamics of natural, irreversible processes in open systems that will be relevant to biological evolution and the coming into existence of living systems. There have been a number of other approaches for quantifying the evolutionary process and these have often been intricately interwoven in discussion of the relation of thermodynamics to the evolutionary process. But it is, in the end, in my view, the concepts of complexity and organization that are the most pertinent, and their relation to the thermodynamic interpretations of evolution are considered in the following section.

3.2 *Thermodynamics and the Evolution of Biological Complexity and Organization*

The biological complexity we actually observe now in the natural world is the product of a long evolutionary history, and different levels in the hierarchy of biological complexity have appeared at successive stages in the history of the earth. Today's biological structures are the result of a long process of development from earlier forms, a fact which is disregarded in the kind of calculations that try to estimate, say, the chance of a molecule of a typical protein being formed *de novo* from its constituent atoms, or even amino acids. Such calculations (*a fortiori* if complete organisms are considered) usually result, as we saw, in the conclusion that this probability is so low that the planet earth has not existed long enough for such a complex assembly to appear by the random motion of atoms and molecules—whether this period be that of the total 4-5 x 10^9 years of the earth's life, or the 1.1 x 10^9 years between the formation of the earth (4.6 x 10^9 years ago) and the oldest known rocks (3.5 x 10^9 years ago) containing the remains of living cells (blue-green algae) found in rocks at Warrawoona, Australia. The fallacy of such calculations lies in their ignoring the actual processes whereby complex self-reproducing (initially molecular and macromolecular) systems might self-organize themselves entirely consistently with currently known thermodynamics and chemical kinetics;[13] in ignoring the role of selection of organizations of macromolecules that have favored reproduction rates and, once established, irreversibly channel the evolutionary process in one particular direction;[14] and in ignoring the fundamental analyses of the architecture and evolution of complexity made by many authors, in particular by H.A. Simon.[15] He showed that complex systems will

[13]Arthur R. Peacocke, *An Introduction to the Physical Chemistry of Biological Organization* (Oxford: Clarendon Press, 1983).

[14]Eigen, "Selforganization of Matter"; and *idem*, "Molecular Organization and the Early Stages of Evolution," *Quarterly Reviews of Biophysics* 4 (1971): 149-212.

[15]H.A. Simon, "The Architecture of Complexity," *Proceedings of the American Philosophical Society* 106 (1962): 467-82.

evolve from simple systems much more rapidly if there are stable intermediate forms than if there are not, and that the resulting complex forms will then be hierarchic in organization. The requirement for stability of an atomic or molecular structure reduces to the requirement that the free energy of the structure be less by virtue of its structure than that of its component atoms or molecules. The complex structures that emerge in the evolution of the first living forms, and of subsequent forms, may be presumed to have a stability of this kind, though, since living systems are open, the stability attributed to them must not be identified directly with the net free energy of formation of their structures. The point that is essential to make here, in relation to the possibility of formation of complex forms, is that natural selection speeds up the establishment of each new stratum of stability in the succession of forms of life. Each stratification of stability is only a temporary resting place before random mutation and natural selection open up further new possibilities and so new levels of (temporary) stability—rather as if the free energy profile were a switchback with any given form stable only to the immediate environment of the controlling parameters and always with a finite chance of mounting the next barrier to settle (again temporarily) into a new minimum.

Bronowski has also pointed out that this "stratified stability," which is so fundamental in living systems, gives evolution a consistent direction in time towards increased complexity.[16] He used the metaphor of evolutionary time as a "barbed arrow," because random change will tend in the direction of increasing complexity, this being the only degree of freedom for change—and Saunders and Ho have also produced a neo-Darwinian argument along similar lines for the inevitability of increases in complexity in evolution.[17]

3.3 *The Origin of Living Systems from the Non-Living and Dissipative Structures*

One of the particular achievements of thermodynamic theory in the last few decades has been its provision of a basis for understanding the spontaneous *coming into existence*, as distinct from the maintenance, of organized structures (which may be as much kinetically as topologically organized) in open (and closed) systems far from equilibrium when flux-force relationships are non-linear. This analysis by the Brussels school, under the leadership of Prigogine, has significance especially for understanding the emergence of functionally and structurally organized living matter from non-living, that is, non-self-copying, matter.[18] They have been able to show how, on a strictly thermodynamic basis, new organized forms of systems can (but not necessarily will) come into existence and be stable, if matter and energy are flowing through to maintain them. These new ordered

---

[16]J. Bronowski, "New Concepts in the Evolution of Complexity: Stratified Stability and Unbounded Plans," *Zygon* 5 (March, 1970): 18-35.

[17]P.T. Saunders and M.W. Ho, "On the Increase in Complexity in Evolution," *Journal of Theoretical Biology* 63 (1976): 375-84.

[18]For references, see Peacocke, *Biological Organization*, 17-72.

forms are called *dissipative structures* and are radically different from the "equilibrium structures" studied in classical thermodynamics, the "order" of which is attained only at low enough temperatures in accordance with Boltzmann's ordering principle (as described above, and coupled with $\exp(-E/k_B T)$ being proportional to the probability of occupation of a state energy E at temperature T). In this non-linear range, non-equilibrium can indeed be the source of an order that would not be predictable by the application of Boltzmann's principle. In such states there can occur, under the right conditions, fluctuations that are no longer damped and that are amplified so that the system changes its whole structure to a *new* ordered state in which it can again become steady and imbibe energy and matter from the outside and maintain its new structured form. This instability of dissipative structures has been studied by these workers who have set out more precisely the thermodynamic conditions for a dissipative system to move from one state to a new state which is more ordered than previously, that is, for "order through fluctuations" to occur.

It turns out that these conditions (described below) are not so restrictive that no system can even possibly obey them. Indeed a very large number of systems, such as those of the first living forms of matter, which must have involved complex networks of chemical reactions, are very likely to do so, since they are non-linear in their relationship between the forces and fluxes involved. The ordered configurations that emerge beyond such an instability of the thermody-namic branch of non-linear systems were called dissipative structures, because they are created and maintained by the entropy-producing "dissipative" processes occurring inside systems through which, being open, there is a continuous flux of matter and energy.

Model physical systems undergoing such transitions are now well-known, for example, the famous Bénard phenomenon wherein a hexagonal organization at right angles to the vertical heat flow is observed, at a certain critical point, in a column of liquid heated from below—and others cited by H. Haken who, because the awareness of them has now become so widespread, and because they share common features in the "bifurcation" of the solutions of non-linear differential equations controlling these phenomena, has invented a new name for the study of such systems, namely "Synergetics."[19] Even more pertinent to biological systems is the observation of order-through-fluctuations in chemical systems. Chemical networks can be of a very high degree of complexity through incorporating one or more autocatalytic steps and they are often non-linear (in the sense above) when not close to equilibrium. Then various kinds of oscillating reactions and other features can occur. One of the most striking of these is the so-called Belousov-Zhabotinsky reaction, the oxidation of malonic acid by bromate in the presence of

[19]Hermann Haken, *Synergetics—An Introduction: Non-equilibrium Phase Transitions and Self-organization in Physics, Chemistry and Biology*, 2d ed., enlarged (Berlin: Springer-Verlag, 1978).

cerium ions in solution.[20] With the right combination of solution conditions, and at constant temperature, the original homogeneous reaction mixture changes into a series of pulsing waves of concentration of cerium ions, moving up and down the tube, until eventually a steady state is reached in which there are static, banded layers of alternating high and low concentrations of ceric ions. From an originally homogeneous system, a highly ordered structure has appeared through the fluctuations that are possible in a non-linear system far removed from equilibrium. What has happened is that fluctuations in such a system have been amplified and, consistently with the laws of chemical kinetics, a new structure has appeared that is ordered, at first in time and then finally in space—representative of an alliance of chance and law. Under the conditions of this reaction the structural formation has a probability of unity, provided that the initial fluctuation arises from within the system. The causal chain leading to this fluctuation, although it cannot be discerned by ourselves, must itself be the result of law-like processes occurring at the micro-level, some of which may be subject to "Heisenberg uncertainty." Because of the discovery of these dissipative systems, and the possibility of "order-through-fluctuations," it is now possible, on the basis of these physico-chemical considerations, to regard not merely as highly probable, but as inevitable the emergence from non-living matter of those ordered and articulated molecular structures that are living. Instead of them having only an inconceivably small chance of emerging in the "primeval soup" on the surface of the earth before life appeared, we now see that ordered dissipative structures of some kind will appear in due course.

As Eigen puts it, the emergence of self-copying patterns of matter, that is, living systems is, through this new physico-chemical analysis, seen as "inevitable," but the *actual* course it will take is "indeterminate."[21]

One has to presume that before life had evolved there existed in this prebiotic stage a system containing replicating macro-molecules that could both maintain itself by means of some simple copying mechanism and have a potentiality of change incorporated into its very structure and function so as to facilitate its multistage evolution to more complex forms more efficient in survival and reproduction. Some of the kinetic and stochastic problems associated with such prebiotic systems have been very fully investigated by Eigen and his colleagues at Göttingen.[22]

In the treatment of the Brussels school it is not only the instability but also a succession of instabilities of dissipative systems, now appearing under the aegis of thermodynamic laws, that bridges the gap between the non-living and the living. This succession of instabilities can be a process of self-organization, and of

---

[20]John J. Tyson, "What Everyone Should Know about the Belousov-Zhabotinsky Reaction," in S.A. Levin, ed., *Frontiers in Mathematical Biology*, Lecture Notes in Biomathematics, 100 (Berlin: Springer-Verlag, 1994), 569-87.

[21]Eigen, "Selforganization of Matter," 519, n. 9.

[22]Ibid., nn. 9 and 11; and Eigen and P. Schuster, *The Hypercycle: A Principle of Natural Self-Organization* (Berlin: Springer-Verlag, 1979).

evolution, provided there are fulfilled the same conditions as for the initial emergence of dissipative structures. They are that a process of self-organization can occur in a system if: (i) the system is *open* to the flux of matter and energy; (ii) the system is not at equilibrium and preferably *far from equilibrium*; and (iii) the system must be *non-linear* in its flux-force relationships, that is, there must be strong coupling between its processes.

The evolution to order through an instability induced by fluctuations, is only a possible, not a certain, development and in fact requires, along with the subsequent stabilization to a dissipative structure, that some other very stringent conditions be fulfilled.

3.4 *Thermodynamics and Kinetic Mechanisms*

The existence of these other conditions reminds us that, like patriotism, thermodynamics "is not enough." For thermodynamics is always the science of the possible: it can allow, it can forbid, but it cannot prescribe. What actually occurs in any system obeying the conditions described above for it to become (more) self-organizing depends on the ability of the actual molecules and higher structures present, on their spatial and temporal arrangements and on the numerical parameters specific to these features. In a molecular system, this means the patterns generated will depend principally on chemical reaction rate constants and on diffusion constants, controlling movement across space. Similar parameters characteristic of higher order structures will also operate at other levels (e.g., in the patterning in predator-prey, herbivore-plant, and host-parasite systems[23]). Ever since the seminal paper of A.M. Turing (1952) on "The Chemical Basis of Morphogenesis," it has become increasingly clear that suitable combinations of rates of chemical reactions and diffusion processes can spontaneously generate patterns in space and time, both permanent and oscillating according to conditions.[24] This separate development of dynamical theory, of kinetics and of fluctuation theory opens up new vistas of interpretation of pattern formation at many levels of biology. These interpretations cannot be described here,[25] but the point is that it is now possible to understand the detailed mechanisms in actual dissipative systems that lead to new ordered ("self organized") forms appearing.

Particularly striking biological examples of such forms are provided by the oscillations in time and the patterning in space of the concentration of intermediates in glycolysis that have been observed by Benno Hess and his colleagues, and the quite extraordinary ability of individual unicellular slime mold organisms (in particular *Dictyostelium discoideum*) to come together, under conditions of

---

[23]Alfred J. Lotka, *Elements of Physical Biology* (New York: Dover, 1924; reprinted as *Elements of Mathematical Biology*, New York: Dover, 1956), reference is to reprint edition.

[24]A.M. Turing, "The Chemical Basis of Morphogenesis," *Philosophical Transactions of the Royal Society of London* B237 (1952): 37-72.

[25]See Peacocke, *Biological Organization*, 111-213.

starvation, in organized spiral and annular patterns in a colony that then behaves, temporarily, as an organized whole until new food sources are found.[26]

So, one may well ask, what is gained from the application of irreversible thermodynamic concepts and criteria to biological systems? The primary and overriding gain is undoubtedly in the ability of thermodynamics to provide, as it were, an architectonic framework which limits but does not in detail prescribe. One can then build on this framework by using other resources of dynamical theory, of kinetics, of fluctuation theory—and of precise experimental information and new knowledge of modes of control and regulation at all levels in biology. Structural order comes from the existence of constraints and the macroscopic and phenomenological approach of thermodynamics is uniquely fitted to handle such factors.

Thermodynamics can never work in isolation from other approaches based on the theory of fluctuations, of stability, of stochastic processes, and of non-linear differential equations. However, it has its own unique insights which serve to link reflection on biological systems with the whole corpus of physico-chemical theory. For the new concepts in irreversible thermodynamics of non-equilibrium as the source of order, of "order through fluctuations," of the decisive role of non-equilibrium constraints and of dissipative structures (spatial, temporal, or both) in open systems, broaden and deepen immeasurably our perspective on biological systems and whole organisms and, indeed, have already proved to be a stimulus for the kind of detailed work that is required to give them a "local habitation and a name."

4 *Theological Reflections*

4.1 *God and the World*

The postulate of God as creator of all-that-is is not, in its most profound form, a statement about what happened at a particular point in time. To speak of God as creator is a postulate about a perennial or "eternal"—that is to say, timeless—relation of God to the world, a relation which involves both differentiation and interaction. God is differentiated from the world in that he is totally other than it (indeed *this* dualism—of God and the world—is the only one that is foundational to Christian thought). God is postulated in answer to the question: Why is there anything at all? He is the "Ground of Being" of the world; or for theists, that without which we could neither make sense of the world having existence at all nor of its having that kind of intellectually coherent and explorable existence which science continuously unveils. But this affirmation of what is termed "transcendence" has to be held in tension with the sense of God's immanence in the world.

For the process of evolution is continuous with that of inorganic and cosmic evolution—life has emerged as a form of living matter and develops by its own inherent laws, both physico-chemical and biological, to produce new forms with

---

[26]See ibid., 203-12.

new emergent qualities requiring new modes of study (the various sciences) as well as new concepts and languages to describe and explicate them. So the stuff of the world has a continuous, inbuilt creativity—such that, whatever "creation" is, it is not confined to a restricted period of time but is going on all the time (and indeed modern physics would support seeing time itself as an aspect of the created order). So, if we identify the creativity of the world with that of its creator, we must emphasize that God is *semper creator*, all the time creating—God's relation to the world is perennially and eternally that of creator. But to speak thus is to recognize also that God *is creating* now and continuously in and through the inherent, inbuilt creativity of the natural order, both physical and biological—a creativity that is itself God in the process of creating. So we have to see God's action as being in the processes themselves, as they are revealed by the physical and biological sciences, and this means we must stress more than ever before God's *immanence* in the world. If the world is in any sense what God has created and that through which he acts and expresses his own inner being, then there is a sense in which God is never absent from his world and he is as much in his world as, say, Beethoven is in his Seventh Symphony during a performance of it.

The processes of evolution, initially the physical and cosmological, and then more strikingly the biological, are characterized by the *emergence* of new forms within and by means of continuous developments subject to their own inherent, regular, law-like behavior that is studied by the sciences. What emerges is usually more complex and, along certain branches of the evolutionary tree, more and more conscious, culminating in the self-consciousness and the sense of being a person that characterizes humanity. In theological terms, God's immanent creative action in the world generates within the created order a being, the human being, who becomes self-aware, morally responsible, and capable of himself being creative and of responding to God's presence. Thus, the inorganic, biological, and human worlds are not just the stage of God's action—they *are* in themselves a mode of God in action, a mode that has traditionally been associated with the designation "Holy Spirit," the creator Spirit. I think that to give due weight to the evolutionary character of God's creative action requires a much stronger emphasis on God's immanent presence in, with, and under the very processes of the natural world from the "hot big bang" to humanity.

If I had to represent on a blackboard the relation of God and the world, including humankind, I would not simply draw three spheres labeled respectively "Nature," "Humanity," and "God" and draw arrows between them to represent their inter-relation. Rather, I would denote an area representing nature and place that entirely within another area representing God, which would have to extend to the edges of the blackboard and, indeed, point beyond it (to infinity). When I came to depict human beings, I would have to place them with their feet placed firmly in nature but with their self-consciousness (perhaps represented by their brains?) protruding beyond the boundary of nature and into the area that attempts to "depict" God, or at least refer to him. The basic affirmation here is that all-that-is, both nature and humanity, is in some sense *in* God, but that God is, profoundly and ultimately, "more" than nature and humanity—there is more to God, who is, profoundly and ultimately, "more" than nature and humanity. God in his being transcends, goes beyond, both humanity and nature, for God is either in everything created from the beginning to the end, at all times and in all places, or he is not there at all. What we see in the world is the mode of God's creativity in the world. The analogy with Beethoven's Seventh Symphony as an expression of Beethoven's own inner creative being is, I think, a fair one. In the actual processes of the world, and supremely in human self-consciousness, God is involving himself and expressing himself as creator. However, since human beings have free will we have also to recognize that God put himself "at risk," as it were, in creatively evoking in the natural world, beings who have free will and who can transcend their perceived world and shape it in their own way.

4.2 *God and Chance*

How may this understanding of God creating "in, with and under" the ongoing processes of the natural world be held in consonance with the recognition that new, and increasingly complex, forms of both inorganic and eventually living matter emerge by a combination of what we recognize as "chance" and "law"—and that this combination is inherently creative in itself and involves, for sound thermodynamical and chemical reasons, an increase in complexity?[27] How can the assertion of God as creator be interpreted in the light of this new and profound understanding of the natural processes by which new organized forms of matter appear, both non-living and living? In evolution there is an interplay between random chance at the micro- level and the necessity which arises from the stuff of the world having its particular "given" properties and law-like behavior. These potentialities a theist must regard as written into creation by the creator's intention and purpose, and they are gradually actualized by chance exploring their gamut.

I have elsewhere[28] tried to express this situation by seeing God as creator as like a composer who, beginning with an arrangement of notes in an apparently simple tune, elaborates and expands it into a fugue by a variety of devices. Thus, I suggested, does a J.S. Bach create a complex and interlocking harmonious fusion of his original material. The listener to such a fugue experiences, with the luxuriant and profuse growth that emanates from the original simple structure, whole new worlds of emotional experience that are the result of the interplay between an expectation based on past experience ("law") and an openness to the new ("chance" in the sense that the listener cannot predict or control it).

Thus might the creator be imagined to unfold the potentialities of the universe which he himself has given it, selecting and shaping by his redemptive and providential action those that are to come to fruition—an improvisor of unsurpassed ingenuity? He appears to do so by a process in which the creative possibilities, inherent (by his own intention) within the fundamental entities of that

---

[27]Peacocke, "Thermodynamics and Life," *Zygon* 19 (December, 1984): 408-12; and *idem*, *Biological Organization*, chaps. 2 and 4-6.

[28]Peacocke, *Creation and the World of Science* (Oxford: Clarendon Press, 1979), chap. 3, esp. pp. 105-6.

universe and their inter-relation, become actualized within a temporal development shaped and determined by those selfsame inherent potentialities.

The image of creation as an act of composing and of the created order as a musical composition is surprisingly rich and fecund and, since propounding it in my 1978 Bampton Lectures,[29] I have come across a number of other authors resorting to the image of music, as flexible form moving within time, to express what they wish to say about both the created order and the act of creation—authors as diverse as Popper, Čapek and Eigen.[30] One recalls in this connection that the music of creation has also been a constant theme of the religions of India, for example the South Indian representations, in bronze, of the dancing Shiva, the creator-destroyer, as lord of the dance of creation.[31]

Both images, of the writing of a fugue and of the execution of a dance, serve to express the idea of God enjoying, of playing in, creation. This is not an idea new to Christian thought. The Greek fathers, so Harvey Cox argues, contended that the creation of the world was a form of play. "God did it they insisted out of freedom, not because he had to, spontaneously and not in obedience to some inexorable law of necessity."[32]

H. Montefiore has criticized this wholehearted acceptance of the creative interplay of chance and law as a sufficient account of the observed growth of complexity as leading to the "God of the deists"—that is, to a distant "absentee-landlord" God who, as it were, sets the universe going and leaves it to get on with it—"a remote, unmoved, unloving" God. He asserts this because he believes "chance and necessity may produce creativity, but they cannot produce purpose."[33] But what these new developments actually show is that the interplay of chance and law bring about that increase in complexity which is both inbuilt according to the laws of thermodynamics and chemical kinetics[34] and is also the basis of that increase of sentience and freedom of the individual organism which is the condition for the appearance of all those qualities in and values of humankind, the eliciting of which in the created order Montefiore quite properly wishes to attribute to the purpose of the creator God of Christian theism. D.J. Bartholomew, a professor of statistical and mathematical science, has made a thoroughgoing study of the role of chance in the natural order in relation to the concept of God.[35] He

---

[29]Ibid.

[30]See, Karl Popper, *The Unended Quest: An Intellectual Autobiography*, rev. ed. (London: Fontana, 1976); Milič Čapek, *The Philosophical Impact of Contemporary Physics* (Princeton, NJ: Van Nostrand, 1961); and Eigen and Ruthild Winkler, *The Laws of the Game: How the Principles of Nature Govern Chance*, trans. Robert and Rita Kimber (London: Allen Lane, 1982).

[31]Ananda Kentish Coomaraswamy, *The Dance of Shiva* (London: Peter Owen, 1958).

[32]Harvey Cox, *The Feast of Fools: A Theological Essay on Festivity and Fantasy* (Cambridge, MA: Harvard University Press, 1969), 151.

[33]Hugh Montefiore, *The Probability of God* (London: SCM Press, 1985), 98.

[34]See n. 26 above.

[35]David J. Bartholomew, *God of Chance* (London: SCM Press, 1984).

contends that chance is actually conducive to the kind of world which we would expect a God, such as Christians believe in, to create, and that God uses chance to ensure the variety, resilience, and freedom necessary to achieve his purposes. His basic hypothesis is that God uses chance, which "offers the potential Creator many advantages which it is difficult to envisage being obtained in any other way."[36] Even more strongly, he believes "God chose to make a world of chance because it would have the properties necessary for producing beings fit for fellowship with himself."[37] His whole position merits careful study although, in my opinion,[38] it takes too externalist a view of God's mode of action in the world and does not emphasize sufficiently, or develop any understanding of, God's immanence in the natural creative processes. Like Montefiore, whose views are otherwise contrary to his, Bartholomew fails to recognize[39] that the joint emphasis *both* on the role of chance in natural creativity *and* on the immanence of God in these same natural processes which I, for one, have been making, leads not to deism but to that integration of immanence and transcendence to which I have already referred, namely that "the Being of God includes and penetrates the whole universe, so that every part of it exists in Him, but . . . that his Being is more than, and is not exhausted by the universe."[40]

4.3 *Living with Chance*

This theoretical acceptance of the role of chance in creation and in the created world also has implication for our attitudes to the role of chance in human life. Rustum Roy, Director of the Materials Research Laboratory at Pennsylvania State University, entitled his first 1979 Hibbert Centenary Lecture, in London, "Living with the Dice-playing God."[41] He urged us to accept that the world displays patterned chance. It is not a chaos, for there is only a "loose coupling" through statistical laws and patterns, which still allows talk of "causes." In human life we must accept, for the stability of our own mental health and of our faith, that reality has a dimension of chance interwoven with a dimension of causality[42]—and that through such interweaving we came to be here and new forms of existence can arise. So we have to accept the interplay of chance and law as the mode of God's creativity. It seems to me to be more consistent with the fundamental creativity of reality than the belief—stemming from a Newtonian, mechanistic, determinist view of the universe with a wholly transcendent God as the great lawgiver—that God

---

[36]Ibid., 97.

[37]Ibid., 138.

[38]Peacocke, review of *God of Chance*, by D.J. Bartholomew, In *Modern Theology* 2 (January 1986): 157-61.

[39]Bartholomew, *God of Chance*, 97.

[40]*The Oxford Dictionary of the Christian Church*, 2d ed., s.v. "Panentheism."

[41]See Rustum Roy, *Experimenting with Truth: The Fusion of Religion with Technology Needed for Humanity's Survival* (Oxford: Pergamon, 1981), 188.

[42]Cf. Paul Burrough, *God and Human Chance* (Lewes: Book Guild, 1985).

intervenes in the natural nexus for the good or ill of individuals and societies. We have to learn to accept these conditions of creation and of creativity in the world—"the changes and chances of this fleeting world." Such an attitude can rightly be urged not simply as a psychological necessity but also as the outcome of the recognition that we have just been developing, that the creation of life itself, and the creativity of the living, inevitably involves an interplay of chance and law.

5 *Conclusion*

Norbert Wiener suggested that a thermodynamic perspective provoked the feeling that the universe was somehow against the experiment of life and that in nurturing our little enclave of organized existence, as biological and as social systems, we were swimming against the entropic stream that was in fact sweeping all to randomness and "dark night."[43] But the picture that is emerging in more recent thermodynamic analyses of dissipative systems and of living organisms has a different tenor. Certainly the stream as a whole moves in a certain general, overall direction which is that of increasing entropy and increasing disorder, in the specific sense I have defined. However, the movement of the stream *itself* inevitably generates, as it were, very large eddies *within* itself in which, far from there being a decrease in order, there is an increase first in complexity and then in something more subtle— functional organization. Now there could be no eddies without the stream in which they are located and so may it not be legitimate to regard this inbuilt potentiality for living organization that the entropic stream manifests as being its actual point— namely, why it *is* at all? There could be no self-conscious ness and human creativity without living organization, and there could be no such living dissipative systems unless the entropic stream followed its general, irreversible course in time. Thus does the apparently decaying, randomizing tendency of the universe provide the necessary and essential matrix (*mot juste*!) for the birth of new forms—new life through death and decay of the old.

---

[43]Norbert Wiener, *I am a Mathematician: The Later Life of a Prodigy* (Cambridge, MA: MIT Press, 1964), 325.

III

CHAOS, COMPLEXITY, AND DIVINE ACTION

THE METAPHYSICS OF DIVINE ACTION

John Polkinghorne

1 *Introduction*

'Metaphysics' is not a popular word in contemporary culture but, in fact, no one can live a reflective life without adopting some broad view of the nature of reality, however tentative and subject to possible revision it might need to be. Even militant scientific reductionists, for whom "physics is all," are metaphysicians. They claim to be able to extend the insights and laws of physics into regimes, such as human behavior, in which their total adequacy is an untested hypothesis. They are certainly going beyond (*meta*) physics.

Anyone who wishes to speak of agency, whether human or divine, will have to adopt a metaphysical point of view within which to conduct the discourse. The conceptual edifice thus constructed must be consonant with its physical base, but it will no more be determined by it than the foundations of a house completely determine the character of the building. In each case, there is constraint but not entailment. Metaphysical endeavor in general, and talk of agency in particular, will inevitably require a certain boldness of conjecture as part of the heuristic exploration of possibility. In our present state of ignorance, no one has access to a final and definitive proposal. The test of the enterprise will be the degree to which it can attain comprehensiveness of explanation and overall coherence, including an adequate degree of consonance with human experience. The principal strategy of nearly all writers on divine agency has been to appeal in some way to an analogy with human agency, though our ignorance about the latter makes this a precarious undertaking.

2 *Epistemology and Ontology*

Metaphysical theories are ontologically serious. They seek to describe what is the case. It is a central philosophical question how what is the case is related to our knowledge of the world. There is clearly no certain and simple way in which to make the connection. There has been a strong tradition since Immanuel Kant which emphasizes the unknowability of "things in themselves." The spectacles we wear behind the eyes (the presuppositions we bring to our interpretation of the world) and the epistemic blinkers imposed by our having to view reality from the limitations of a human perspective are held so to refract and limit our perceptions of the way things are that reality is inaccessible to us.

It is not necessary to give way to such metaphysical despair. Of course there is no *deductive* way of going from epistemology to ontology. In fact, an important aspect of the connection is precisely the problem of *induction*: what degree of knowledge could lead to an ontological conclusion? Yet almost all scientists believe that they are learning about the actual nature of the physical world that they investigate. Consciously or unconsciously, they are critical realists. One could define the program of critical realism as the strategy of seeking the maximum correlation between epistemology and ontology, subject to careful acknowledgment that we view reality from a perspective and subject to pushing the search for knowledge to any natural limits it may possess. Its motto is "epistemology models ontology"; the totality of what we can know is a reliable guide to what is the case. It has to be a *critical* realism because in some regimes (such as the quantum world) what is the case is so counterintuitive in terms of common sense expectation that it cannot be reduced to a simple-minded objectivity. We have to respect its idiosyncrasy, but that does not prejudice its reality. One can see how natural this strategy is for a scientist by considering the interpretation of the uncertainty principle in quantum theory. Heisenberg's original discovery was epistemological; he showed there were intrinsic limitations on what could be *measured.* Very shortly, he and almost all other physicists were giving the principle an ontological interpretation. It was treated as a principle of actual indeterminacy, not mere ignorance.

There was no logical necessity to make this transition. It could not be deduced. This is clearly established by the existence of alternative interpretations in which there is complete determinacy, but in ways that are hidden from human knowledge. One such interpretation is Bohm's version of quantum theory,[1] where a hidden wave guides the perfectly determined motion of purely classical particles. Another is many-worlds quantum theory,[2] in which the perfectly deterministic Schrödinger equation controls all that is, but its consequences are spread between parallel universes, not simultaneously open to human observation.

Neither of these interpretations has commended itself to the majority of physicists. They have freely (and in my view rightly) made the metaphysical decision to interpret quantum theory as indicating an intrinsic indeterminacy in physical reality. I have been arguing[3] that it is a rational and attractive option to pursue the same strategy in relation to other intrinsic unpredictabilities which we

---

[1]David Bohm, *Wholeness and the Implicate Order* (London: Routledge & Kegan Paul, 1980); Bohm and B.J. Hiley, *The Undivided Universe: An Ontological Interpretation of Quantum Theory* (London: Routledge, 1993).

[2]H. Everett, *Reviews of Modern Physics* 29 (1957): 454. See also, Alastair Rae, *Quantum Physics: Illusion or Reality?* (Cambridge: Cambridge University Press, 1986), chap. 6.

[3]John Polkinghorne, *Science and Creation: The Search for Understanding* (London: SPCK Press, 1988) chaps. 3 and 5; *idem*, *Science and Providence: God's Interaction with the World* (London: SPCK Press, 1989), chap. 2; *idem*, *Reason and Reality: The Relationship Between Science and Theology* (London: SPCK Press, 1991), chap. 3; and *idem*, "The Laws of Nature and the Laws of Physics" in *Quantum Cosmology and the Laws of Nature: Scientific Perspectives on Divine Action*, ed. Robert John Russell, Nancey Murphy, and C.J. Isham (Vatican City State: Vatican Observatory, 1993; Berkeley, CA: Center for Theology and the Natural Sciences, 1993), 437-448.

discover in nature. We should treat these epistemic limitations as being ontological opportunities for fruitful metaphysical conjecture.

3 *Some Questionable Metaphysical Strategies*

3.1 *Primarily Science-Based: Physicalism*

Our growing recognition of the remarkable powers of self-organization displayed by complex physical systems far from equilibrium has encouraged some to adopt a refined form of physicalism. They suppose that this will enable the completion of an adequate descriptive program of human experience on the basis of natural science alone.[4] I have already stressed that such a claim is metaphysical in character, however much it may seek to hide that fact behind the language of physics.

Such a strategy may be defended on the grounds that science has already explained much which was not understood by previous generations and why should we set limits to its eventual successes? Indeed it can be argued that the lessons of history encourage this point of view. The boundaries between organic and inorganic matter, between living entities and inanimate objects, are no longer perceived as total barriers to the advance of scientific explanation. Why should consciousness or human agency be thought to be different?

I would respond by pointing out that the lessons of history are more ambiguous than this argument acknowledges. Even within physical science itself there are many phenomena (the stability of atoms, superconductivity, the energy sources of stars) which only proved intelligible in terms of an extremely radical revision of then currently accepted physical principles, represented by the advent of quantum theory and relativity. When one considers the big ugly ditch which seems to intervene between physical talk (however complex and sophisticated in terms of neural networking or whatever) and mental talk (even at the most elementary level of perceiving a patch of pink), there seems no reason to suppose that its bridging will not require the most drastic revision, in unforeseeable ways, of our understanding of the nature of reality. In the words of the sharp-tongued theoretical physicist, Wolfgang Pauli, it is no use simply claiming "credits for the future," waving one's hands and hoping that one day present understanding will turn out this way.

Physical science seems light-years distant from the unaided understanding of the mental or the intentional, an indispensable requirement for an adequate metaphysical strategy. Moreover, the reductionist program that underlies physicalism is threatened by developments in physical science itself. The non-locality found in quantum theory shows that the subatomic world is one which cannot be treated atomistically.[5] The vulnerability of chaotic systems to the smallest

---

[4]See Bernd-Olaf Küppers, "Understanding Complexity" (in this volume).

[5]See, e.g., Polkinghorne, *The Quantum World* (London: Longman, 1984), chap. 7.

influences from their environment, consequent upon the exquisite sensitivity of such systems to fine details of their circumstance,[6] shows that they are never truly isolable. Physics is taking a holistic turn. The possibility of the existence of holistic laws of nature is one which should not be discounted. Certainly such laws would be more difficult to discover than the familiar laws governing the behavior of parts, and their form would surely be different from that of the differential equations which are the staple of current localized mathematical physics. Yet it would be a Procrustean imposition on science to deny that it could have access to such laws. It is clearly worthwhile to pursue the program of reductionist explanation as far as it can legitimately be pursued, but that is a methodological strategy for investigation, not a metaphysical strategy determining the total nature of reality. The dawning holism of physics points in a more hopeful direction if science is eventually to find a satisfactory integration into a comprehensive and adequate metaphysical scheme.

One final criticism of too great a reliance on the principle of self-organization needs to be made. The insights of non-equilibrium thermodynamics seem helpful in relation to the generation of structure and long-range order. Agency, however, seems to correspond to an altogether more flexible and open kind of time-development than that corresponding to typical self-organizing patterns, such as convection columns or chemical clocks.

3.2 Primarily Theology-Based: Primary Causality

At least since Thomas Aquinas, there has been a tradition of theological thinking which seeks to explain divine agency by appeal to the distinction between primary and secondary causality. A notable modern exponent of this point of view has been Austin Farrer with his idea of double agency.[7] The secondary web of created causality is treated as being complete and unriven. Yet the primary causality of God is supposed nevertheless to be ineffably at work in and through these created causalities. How this is so is not explained. Indeed Farrer would regard it as risking monstrosity and confusion if one were to attempt to discern the "causal joint" by which divine providence acts.

It is not clear to me what is gained by so apophatic an account of God's action. In the end, the answer seems to be "God only knows." I agree with Arthur Peacocke's judgment on the paradox of double agency that it "comes perilously close to that mere assertion of its truth . . . since Farrer on his own admission can give no account of the 'causal joint' between the agency of the Creator and even human action."[8] This seems to me to be a strategy of absolutely last resort, only to

---

[6]See, e.g., James Gleick, *Chaos: Making a New Science* (London: Heinemann, 1988), chap. 1.

[7]Austin Farrer, *Faith and Speculation: An Essay in Philosophical Theology* (London: A&C Black, 1967).

[8]Arthur Peacocke, *Theology for a Scientific Age: Being and Becoming—Natural and Divine* (Oxford: Blackwell, 1990), 149.

be undertaken if it proves impossible to make any satisfactory conjecture about the causal joint of God's agency. I do not believe we are in so desperate a case, and I make my own suggestion in the course of what follows.

3.3 *Top-Down Causality*

The causality which physics most readily describes is a bottom-up causality, generated by the energetic interaction of the constituent parts of a system. The experience of human agency seems totally different. It is the action of the whole person and so it would seem most appropriately to be described as top-down causality, the influence of the whole bringing about coherent activity of the parts. May not similar forms of top-down causality be found elsewhere, including God's causal influence on the whole of creation?

It is an attractive proposal, but it is important to recognize that without further explanation top-down causality is a far from unproblematic concept. Its uncritical use would amount to no more than sloganizing. It seems to me that two important difficulties have to be faced and discussed.

The first is one I have already referred to in discussing the limitations on the insights provided by the principles of self-organization. If one is to give an account of intentional agency, it will require something much more open and dynamic than simply the generation of long-range order or the propagation of boundary effects. Striking as instances of this kind can be (involving the coherent motion of billions of molecules), they are often fully explicable in terms of a bottom-up approach, generating long-range correlations between localized constituents (phase transitions in physics are good examples of this kind of phenomenon). True top-down causality will have to be more open and more non-local than that. I believe that chaotic dynamics, with its picture of the open exploration of proliferating possibilities within the confines of a strange attractor, may offer an important clue to how this might come about. Self-organization offers the prospect of the generation of different patterns of spatial order; chaotic dynamics offers the prospect of the generation of different temporal patterns of dynamical history. The latter seems much closer to notions of agency than the former.

The second point, closely related to the first, is that if there is to be room for the operation of true top-down causality, then there will have to be intrinsic gaps, a degree of underdetermination in the account of the bottom-up description alone, in order to make this possible.[9] It is to the possible identification of the source of this intrinsic openness that I now turn.

4 *Ontological Gaps*

It seems to me that our experience of human agency is basic and by itself sufficient to indicate that a metaphysical scheme affording no scope for top-down causality

[9]See Thomas Tracy, "Particular Providence and the God of the Gaps" (in this volume).

would be seriously defective. Yet metaphysics must be consonant with its physical basis and so it is necessary to consider whether there are appropriate intrinsic gaps already known to us in the bottom-up description of the physical world. There seem to be two broad possibilities:

4.1 *Quantum Theory*

May not agents, human or divine, act in the physical world by a power to determine the outcomes of individual indeterminate quantum events, even if the overall statistical pattern of many such events may still be expected to lie within the limits of probabilistic quantum laws?[10]

This form of causality would actually be effected in the basement of subatomic processes. The proposal requires, of course, the adoption of the metaphysical strategy of interpreting quantum theory as involving intrinsic indeterminacies, but that is a strategy consciously or unconsciously endorsed by the great majority of physicists. For agency thus exercised, these microscopic determinations would have to have their consequences amplified up to the macroscopic level.

There are a number of difficulties about this proposal in relation to human and divine agency. One relates to the amplification effect. Exactly how the quantum world interlocks with the everyday world is still a question of unresolved dispute. In essence, this is the measurement problem in quantum theory.[11] Until this question is settled, the micro-macro boundary is a difficult barrier to cross with confidence. One might hope that a way around this might result from the sensitivity of chaotic systems to small triggers. Very quickly, there seems to be established a dependence of the behavior of such systems on details of what is going on at the level of quantum indeterminacy. Yet the grave and unresolved difficulties of relating quantum theory to chaos theory,[12] or of what is often called "quantum chaos," makes this a perilous strategy to pursue.

There is a particular difficulty in using quantum indeterminacy to describe divine action. Conventional quantum theory contains much continuity and determinism in addition to its well-known discontinuities and indeterminacies. The latter refer, not to all quantum behavior, but only to those particular events which qualify, by the irreversible registration of their effects in the macro-world, to be described as measurements. In between measurements, the continuous determinism of the Schrödinger equation applies. Occasions of measurement only occur from time to time and a God who acted through being their determinator would also

---

[10]William Pollard, *Chance and Providence: God's Action in a World Governed by Scientific Law* (London: Faber, 1958); see also Nancey Murphy, "Divine Action in the Natural Order: Buridan's Ass and Schrödinger's Cat" (in this volume); and Tracy, "Particular Providence."

[11]See, e.g., Polkinghorne, *Quantum World*, chap. 6.

[12]Joseph Ford, "What is Chaos, that we should be mindful of it?" in *The New Physics*, ed. Paul Davies (Cambridge: Cambridge University Press, 1989).

only be acting from time to time. Such an episodic account of providential agency does not seem altogether satisfactory theologically.

4.2 *Chaos Theory*

The exquisite sensitivity of chaotic systems certainly means that they are intrinsically unpredictable and unisolable in character. In accordance with the realist strategy already discussed, I propose[13] that this should lead us to the metaphysical conjecture that these epistemological properties signal that ontologically much of the physical world is open and integrated in character. By 'open' is meant that the causal principles that determine the exchange of energy among the constituent parts (bottom-up causality) are not by themselves exhaustively determinative of future behavior. There is scope for the activity of further causal principles. By 'integrated' is meant that these additional principles will have a holistic character (top-down causality).

The deterministic equations from which classical chaos theory developed are then to be interpreted as downward emergent approximations to a more subtle and supple physical reality. They are valid only in the limiting and special cases where bits and pieces are effectively insulated from the effects of their environment. In the general case, the effect of total context on the behavior of parts cannot be neglected.

Of course, with present ignorance, it is no more possible for me to spell out the details of the subtle and supple physical reality I propose than it is for the physical reductionist to spell out how neural networks generate consciousness, or for those who rely on quantum indeterminacy to spell out how it generates macroscopic agency, or for those who rely on an unanalyzed notion of top-down causality through "boundary conditions" to spell out how it actually operates. We are all necessarily whistling in the dark. I prefer the tune I have chosen because it has a natural anchorage in what we know about macroscopic physical process and because it exhibits certain promising features which I will now discuss.

For a chaotic system, its strange attractor represents the envelope of possibility within which its future motion will be contained. The infinitely variable paths of exploration of this strange attractor are not discriminated from each other by differences of energy. They represent different patterns of behavior, different unfoldings of temporal development. In a conventional interpretation of classical chaos theory, these different patterns of possibility are brought about by sensitive responses to infinitesimal disturbances of the system. Our metaphysical proposal replaces these physical nudges by a causal agency operating in the openness represented by the range of possible behaviors contained within the monoenergetic

---

[13]See n. 3 above. See also Polkinghorne, *Science and Christian Belief: Theological Reflections of a Bottom-up Thinker* (London: SPCK Press, 1994; printed in the United States as *The Faith of a Physicist: Reflections of a Bottom-up Thinker* [Princeton, NJ: Princeton University Press, 1994]), chap. 1.

strange attractor. What was previously seen as the limit of predictability now represents a "gap" within which other forms of causality can be at work.

Because of the unisolability of chaotic systems, this new agency will have a holistic top-down character. It will be concerned with the formation of dynamic pattern, rather than with transactions of energy. In a vague but suggestive phrase I have proposed that it might best be thought of as "active information." There seems a hope that here we might discern a glimmer of how it comes about that intentional agency is exercised, either by our minds upon our bodies or by God upon creation.

It is important to recognize that, in this scheme, the significance of the sensitivity of chaotic systems to the effect of small triggers is *diagnostic* of their requiring to be treated in holistic terms and of their being open to top-down causality through the input of active information. It is not proposed that this is the localized mechanism by which agency is exercised. I do not suppose that either we or God interact with the world by the carefully calculated adjustment of the infinitesimal details of initial conditions so as to bring about a desired result. The whole thrust of the proposal is expressed in terms of the complete holistic situation, not in terms of clever manipulation of bits and pieces.[14] It is, therefore, a proposal for realizing a true kind of top-down causality. It may fittingly be called *contextualism*, for it supposes the behavior of parts to be influenced by their overall context. This implies a strong form of anti-reductionism in which processes are capable of being modified by the context in which they take place. This will be so for "cloudy" chaotic systems, but there will also be some "clockwork" systems, insensitive to details of circumstance, in which the behavior of the parts will be unmodified. Thus, one can understand the successes of molecular genetics in describing the (mechanical) behavior of DNA, without having to suppose that this justifies a claim that all aspects of living systems are adequately described in this reductionist fashion.

5 *A Metaphysical Proposal*

The classical metaphysical options were materialism, idealism, and dualism. None seems satisfactory. Materialism implausibly devalues the mental. Idealism implausibly devalues the physical. Dualism has never succeeded in satisfactorily integrating the disjoint realms of matter and mind and it faces the problem of how to account for the apparent continuity of evolutionary history, in which a world which was once a hot quark soup (apparently purely material) has turned into the home of human beings.

In consequence, in the twentieth century some have felt encouraged to explore the possibility of a dual-aspect monism, in which the mental and material

---

[14]The discussion of Peacocke in *Theology for a Scientific Age*, p.154, does not correctly represent my view. I have never supposed agency to be exercised through (calculated!) manipulations of individual atoms and molecules. See n. 3 above.

are conceived of as being opposite poles (or phases, as a physicist might say) of a single (created) reality. A key idea may well be that of complementarity. Quantum theory discovered that the apparently qualitatively different characters of wave and particle were present in the nature of a single entity, light. This proved possible to understand when quantum field theory identified the feasibility of reconciling these complementary descriptions as due to the presence of an *intrinsic* indefiniteness. (A wavelike state is associated with the presence of an indefinite number of photons.) The essence of complementarity is its ability to hold together apparently irreconcilable characteristics (spread out wave and point-like particle) in a simple reconciling account. We experience the apparently qualitatively different realms of the material and the mental. May not the understanding of this duality be found in the intrinsic indefiniteness associated by our hypothesis with the behavior of chaotic systems, influenced by both energetic transactions and by active information? Of course consciousness is a much more profound and mysterious property than history formation by active information, but at least the latter seems to point in a mildly hopeful direction.

In common with all the other metaphysical proposals here discussed, a dual-aspect monism based on a complementary mind/matter metaphysic, is largely conjectural and heuristic. We do not have the knowledge to produce definitive proposals of a fully articulate kind. Nevertheless, I believe this is a sensible and hopeful direction in which to look for an understanding consistent with our knowledge of physical process and with our experience of human agency. It would afford a picture of reality which would also be hospitable to the theological concept of divine providential interaction with creation. Motivation for belief in divine providence is found in the religious experiences of prayer and of trust in a God who guides.

6 *Some Comments*

There are well-known relationships, due to Leon Brillouin and Leo Szilard, which connect the transfer of units of information (in a communications-theory sense) with minimal transfers of energy. This might seem to imply that for a physical system there could not be a totally pure distinction between energetic action and active information. Careful analysis would be required before such a conclusion was firmly established. It is not clear that active information is subject to exactly the same constraints as communications theory imposes on the storage of elements of passive information.[15] Even if that were so, it would simply reflect the embodied character of human beings. We are mind/matter amphibians and are never in the state of being pure spirits.

God, in any case, is not embodied in the universe and there does not seem to be any reason why God's interaction with creation should not be purely in the form of active information. This would correspond to the divine nature being pure spirit

---

[15]See the discussion in Bohm and Hiley, *Undivided Universe*, 35-38.

and it would give a unique character to divine agency in a way that theologians have often asserted to be necessary. (God is not just an invisible cause among other causes.)

A world open to both bottom-up and top-down causality is a world released from the dead hand of physical determinism. It is a world of true becoming, in which the future has novel aspects not predictable from the past. It is a world of true temporality.[16] God knows things as they really are and this surely implies that God knows the temporal in its temporality. Divine knowledge of temporal events must be knowledge of them in their succession, not just that they are successive. This implies, I believe, that the God who is the creator of a world of becoming must be a God who possesses a temporal pole as well as an eternal pole.[17] Because the future of such a world is not yet formed, even God does not yet know it. This is no imperfection in the divine nature. God knows all that can be known, but the future is still inherently unknowable.

---

[16]Cf. C.J. Isham and J.C. Polkinghorne, "The Debate over the Block Universe" in *Quantum Cosmology*, 135-144.

[17]Polkinghorne, *Science and Providence*, chap. 7.

THE DISCOVERY OF CHAOS AND THE RETRIEVAL OF THE TRINITY

Denis Edwards

The triumph of Newtonian science led to a view of the world that was mechanistic, deterministic, and predictable. Recent theories of chaos and complexity, coming on top of the insights of quantum mechanics, have led to a radically different view of the world, as unpredictable and open to the new. Chaotic systems are understood as unpredictable in practice, while quantum systems are seen as unpredictable in principle and therefore as indeterministic.

Essays in this volume consider divine action in the light of this new worldview. I argue that it is a matter of great importance that we consider the view of God that we bring to this inquiry.

Classical science, with its methodological reductionism, tended to objectify physical reality and to lend support to an Enlightenment concept of the human being as the individual subject, the detached observer of the objects of science. This individualistic concept of the subject has led to well-known problems in understanding human social relations and human relationship with other creatures. It has also tended to support a view of God as the great Subject, an individualistic creator set over against creation.

When we ask about divine action we need to bring to our task not just a new scientific paradigm, but also a new, or retrieved, theological paradigm. The old scientific worldview is giving way to a new paradigm of an open and self-organizing universe. It is equally important for our purposes to note that in systematic theology the old concept of God as the individual Subject is giving way to a relational, dynamic trinitarian theology. We need to attend not only to a renewed understanding of the complexity and dynamism of physical processes, but also to a renewed view of a relational God.

In this paper, I will begin by exploring the general concept of divine action from the perspective of a retrieved trinitarian theology. Then, in the second part of the paper, I will bring this trinitarian theology of divine action into dialogue with the work of both John Polkinghorne and Arthur Peacocke, including their reflections on divine action in the light of chaos theory. In this part I will be particularly concerned with the question of what can be said about particular divine acts.

1 *Towards a Theology of Trinitarian Divine Action*

As the scientific community is coming to a new understanding of physical process, so, it seems to me, the theological community is at the beginning of a new appropriation of its central doctrine of the Trinity. The trinitarian theology which we have inherited in the West, under the influence of Augustine and Aquinas, has much to recommend it, but it has also had negative effects. It has emphasized an individual psychological model of the Trinity rather than a communitarian one, the divine unity rather than the persons, and the divine being rather than divine love.

In what follows, I will outline a theology of trinitarian divine action in which the Trinity is understood (1) as Persons-in-Mutual-Communion and (2) as dynamic, ecstatic, and fecund. I will argue (3) that the universe is God's trinitarian self-expression, (4) that there are "proper" roles for the trinitarian persons in creation, and (5) that divine interaction with creation is characterized by the vulnerability and liberating power of love.

1.1 *Richard of St. Victor: The Trinity as Persons-in-Mutual-Communion*

Great theologians such as Karl Barth and Karl Rahner not only built on the Augustinian tradition, which stressed the one divine nature of the trinitarian God; they transformed it in terms of the nineteenth-century emphasis on the individual subject. In the light of their work, there has been a tendency for God's unity to be stressed even more—God is seen as the one self-conscious subject, in three modes of being or subsistence. By contrast, I would suggest that the contemporary recovery of the Trinity is more focussed on a relational and communal model of the Trinity—as a dynamic communion of persons in mutual love. I am thinking of the work of Jürgen Moltmann, Elizabeth Johnson, and Ted Peters, as well the work of a number of other scholars who come from diverse theological perspectives, including Walter Kasper, John Zizioulas, Leonardo Boff, Joseph Bracken, Wolfhart Pannenberg, and Catherine Mowry LaCugna.[1] In light of this contemporary recovery of the communal and relational notion of the Trinity, I have found it helpful to explore the line of trinitarian theology represented by Richard of St. Victor and Bonaventure.

Richard of St. Victor (d. 1173) was perhaps the supreme Western advocate of a relational theology of the Trinity. He built his approach on the model of mutual human friendship. Richard lived in a century that was marked by the discovery of romantic love and by an intense interest in friendship in the new

---

[1]Jürgen Moltmann, *The Trinity and the Kingdom of God* (London: SCM Press, 1981); Elizabeth Johnson, *She Who Is: the Mystery of God in Feminist Theological Discourse* (New York: Crossroad, 1992); Ted Peters, *God as Trinity: Relationality and Temporality in Divine Life* (Louisville, KY: Westminster/John Knox, 1993); Walter Kasper, *The God of Jesus Christ* (London: SCM Press, 1983); Leonardo Boff, *Trinity and Society* (Maryknoll, NY: Orbis Books, 1988); Joseph Bracken, *Society and Spirit: A Trinitarian Cosmology* (Cranbury, NY: Associated University Presses, 1991); Wolfhart Pannenberg, *Systematic Theology*, vol. 1 (Grand Rapids, MI: William B. Eerdmans, 1991); John Zizioulas, *Being as Communion* (Crestwood, NY: St. Vladimir's Seminary Press, 1985); Catherine Mowry LaCugna, *God For Us: The Trinity and Christian Life* (San Francisco: HarperSanFrancisco, 1991).

monastic movements.[2] Richard's unique contribution was in his application of reflection on Christian friendship to the central mystery of the Trinity.

He begins with an understanding of God as supreme goodness. Such goodness, he argues, must involve full and perfect love. But love which is centered on the self cannot be considered to be the fullness of love, or charity. If there is the fullness of love in God, then it cannot be self-love or private love (*amor privatus*), but only love which involves more than one person (*amor mutuus* or *caritas*).[3] Richard's thought moves from the fullness of goodness in God, to self-transcending love, to the plurality of persons.

Richard sees the limited and imperfect human experience of self-transcendence in love as pointing to the infinitely more profound experience of trinitarian communion. The Trinity is the "ideal of interpersonal relations because here there is infinite self-giving and infinite reception of love."[4] In the fullness of divine love the trinitarian persons give of themselves infinitely without fear of losing themselves or of being rejected by the other.

Having argued towards an understanding of the plurality of persons in God, Richard's second step is to insist that the love between them must be completely mutual and equal. Since "nothing is more glorious, more magnificent than to wish to have nothing that you do not wish to share," we can conceive of the persons in God only as sharing equally from all eternity.[5] Supreme charity demands complete mutuality and equality of persons. For supreme charity, the beloved must be loved supremely, and return love in the same way.

The next step in Richard's argument is to establish the meaning of the third person in the Trinity. He argues from a subtle psychological reflection on Christian friendship that mutual love demands a third. Love between two can exclude others, but this would not be mature or Christian love. In his thought, "the very dynamism that led to self-transcendence in uniting the two, now leads to a further transcendence from the sphere of the two alone to the third."[6] For perfect love we look for one who can share in love for the beloved. Richard speaks of this person as the "*condilectus*."[7] This word seems to be his own invention. It refers to one who is "loved with," and to one who shares in love for a third.

This model of the Trinity as mutual love reflects the centrality of love revealed in the "good news" of Jesus of Nazareth—above all in the divine "foolishness" of the cross. It can lead to a radically different worldview. It offers the basis for moving from a worldview, and a metaphysics, centered on being or substance to one centered on dynamic relationships. In this kind of theology, as

---

[2]Ewert Cousins, "A Theology of Interpersonal Relations," *Thought* 45 (1970): 59-60.

[3]See Richard of St. Victor, *De Trinitate* (Book III has been translated by Grover A. Zinn, *Richard of St. Victor: The Twelve Patriarchs, the Mystical Ark, Book Three of the Trinity* [New York: Paulist Press, 1979]), 3.2 (*PL* 196, 916-917).

[4]Cousins, "A Theology of Interpersonal Relations," 69-70.

[5]*De Trinitate*, 3.6 (Zinn, 379).

[6]Ibid.

[7]Ibid., 3.15.

Kasper comments, "the ultimate and highest reality is not substance but relation."[8] Or as LaCugna puts it, person, not substance, becomes the ultimate ontological category.[9]

Today we are increasingly aware that Big Bang cosmology reveals our common history and our interconnectedness with all other creatures of the universe. In addition, quantum mechanics tells us that we live in a reality which is more a network of relationships than a world of billiard-ball-like atoms. Finally, the ecological crisis forces us to begin to think relationally, and to begin to see our own interconnectedness in the delicate web of life. Richard's trinitarian theology suggests that relationships of mutual love are the foundation of all reality. It argues that all creation springs from this dynamism of mutual love. Relationality is the source of creaturehood.

1.2 *Bonaventure: The Trinity as Dynamic, Ecstatic, and Fecund*

Many of those who offer a critique of the God of classical theism have not taken into account the fecundity tradition of the Middle Ages, which proposes a trinitarian theology which is dynamic and related to the world.[10] The key figure in this tradition is Bonaventure (1221-1274). Bonaventure joined the Franciscans at an early age, was accepted as a Doctor of the University of Paris, and became Minister General of the Franciscan order in 1257. As Zachary Hayes notes, Bonaventure's work reflects the interest of the early Franciscan movement in the theological implications of Francis' religious experience, above all in the experience of the goodness of God, and the experience of creation as expressive of this divine goodness.[11]

Although he stands in the Western Latin tradition, Bonaventure's thought owes to the East the concept of the dynamic fruitfulness of divine goodness. He inherits from Pseudo-Dionysius the axiom that goodness is self-diffusive (*bonum diffusivum sui*), and this becomes one of the basic metaphysical principles underlying his theology of the Trinity and creation. The first name of God is 'the Good' (Luke 18:19), and this divine goodness is radically self-communicative.[12]

---

[8]Kasper, *The God of Jesus Christ*, 156.

[9]LaCugna, *God for Us*, 248-49.

[10]See Cousins, "St. Bonaventure, St. Thomas, and the Movement of Thought in the 13th Century," in *Bonaventure and Aquinas: Enduring Philosophers*, ed. Robert W. Shahan and Francis J. Kovach (Norman, OK: University of Oklahoma Press, 1976), 16.

[11]See Zachary Hayes, *Saint Bonaventure's Disputed Questions on the Mystery of the Trinity: An Introduction and a Translation* (St. Bonaventure, NY: The Franciscan University Press, 1979), 32-33.

[12]St. Bonaventure, *Itinerarium Mentis in Deum* 5.2. For the Latin text with translation, see *Saint Bonaventure's Itinerarium Mentis in Deum: With an Introduction, Translation and Commentary*, trans. Philotheus Boehner (St. Bonaventure, NY: The Franciscan Institute Press, 1956).

For Bonaventure, the life of the Trinity originates eternally from the one who is Fountain Fullness (*fontalis plenitudo*).[13] This Fountain Fullness expresses itself perfectly in the one who is Image and Word. This process reaches its consummation in the love between them, which is the Spirit.

The procession of the Word arises by way of nature from the fecund, ecstatic divine goodness. Goodness needs to communicate itself. The Word is the full expression of God's primal fruitfulness. The infinite variety of things is given unified expression in the Word. The Word is the locus of the divine ideas, and is thus called the eternal Exemplar.[14] These divine ideas give rise to the created world, in God's free choice to create. They are dynamic causes of creation. Individual things in their distinctness, not just Platonic universals, really are in God, in the divine Exemplar.[15] Here, in the eternal Word, is the "Eternal Art" (*ars eterna*) of the trinitarian God.[16]

Cousins observes: "[B]y placing absolute fecundity in the Trinity, Bonaventure has saved both God and the world."[17] He has rescued the transcendence of divine fecundity, which is such that it cannot be exhausted by creation. God is not swallowed up in creation. On the other hand, creatures are free to be their specific and limited selves, not overwhelmed by the divine power and immensity.

Bonaventure may have been the first in the West to speak of the *circumincessio* of the Trinity. It is an equivalent of the Greek word *perichoresis*, which expresses the idea of persons in a profound and dynamic communion of interdependence and mutuality. *Circumincessio* comes from *circum-incedere*, to "move around one another." It can bring to mind the image of the divine dance, a dance of unthinkable intimacy and mutual love, a dance which freely overflows in creation and becomes expressed in the dance of the universe.

1.3 *The Universe as Divine Self-Expression*

In Bonaventure's thought, the dynamic goodness of the Fountain Fullness finds expression in the Trinity, and freely "explodes into a thousand forms" in the world of creation.[18] In the free divine action of creation, the trinitarian fullness of divine goodness "explodes" into what is not God, into creatures which are not God, but

---

[13]See St. Bonaventure, *Sent.* 1.31, p.2, dub.7.

[14]Ibid., 1.6, a.u., q.1-3.

[15]Ibid., 1.35, a.u., q.4, conc; *idem, Breviloquium* 1.8.7. See Leonard Bowman, "The Cosmic Exemplarism of Bonaventure," *The Journal of Religion* 55 (1975): 182-83.

[16]*Sent.* 1.6, a.u., q.3, resp.; *Breviloquium* 1.8; *idem, Hexaemeron* 1. Augustine had used this expression in his *De Trinitate* 6.10.11.

[17]Cousins, "Movement of Thought," 18.

[18]Bowman uses this expression in "Exemplarism," 183. He borrows it from Alexander Gerkin, *La Theologie du verbe: La relation entre l'Incarnation et la Creation selon S. Bonaventure*, trans. Jacqueline Greal (Paris: Editions Franciscaines, 1970), 132.

are God's self-expression. Hayes comments that creation can be understood as the "free overflow of God's necessary, inner-divine fruitfulness."[19]

Bonaventure's view of creatures is that they exist by way of exemplary causality. An exemplary cause is the pattern, model, or exemplar according to which something is made. The Word is the Exemplar for all things. Bonaventure writes that creatures seem to be "nothing less than a kind of representation of the wisdom of God, and a kind of sculpture."[20] They are the work of art produced by divine Wisdom. He tells us that "every creature is of its very nature a likeness and resemblance to eternal wisdom."[21] It is an intrinsic characteristic of creatures that they represent and give expression to the Wisdom of God.[22]

For Bonaventure, then, the universe is a book which can be read, a book whose words reveal the creator:

> From this we may gather that the universe is like a book *reflecting*, *representing* and *describing* its Maker, the Trinity, at three different levels of expression: as a *trace* (*vestigium*), an *image*, and a *likeness*. The aspect of trace is found in every creature; the aspect of image, in the intellectual creatures or rational spirits; the aspect of likeness, only in those who are God-conformed.[23]

What is this vestige of the Trinity that we find in all creatures? It is the reflection in a creature of the Trinity as efficient, exemplary, and final cause of the creature's inner structure. The first person of the Trinity is reflected as the Power that holds the creature in being (efficient causality). The second person is reflected as the Wisdom, or the Exemplar, by which it is created (exemplary causality). The third person is reflected as the Goodness which will bring the creature to its consummation (final causality).

The center of Bonaventure's thought about creation has to do with exemplary causality, because here we see, in the very form of a creature, its reference to the Exemplar. From exemplary causality comes the truth, form, species (intelligibility),

---

[19]Hayes, "Incarnation and Creation in the Theology of St. Bonaventure," in Romano Stephen Almagno and Conrad L. Hawkins, eds., *Studies Honoring Ignatius Charles Brady, Friar Minor* (St. Bonaventure, NY: The Franciscan Institute Press, 1976), 315.

[20]"Unde creatura non est nisi quoddam simulacrum sapientiae Dei, et quoddam sculptile" (*Hexaemeron*, 12).

[21]"Omnis enim creatura ex natura est illius aeternae sapientiae quaedam effigies et simultudo" (*Itinerarium*, 2.12).

[22]See Etienne Gilson, *The Philosophy of St. Bonaventure* (London: Sheed & Ward, 1938), 215. See *Sent*. 2.16, a.u., q.2, where Bonaventure says that it cannot be an accident in a creature to be a vestige.

[23]*Breviloquium* (trans. Jose de Vinck in *The Works of Bonaventure II: The Breviloquium* [Paterson, NJ: St. Anthony Guild, 1963], 104.), 2.12. My italics. Bonaventure can also mention the shadow (*umbra*) of God, which is the most elementary and general reflection of God in creation. It reflects God as one, whereas the trace or vestige reflects the three persons.

number, and beauty of a creature.[24] Bonaventure has a special word, "contuition," to describe a way of seeing creatures in their relationship to God. It implies a simultaneous awareness of the reality of, for example, a giant sequoia tree, and of its eternal Exemplar. In this life, the apprehension of the divine Exemplar is veiled, distorted by sin, and indirect, but it is real.

Bonaventure's trinitarian theology has much to offer a contemporary theology of creation. Every creature is understood as an aspect of God's self-expression in the world. Every creature in its form, proportion, and beauty, reflects the Word and Wisdom of God, the divine Exemplar. Every creature is a revelatory word written in the great Book of Creation. Every species and every ecosystem, every grain of sand and every galaxy, are the self-expressions of the eternal art of divine Wisdom. For Bonaventure there is no simple identification between God and the world, no hint of pantheism. But his teaching could be described as a kind of panentheism. As Ewert Cousins writes, "in a most emphatic way for Bonaventure, God is in the world and the world is in God."[25] God is profoundly present to all things, and God is expressed in all things, so that each creature is a symbol and a sacrament of God's presence and trinitarian life.

1.4 *Creation is the Action of the Whole Trinity, but Involves Distinct and "Proper" Roles of the Trinitarian Persons*

In the previous three sections, I have been outlining some of the insights of Richard of St. Victor and Bonaventure which I think are important to recover for today. In what follows I will enter into areas where a contemporary theology needs to move beyond medieval theology. The first of these is the argument that the self-expression of God in creation is "undivided" but properly trinitarian. The Augustinian teaching that "*opera trinitatis ad extra indivisa sunt*" (the Trinity's acts *ad extra* are one) has led to the conviction that since creation is an action of the essence of God, there is no proper role for each person. This means that when we attribute a particular role in creation to Wisdom or to the Spirit, we do so only by "appropriation." We attribute to one person what really applies to all three.

Contemporary theology has recovered the understanding that we must attribute a proper role to the trinitarian persons in the salvific missions of the Word and the Holy Spirit.[26] This needs to be extended to include a properly trinitarian notion of creation. This does not deny the unity of the divine action in creation, but points to distinctions proper to the persons within this common action. It suggests that the universe is the self-expression of the trinitarian God: the Fountain Fullness is the Source of the existence of each creature; the Divine Wisdom is the Exemplar

---

[24]See Bowman, "Exemplarism," 190-94.

[25]Cousins, "Movement of Thought," 19.

[26]On the proper role of the persons in the economy of salvation see Karl Rahner, *The Trinity* (New York: Herder & Herder, 1970), 27. On the Holy Spirit, see David Coffey, "A Proper Mission for the Holy Spirit," *Theological Studies* 47 (1986): 227-50; and *idem*, "The 'Incarnation' of the Holy Spirit in Christ," *Theological Studies* 45 (1984): 466-80.

for the unique identity of each creature; the Spirit is the immanent Presence of God in all creatures and the Bond of Love who brings all things into unity.

Three reasons support attributing proper roles to the trinitarian persons in the act of creation. First, the inner logic of the system of Bonaventure, which I am following, suggests that the one trinitarian work of creation be understood as proper to each person, since for Bonaventure what is distinctive of the persons (Absolute Source, Divine Exemplar, and the Presence and Bond of Love) is what comes into play in the one work of creation.[27] Second, this seems faithful to the biblical texts, including the texts which identify Christ and cosmic Wisdom and stress creation "in" and "through" Christ, and the biblical teaching which suggests a real and "proper" engagement of the Spirit in creation.[28] Third, I would argue that once theology posits proper missions of the Word and Spirit in salvation history, the inner theological connection between creation and redemption requires that creation too be understood as properly trinitarian.[29]

1.5 *The Trinity's Interaction with Creatures is Characterized by the Vulnerability and Liberating Power of Love which Respects Human Freedom and Natural Processes*

Christian theology, as it engaged with Greek philosophical thought, removed change and feeling from the biblical God. Theologians sought to preserve divine freedom, and the classical ideal of divine freedom meant self-sufficiency and freedom from dependency on creation. Hence, it seemed, God must be understood as immutable. Even the cross of Jesus was no exception because, it was argued, the suffering of Jesus touched only the human nature and not the divine. Later, God was thought of as "pure act" (*actus purus*), and if the divine being was pure act, there was no movement from potency to act in God, and hence no change in God.

This theology has proved radically inadequate in the century of the holocaust and of Hiroshima. Feminist theologians have analyzed the traditions of God's invulnerability and omnipotence, and shown their relation with patriarchal culture and its horrific effects.[30] There is widespread agreement in contemporary theology

---

[27]Lateran IV (1215) affirmed the Augustinian view that the world was created by the divine essence. This teaching can co-exist with a proper role of the trinitarian persons in creation.

[28]On the biblical view of the work of the Holy Spirit in creation see Jürgen Moltmann, *The Spirit of Life: A Universal Affirmation* (Minneapolis, MN: Fortress Press, 1992).

[29]See Moltmann, *God in Creation: An Ecological Doctrine of Creation* (London: SCM Press, 1985), 94-103.

[30]For a strong analysis and theological argument along these lines see particularly the last chapter of Johnson's *She Who Is*, "Suffering God: Compassion Poured Out" (246-72). See also Anna Case-Winters, *God's Power: Traditional Understandings and Contemporary Challenges* (Louisville, KY: Westminster/John Knox, 1990).

that we need to return to a way of speaking which is more faithful to the biblical God.[31]

Central to Pauline Christology is the fact that Divine Wisdom is revealed in the shocking image of the crucified (1 Cor. 1-2). This cannot be explained away as the suffering of only the human nature of Jesus. Paul is surely saying more than that. He is pointing to the cross and saying this is the revelation of God. This is the Wisdom of God. God suffers in the death of Jesus and God suffers in the sufferings of our world. But God not only shares the suffering of the world, not only stands with a suffering world radically in the cross, but God offers a promise that resurrection, liberation, and life are at work in our universe.

In response to the concerns of classical theology it can be said, first, that when God is seen in terms of persons-in-mutual-love then the ideal of freedom is not one of isolation and self-sufficiency. Real freedom is the freedom to enter into love, to risk oneself with another, to enter into love in openness to the other. This kind of freedom demands both self-possession and self-giving in vulnerability. And there is every reason to assume that the trinitarian God is supreme in personal freedom, and free beyond comprehension to enter into the vulnerability of loving communion. This assumption is verified in a staggering way in the love revealed in the cross of Jesus. And since the basic metaphor for the trinitarian God, I have been arguing, is not *actus purus* but persons-in-communion, then this view of God removes the force of the theological argument that the one who is pure act cannot suffer.[32] The trinitarian God is now understood as the one in whom self-possession and self-giving, freedom and vulnerability, exist in a way beyond comprehension.

This suggests, in contrast with Aquinas, that God must be understood as having a real relationship with the world. From the side of the creature, Aquinas saw creation as a *real* relation of dependence between the creature and God, by which God enables the creature to be. But, from God's side, Aquinas allowed only a *logical* relation between God and the creature, since in his philosophical framework a real relation would seem to involve a necessity of nature.[33] He could not allow that creation belongs by way of necessity to God's nature. God's relation to the creature does not constitute God's being. By contrast, in the relational metaphysics used here, I believe that it is possible and essential to affirm both (1) that creation is a *free* act of God's love, and (2) that through this freely chosen love, God enters into a *real* relationship with creation, which means that God freely accepts the limitation and vulnerability of such relationship (Phil. 2:6-11). God's supreme goodness and capacity for love is of such a kind as to be able to make

---

[31]Peters, *God—The World's Future: Systematic Theology for a Postmodern Era* (Minneapolis, MN: Fortress Press, 1992), 201. Moltmann has been at the forefront of this reinterpretation for many years. See, for example, *The Crucified God* (London: SCM Press, 1974).

[32]See Johnson, *She Who Is*, 265.

[33]Thomas Aquinas, *Summa Theologiae*, 1.13.7; 1.28.1; 1.45.3, ad 1. A key problem for Aquinas was that in Aristotelian thought, relation is classified as an accident, and no accident can inhere in the divine being.

space for others.[34] As Kasper notes: "For the Bible, then, the revelation of God's omnipotence and the revelation of God's love are not contraries." Rather God's omnipotence is the supreme capacity to love: "It requires omnipotence to be able to surrender oneself and give oneself away; and it requires omnipotence to be able to take oneself back in the giving and to preserve the independence and freedom of the recipient."[35]

Polkinghorne points out that God's love and faithfulness apply not just to human beings, but also to the physical universe. God is the great respecter and allower of freedom not only of human creatures, and other living creatures, but also of physical processes.[36] This suggests that we need to think of Divine Wisdom as responding creatively to the universe, as "improvising" on the theme of creation in the light of the interplay of chance and necessity. God's ongoing creative action (*creatio continua*) is through loving, faithful, and fruitful interaction with both chance and lawfulness. Peacocke has done much to show the place of chance and law in Christian theology.[37] He speaks of God's action in creation under the beautiful image of God as the Great Improvisor.

2 *Reflections from a Systematic Trinitarian Theology on Particular Divine Acts*

I have been arguing that these key insights of trinitarian theology can illuminate what is meant by the word 'divine' in the concept of divine action. They are the basis for a general trinitarian theology of divine action through continuous creation, as well as for an approach to the issue of particular divine acts. It is to this issue that I turn now. What does this theology have to say about particular divine actions, like the incarnation, the experience of the Holy Spirit, and the experience of divine providence? I will attempt to explore this question through dialogue with the works of Polkinghorne and Peacocke on divine action.

Polkinghorne and Peacocke are both, in their own distinctive ways, outstanding contributors to the dialogue between science and theology. In particular, both have articulated new approaches to an understanding of divine action in the light of chaos theory. I will attempt a brief summary of their approaches, and in the light of their views, offer my own reflections on the

---

[34]See Moltmann, *Trinity*, 108-11; and William Hill, *The Three-Personed God: The Trinity as a Mystery of Salvation* (Washington, D.C.: Catholic University of America Press, 1982), 76, n.53.

[35]Kasper, *The God of Jesus Christ*, 194-5.

[36]See John Polkinghorne, *Science and Providence: God's Interaction with the World* (London: SPCK Press, 1989); and *idem*, *Reason and Reality: The Relationship between Science and Theology* (London: SPCK Press, 1991).

[37]Arthur Peacocke, *God and the New Biology* (London: J.M. Dent, 1986), 97-98; and *idem*, *Theology for a Scientific Age: Being and Becoming—Natural and Divine* (Oxford: Basil Blackwell, 1990), 175-77.

contribution of the retrieval of trinitarian theology to the discussion of particular divine action.

2.1 *Polkinghorne and Peacocke on Divine Interaction with Creation*

Polkinghorne points out that chaotic systems are so exquisitely sensitive to initial conditions and circumstance that we can never know enough to predict their behavior.[38] This is a statement about our limited knowledge, an epistemological statement. But Polkinghorne argues that this unpredictability suggests a real openness in nature. It can suggest that there is an openness which is not just epistemological but ontological.[39]

Physical processes can be understood as open from below, and as able to be influenced by holistic organizing principles which are still unknown, but open to scientific discovery.[40] Polkinghorne suggests that this openness makes room not only for human free acts, but also for divine interaction with creation. Physical reality will be open from below to a proliferating world of possibilities. The future of the system can be determined by downward causation, through what Polkinghorne describes as information input.

He argues that human beings can be thought of as a matter/mind unity, for which the complementarity of wave/particle in quantum theory is an analogy. What we call the material and the mental are really one "stuff" encountered in different regimes. He sees causality through information input as offering a glimmer of understanding of how a mental decision to lift an arm, say, brings about the physical act of movement.

Polkinghorne suggests that divine agency in the world might be understood by analogy with human agency. God can be thought of as acting through

---

[38]See Polkinghorne, "The Laws of Nature and the Laws of Physics," in *Quantum Cosmology and the Laws of Nature: Scientific Perspectives on Divine Action*, ed. Robert John Russell, Nancey Murphy, and C.J. Isham(Vatican City State: Vatican Observatory, 1993; and Berkeley, CA: Center for Theology and the Natural Sciences, 1993), 437-48; and *idem*, "The Metaphysics of Divine Action" (in this volume). See also *idem, Science and Creation: The Search for Understanding* (London: SPCK Press, 1988); *idem, Science and Providence*; and *idem, Reason and Reality*.

[39]It seems to me that Nancey Murphy is right in her criticism of an argument from epistemological unpredictability to ontological indeterminacy. However, Polkinghorne allows that there is no strict argument from chaos theory to ontology, and admits that he is making a "metaphysical guess" about the nature of reality. In my view this "guess" is worth exploring from a theological perspective. See Murphy, "Divine Action in the Natural Order: Buridan's Ass and Schrödinger's Cat" (in this volume).

[40]Polkinghorne writes that ". . . chaos theory presents us with the possibility of a metaphysically attractive option of openness, a causal grid from below which delineates an envelope of possibility (it is not the case that anything can happen but many things can), within which there remains room for manoeuvre." Polkinghorne, "The Laws of Nature and the Laws of Physics," 443.

information input into the flexible and open process of cosmic history. But Polkinghorne also sees God as valuing and protecting the freedom of creation. This provides the basis for a theodicy which he calls a "free-process defense" in relation to physical evil, patterned after and along with a "free-will defense" in relation to moral evil.

He insists that this theory of divine action is not a return to the discredited "God of the gaps" because the openness of the physical universe is not a gap in scientific information, but an intrinsic openness of the universe to top-down causality. He insists, too, that God is not to be understood as simply one cause amongst other causes, and that God's input is not one of energy, but one of information.

Polkinghorne argues that a God who is involved with physical processes will be radically involved with time. Borrowing from process theology he suggests that God needs to be understood in a dipolar (time/eternity) fashion, and as intimately involved in the world of becoming. He makes the more radical suggestion that God does not know the future, since the future is not there to be known. There is a kenosis of divine omniscience in creation, and God allows the future to be truly open. God's purposes will be brought about by way of "contingent paths"—through interaction with the processes of the universe.

Peacocke, too, points to the fact that many systems which are governed by non-linear dynamical equations become unpredictable and that this unpredictability cannot be eradicated. He makes it clear that he sees this as an unpredictability in practice and not in principle—he is not claiming that it stems from ontological indeterminacy in nature. But he points out that in certain cases, systems can reach a point of bifurcation, an either/or transition, which is completely unpredictable in practice.[41] Open dissipative systems, far from equilibrium, may evolve into new levels of order and organization. Through the amplification of small fluctuations, a system can have access to novelty. In such systems we recognize that alongside the normal description of "bottom-up" causation, there is need for a "top-down" or, more precisely, a "whole-part" description where this whole influences its parts.[42] An example of this is found in the jaws of worker termites, which depend for their effectiveness on particular proteins, whose efficiency has been enhanced through natural selection.

Peacocke suggests, as does Polkinghorne, that this influence can be thought of as a flow of information rather than of energy. It is "information" from the environment which determines the selection of the DNA sequence in the termite jaw. Peacocke speculates that the human brain operates through "top-down" causation, as the brain state as a whole determines what happens at the level of individual neurons. He suggests that this relationship between the brain state and

---

[41]See Peacocke, "God's Interaction with the World: The Implications of Deterministic 'Chaos' and of Interconnected and Interdependent Complexity" (in this volume). See also *idem, Theology for a Scientific Age*.

[42]See Peacocke, "God's Interaction with the World."

individual neurons is better conceived of as a transfer of information than of energy.

Can God be thought of as interacting with the universe in the unpredictability of chaotic systems? Peacocke answers no. He would see this as a new kind of implicit interventionism.[43] God has given a certain autonomy to natural processes. He believes that God's omniscience is freely self-limited, such that God cannot know the outcome of unpredictable events and so cannot manipulate them. While our new awareness of open-endedness does help us see the natural world as a matrix in which openness and human freedom emerge, it does *not* help us understand what Austin Farrer called the "causal joint" of God's action in the world.

Rather, Peacocke sees God interacting with the world as a whole, and through this with individual components of the world. God is then understood as interacting in a "top-down" way, without abrogating the laws, regularities, or unpredictabilities of natural systems. He invokes the analogy of the relation of the human mind/brain to the body. Just as there is top-down causation from the state of the whole brain—in the body—to bodily action, so God might be regarded as exerting continuous "top-down" causative influences on the world as a whole. God is thus the unifying and centering influence on the world's activity. The succession of the states of the world is also a succession in the experience of God. God's interaction with the world as a whole is envisaged as an input of information, rather than as a flow of energy. Since God is personal this can be understood as a "communication" to the world of God's purposes and intentions. Such a mode of influence belongs to God alone, to one who is both transcendent and immanent.

In Peacocke's view, God can influence not only the whole system but also particular events through top-down causative influence, but this would never be scientifically observable, and would not appear as a divine intervention. It would not interfere with the laws, regularities, or unpredictabilities of nature.

2.2 *Reflections on Issues Raised by Polkinghorne and Peacocke*

There is a significant measure of agreement between Polkinghorne and Peacocke. Both believe that the epistemic openness of chaos theory and theories of complexity suggests, and allows for, a concept of the universe as unpredictable in practice. Both use the image of human action and the mind/body unity as a way of understanding divine action. Both suggest that divine action is by way of information input rather that through energy input. And both agree that the future is undetermined and that God does not know the future.

---

[43]Polkinghorne also makes it very clear in his article in this volume that he does not propose that chaos is "the localized mechanism by which agency is exercised." Rather, he writes, "the whole thrust of the proposal is expressed in terms of the complete holistic situation, not in terms of the clever manipulation of bits and pieces." See "The Metaphysics of Divine Action" (in this volume).

But there are real differences between them. First, while both argue that chaos and complexity lead to a position of epistemic openness (unpredictability), Peacocke sees this only as an openness *in practice*, while Polkinghorne sees it as a basis for positing an openness *in principle*. Second, Polkinghorne points particularly to chaotic systems as providing the basis for judging that there are "intrinsic gaps" from below in nature, so that God's action can be understood in terms of holistic top-down causality through information input. Peacocke's emphasis is not on chaotic systems but simply on the whole. He suggests that we understand God as interacting with the world as a whole, by way of top-down influence. This does effect particular events but always respects the laws, regularities, and unpredictabilities of nature. His analogy is the influence of the human brain on the whole human person.

My own reflections on the issues discussed by Polkinghorne and Peacocke, on the basis of a trinitarian theology of creation, can be cast in the form of the following six statements.

1) *The trinitarian God works in and through the process of the universe, through laws and boundary conditions, through regularities and chance, through chaotic systems and the capacity for self-organization.* Divine Wisdom is at work in our universe, in the great Milky Way galaxy with its hundred billion stars, and in the hundred billion or more other galaxies that make up our observable universe, sustaining all things, enabling creatures to break through to new levels of complexity and organization, and enfolding the whole curvature of space within divine creative love. Here I am in agreement with the common position of Polkinghorne and Peacocke concerning the ongoing and general creative action of God, and with their argument that this general divine action must be understood to operate through what we experience as chance and lawfulness. In this Polkinghorne and Peacocke offer an important challenge to traditional theology. As Michael Heller writes, in this new view of divine action randomness is no longer perceived as a competitor of God, but rather as a powerful tool in God's strategy of creating the world. Cosmos and chaos are not understood as antagonistic forces but rather as "two components of the same Logos immanent in the structure of the Universe."[44] In the trinitarian theology developed above, this divine action is understood as a free act of love, an ecstatic overflow of the mutual love of the Trinity, which allows for and respects human freedom and the integrity of natural processes, and which freely shares in the self-limitation and responsive character of love.

2) *The trinitarian God allows for, respects, and is responsive to the freedom of human persons and the contingency of natural processes, but is not necessarily to be denied a knowledge of future contingent events.* Here I differ from both Polkinghorne and Peacocke. A trinitarian theology, which is built upon a theology of the cross, will certainly think, with Polkinghorne and Peacocke, of God's self-limitation and vulnerability with regard not only to human freedom, but also to

[44]Michael Heller, "Chaos, Probability, and the Comprehensibility of the World" (in this volume).

natural processes. In such a theology the future is understood as radically open and not determined by the present. But I do not see that adequate justification has been provided by Polkinghorne or Peacocke for the assertion that God cannot know the future. The argument appears to be that since natural processes are unpredictable, and since God respects natural processes, God cannot foresee their outcome. But this, it seems to me, involves an unwarranted logical leap. In a trinitarian theology which stresses the mystery and incomprehensibility of God, divine knowledge of future contingent events is not based on God's capacity to predict from the present to the future. It is based on the conviction that God's eternity embraces all of time, and on the assumption that future contingent events are present to God.[45] The trinitarian knowledge of these events does not deny their contingent character. The concept of God's eternal presence to the whole of time is not in conflict with the idea that the future is radically unpredictable from within time. The God revealed in Jesus Christ and in the outpouring of the Spirit is not only eternal but also freely and radically involved with time.[46] But the trinitarian perichoresis *is* eternal and does transcend time and all our concepts of eternal life. We cannot see spacetime from an eternal divine perspective.[47] On the basis of trinitarian theology and its respect for negative theology, I would be cautious regarding statements about divine knowledge of the future.

3) *A trinitarian theology points not only to the divine action of continuous creation, but also to particular or special divine acts.* In making this affirmation, I am simply agreeing with other essayists here, including Polkinghorne and Peacocke, that the existence of *particular* divine actions constitutes a datum of Christian tradition, experience, and theology. For Christian believers, these would include, above all, God's trinitarian self-communication in Jesus of Nazareth, and in the mission and presence of the Holy Spirit. Particular divine acts also include great historical events like the Exodus and personal experiences of healing and of grace, as well as the experience of providence in day-to-day events.

4) *If God is acting creatively and responsively at all times, and also in particular ways, then this seems to demand action at the level of the whole system (Peacocke's emphasis) as well as at the everyday level of events* (Polkinghorne's emphasis), *and also at the quantum level* (the emphasis of George Ellis, Nancey Murphy, Thomas Tracy, and Robert Russell).[48] I see every reason to embrace the

---

[45]See *STh*, 1.14.13.

[46]On this, see the argument developed by Peters in *God as Trinity*.

[47]It may be relevant to note that science treats time as irreversible in some contexts and as reversible in others, and there is no general agreement about the epistemological weight of the "block" universe as compared to the "emergent" universe. See C.J. Isham and Polkinghorne, "The Debate over the Block Universe," in *Quantum Cosmology*, 135-44.

[48]See the articles in this volume: George F.R. Ellis, "Ordinary and Extraordinary Divine Action: The Nexus of Intervention"; Murphy, "Divine Action"; and Thomas F. Tracy, "Particular Providence and the God of the Gaps." See also Russell, "Theistic Evolution," in *Genes, Religion, and Society: Theological and Ethical Questions Raised by the Human Genome Project*, ed. Ted Peters (forthcoming, 1996).

argument that God can be understood as interacting with the universe as a whole, but this has to find expression in particular events. I have advocated a thoroughly relational trinitarian theology of creation—every creature exists at every moment from trinitarian relations. In this kind of theology every creature, from a proton to a galaxy, in its being and its becoming, is the expression and the presence of the dynamic overflowing mutual love at the heart of all things. Divine action at the levels of both the whole and of the parts is necessary for a trinitarian theology. It is theologically inconceivable to think of divine action at the level of particular events which is not interconnected and part of the divine action of continuous creation with regard to the whole universe.

5) *Particular divine acts are always experienced as mediated through created realities*. I agree with William Stoeger, that our experience of particular divine action, whether it be of the incarnation, the promptings of the Holy Spirit, or of divine providence, is always mediated and indirect.[49] We have no experience of unmediated or direct divine action. I agree with Stephen Happel that God acts in every event to sustain its existence, but this action is always mediated in and through the structures of the world.[50] Our experience of the trinitarian God is, in the first instance, through the "economy" of salvation. Our Christological theology affirms that God's action in Jesus of Nazareth is mediated through the humanity of Jesus.[51] This is taken in Christian theology to be an indication that God's other actions with regard to us are "incarnational" in structure, mediated through creation.

As liberation theologians like Gustavo Gutierrez have shown, God's action in an event like the Exodus is mediated through the political struggle of a group of slaves.[52] Our experiences of the Spirit and of providence are always mediated by our day-to-day experiences of life in the world, though these experiences may well open out into mystery, wonder, and grace. A critical analysis of experience reveals that even the most transcendent of human experiences must be mediated in consciousness through images, concepts, words, and stories, all of which are socially and historically conditioned.

6) *The unpredictability, openness, and flexibility discovered by contemporary science is significant for talk of particular divine actions because it provides the basis for a worldview in which divine action and scientific explanation are understood as mutually compatible, but it is not possible or appropriate to attempt to identify the "causal joint" between divine action and created causality*. While

---

[49]See William Stoeger, "Describing God's Action in the World in Light of Scientific Knowledge of Reality" (in this volume).

[50]See Stephen Happel, "Divine Providence and Instrumentality: Metaphors for Time in Self-Organizing Systems and Divine Action" (in this volume).

[51]Happel writes in "Divine Providence and Instrumentality" that "it must still be noted that the humanity of Jesus is an *instrument, not* an unmediated intervention of divine action!" (p. 196).

[52]Gustavo Gutierrez, *A Theology of Liberation: History Politics and Salvation*, rev. ed. (Maryknoll, NY: Orbis Books, 1988), 86-91.

I believe with Polkinghorne that we can conceive of God acting in the openness of the universe at least as suggested by chaotic systems, and with Ellis, Murphy, Tracy, and Russell that it makes sense to think of God acting in the openness suggested at the quantum level, I would differ from these thinkers because I would argue that we cannot identify any one specific or exclusive locus of particular divine action or any real "causal joint" between divine action and created causality. A trinitarian theology suggests a dynamic relational presence of God in every quantum event, chaotic system, and human free act, but it suggests a presence and a causality that finally escapes comprehension. However, I agree with the writers mentioned above that the unpredictability discovered by contemporary science is highly significant for talk of particular divine actions because it provides the basis for a worldview in which divine action and scientific explanation are compatible. As Murphy points out, it is an important intellectual gain that contemporary science, unlike its mechanistic precursor, does not positively exclude human or divine free acts. There is an openness or flexibility in the modern scientific worldview which leaves intellectual room for both the insights of contemporary science and a theology of human and divine free acts.[53] As Russell notes, it is significant that the limitation on predictability found in the theory of deterministic chaos shows that science can *never* fully specify whether or not God might be acting in a particular way in a chaotic system:

> The real message of chaos theory is that because of "eventual unpredictability," we know that we can *never* know the future in advance in full detail. No improvement in technology, and no change in scientific theories, will ever overcome the problem entirely. The future will always be, in part, unknown to us, not just because we don't have a good enough physical theory but for a much more fundamental reason, because we really cannot specify the totality of the present. Theologically this suggest that we can never fully specify what nature "on its own" is capable of, or what God the Trinity is doing in nature, let alone which is doing which.[54]

Chaos and quantum theories both suggest a humbler epistemological stance on the part of science than that of mechanism, a stance which can leave room for both scientific explanation and for a theology of divine action.

I do believe that we need to struggle to understand where natural process is open to divine action. This is where I see the value of Polkinghorne's work, not as providing an understanding of a "causal joint" between divine and created causality, but as showing a rationale for moving beyond a closed worldview which excludes human and divine action, to a worldview which has space for human and divine action.[55] It is a worldview where, as Stoeger says, the laws of nature and

---

[53]Murphy, "Does Prayer Make a Difference?" in *Cosmos as Creation: Theology and Science in Consonance*, ed. Ted Peters (Nashville: Abingdon Press, 1989), 235-45.

[54]Russell, "Trinitarian Chaotic Creation" (unpublished manuscript).

[55]As I have shown above, Polkinghorne takes a step beyond the position I am espousing.

nature itself constrain but do not fully determine what develops or occurs. Many possibilities are left open in nature, and it is this flexibi / which enables humans beings and other animals to influence reality.[56] There is, then, great theological value in attempting to show where there is flexibility in nature both at the quantum level and the macro-level.

But as I have said above, I agree with Peacocke, and with Stoeger, that we cannot expect to see directly what Farrer called the "causal joint" between divine action and natural processes.[57] I am cautious about further speculative specification of the locus of divine interaction with creation, because it seems to me that it runs the risk of compromising the transcendence of God on the one hand, and creaturely autonomy and the integrity of scientific explanation on the other.

If we are to follow the creative theological lead of Polkinghorne and Peacocke and speak of divine action in the language of "information input" and "top-down causation," then the analogical nature of this use of language must be noted, so that God does not appear once again as a cause within the world of natural causes, as explaining, for example, what science would attribute to boundary conditions.[58] It seems to me that this kind of language can function theologically to express the conviction that through God's creative (and mediated) action, one of many possible states of the physical universe actually occurs. But such language must be understood as functioning as creative theological metaphor and *not* as alternative scientific explanation.

In the relational ontology I have been suggesting here, creation is essentially a relation of both the whole universe and individual beings to the divine perichoresis of mutual love. In this kind of trinitarian theology divine action and creaturely autonomy are not in opposition. Rather creaturely autonomy would be understood as flourishing vis-à-vis divine action. Again, I believe that the model to which we should appeal is the model of mutual love, where the influence of the beloved does not annihilate autonomy but allows it to flourish. I would suggest that this model should be applied not just to divine grace at work with human persons, but to the work of the Holy Spirit in all creatures. The trinitarian vision of creation is one which involves the kenosis of love, love which winningly allows itself to become vulnerable to human freedom and to the natural processes of the universe.

All of this suggests that the trinitarian will to save, revealed in the death and resurrection of Jesus, will be achieved in an open and (for us) unpredictable universe. This love is infinitely creative and powerful, but with the kind of power

---

[56]Stoeger, "Describing God's Action" (in this volume).

[57]Ibid.

[58]This is certainly part of the message to theologians of Bernd-Olaf Küppers in his article in this volume, "Understanding Complexity." It also brings to mind the important cautions expressed by Willem B. Drees in his article "Gaps for God?" (in this volume).

revealed in the death and resurrection of Jesus. My image is that of the dance of the universe, a dance led by the trinitarian persons in ever new improvisations, which touch each creature and embrace the whole, which respect freedom and the structure of all that is, and which open out onto what is radically new.

DIVINE PROVIDENCE AND INSTRUMENTALITY:
Metaphors for Time in Self-Organizing Systems and Divine Action

Stephen Happel

There is no such thing as a theological
statement about a beetle.
Karl Rahner, *Theology and the Arts*

1 *Introduction*

I find it not a little curious that a premier Catholic theologian assumed that theology could dismiss realities other than human beings.[1] Indeed, I believe that the epigraph indicates flaws in Rahner's ontology; but in this context, I wish to point to the continued split between an anthropocentrically focused theology and an equally one-sided, objectivist science.

So Steven Weinberg, in *Dreams of a Final Theory*, asserts that "we" are not likely to find an "interested God in the final laws of nature." These laws are more likely to lead inexorably toward a "chilling impersonality." It is better to avoid the cheap consolations of religion by courageously embracing the stoic resistance of science.[2]

---

[1]Karl Rahner, "Theology and the Arts," *Thought* 57 (1982): 26. Rahner's theoretic interpretation of the relationships between the natural/social sciences and theology not only developed, but remained more complex than the seeming dismissal in the article cited. Nonetheless, Rahner focused theology as a discipline (with interdisciplinary concerns) upon *human* self-transcendence toward God. See Karl Rahner, "Theology as Engaged in an Interdisciplinary Dialogue with the Sciences," in *Theological Investigations*, vol. 13, trans. David Bourke (London: Darton, Longman & Todd, 1975), 80-93; *idem*, "On the Relationship between Theology and the Contemporary Sciences," in *Theological Investigations*, vol. 13, 94-102; and *idem*, "On the Relationship between Natural Science and Theology," in *Theological Investigations*, vol. 19, trans. Edward Quinn (New York: Crossroad, 1983), 16-23. In a 1967 essay Rahner argues that "sub-human" levels of natural necessity are meaningful only because they are an element in the history of free spirit. See "Theological Observations on the Concept of Time," in *Theological Investigations*, vol. 11, trans. David Bourke (New York: Seabury, 1974), 298-308.

[2]Steven Weinberg, *Dreams of a Final Theory: The Search for the Fundamental Laws of Nature* (New York: Pantheon, 1992), 257-58. Note how this position is restated (and parallels Karl Rahner's) by Philip Hefner: "Religious cosmological myths, then, do not really answer directly the question, 'What is the Universe like?' Rather, they speak to the

2 *An Intellectual Map*

My particular method for understanding the relationship of science and theology studies the language that both of these disciplines use. Religious thinkers often observe that there is a *first-level discourse* (a poetics and rhetoric) enshrined in the symbols, metaphors, stories, and experiences of revelation, though theologians often shed that language rather quickly. There is also an *intermediate level* of doctrines, creeds, and dogmas (an ecclesial pragmatics) that has a complex relationship to the first-level languages. But in addition, there is theology, a *reflective discourse* (dialectics) that primarily (until the rise of science and historical consciousness) made use of philosophy as a dialogue partner (for the first one thousand years, primarily Plato or neo-Platonism; then Aristotle). I use the hermeneutics of language as my primary dialogue partner. *Religion* is faith expressed in particular first-level languages (the scriptures, popular piety, or devotion) and must be distinguished from, but not divided from, *theology*, a reflection upon those expressions (a linguistic version of *fides quaerens intellectum*).

Science and theology can be related as disciplines about their respective conceptual content or about their methods of data manipulation. They can also be related through the linguistic performance of theologians and scientists. Scientists and theologians may *say* they are doing one thing thematically, and in fact *perform* quite another language. What are theologians and scientists really doing when they use language within these disciplines?

Christian theologians maintain a dependence upon revelation and faith. But neither faith nor revelation is a univocal concept.[3] "Objectively," theologians depend upon community traditions of worship, oral preaching, and religious experiences; genres of written expression (gospels are not visions; nor are they letters or apocalypses); and institutional mediations (such as office or authorities) among others. "Subjectively," faith is a condition for apprehending the meaning of these revelatory discourses. But faith itself has many levels—from a willing suspension of disbelief to the highest levels of contemplative participation in the divine life.[4] Theology, therefore, is never a totally *independent* discipline.[5] Though

---

question, 'What must we believe about the Universe—as we know it to be from our science—in order for us to live optimally?" (Hefner, "Christian Assumptions about the Cosmos," in *Cosmology, History, and Theology*, ed. W. Yourgrau and A.D. Breck [New York: Plenum Press, 1977], 351.)

[3]See Paul Ricoeur, "Toward a Hermeneutic of the Idea of Revelation," in *Essays in Biblical Interpretation*, ed. Lewis S. Mudge (Philadelphia: Fortress Press, 1980), esp. 73-95.

[4]For a nineteenth-century version of this multi-leveled understanding of faith, see Samuel Taylor Coleridge, *Confessions of an Inquiring Spirit*, ed. Henry Nelson Coleridge (London: William Pickering, 1840; reprint, Menston: Scholar Press, 1971).

[5]It is a *subalternated* discipline, as Aquinas stated. *STh*. 1.1.2, *corp*. The "higher science" or knowledge in this case is God's own knowledge of the divine identity, which is disclosed to human beings out of love.

theology has an inner logic and dialectic, it also operates from a rhetoric which precedes its inception and to which it returns.[6] Theology depends for its language upon prior discourse.

Contemporary scientists also have such a rhetoric.[7] Their performance as scientists is not simply involved in mathematical concepts (i.e., dialectical thinking). It also involves a rhetoric, a public community of discourse (in the politics of common speech and grant-getting) and the communication to non-scientists of their results. Moreover, the metaphors of such rhetoric and pragmatics perdure *during* scientific activity, guiding investigations and expressing results. Nonetheless, scientists often maintain that the definition of science is to be limited to the experimental and/or mathematical dimensions of their language. I deny that assertion as a gratuitous assumption about the language of science.[8]

This essay focuses upon *time* because the historical consciousness that entered with modernity in its successive phases has affected both science and religion. "Science is rediscovering time."[9] Temporality is no longer simply a given, but a question. Prigogine states that "time becomes an operator or agent within

---

[6]For my own argument that foundational theology is a rhetoric, see Stephen Happel, "Religious Rhetoric and the Language of Theological Foundations," in *Religion and Culture: Essays in Honor of Bernard Lonergan, S.J.*, ed. Timothy P. Fallon and Philip Boo Riley (Albany, NY: SUNY Press, 1987), 191-203.

[7]Marcello Pera and William R. Shea, ed., *Persuading Science: The Art of Scientific Rhetoric* (Canton, MA: Science History Publ., 1991); Herbert W. Simon, ed., *The Rhetorical Turn: Conviction and Persuasion in the Conduct of Inquiry* (Chicago: University of Chicago Press, 1990).

[8]The classic collection of essays on metalinguistic issues in philosophy is *The Linguistic Turn: Recent Essays in Philosophical Method*, ed. Richard Rorty (Chicago: University of Chicago Press, 1967). My own positions on language are more indebted to continental hermeneutical phenomenology (H.G. Gadamer, Paul Ricoeur, Jürgen Habermas) than to linguistic analysis. Though both have had their effect on the philosophy of science, linguistic analysis, particularly in its English-language tradition (especially in England, where the use of continental hermeneutics continues to be suspect), has more affected issues in the "hard sciences." For a brief discussion of the ways in which linguistic analysis has affected science-and-religion debates, see Ian G. Barbour, *Religion in an Age of Science*, The Gifford Lectures, 1989-1991, vol. 1 (San Francisco: HarperCollins, 1990), 13-16. For an argument about the role of hermeneutics in science more closely related to my own, see Marjorie Grene, "Perception, Interpretation, and the Sciences: Toward a New Philosophy of Science," in *Evolution at a Crossroads: The New Biology and the New Philosophy of Science*, ed. David J. Depew and Bruce H. Weber (Cambridge, MA: MIT Press, 1985), 1-20; and in an application to neurobiology, see Gunther S. Stent, "Hermeneutics and Complex Biological Systems," in *Evolution at Crossroads*, 209-25.

[9]Ilya Prigogine, "The Rediscovery of Time," *Zygon* 19, no. 4 (1984): 434.

these systems."[10] (See section 3 below.) Ultimately, the discussion about divine action in the world must ask if and how divine temporality relates in a non-conflictual way with non-human and human time-consciousness.

The large-scale meaning of the universe seems to be involved with the notions of whether or not the universe has a narrative structure. The stories we tell elicit meaning within or from time. Can we say that these narratives emerge from the structure of the universe itself as we know it through science? Or must we say that they are impositions by a feckless intelligence, busy consumerizing and colonizing time as a function of our need to survive?[11] Whether there is some *inner dialectical or mutually implicative relationship* between the conditions and potential of the universe and the spiritual and ethical imperatives of particular religious traditions is an important question. If philosophy and science are simply *imposing* order upon disorder, if religion is simply *proposing* commands by fiat, creating historical order upon reversible physical processes, then the Transcendent can only appear as an intrusion, an intolerable burden upon nature and history.[12] Part of what Christianity (at least) asks is whether there is a potential (indeed even more, an actualized potential) for its stories and their consequences in the cosmos? What I will examine here is whether the languages of science and religion or theology share common characteristics with regard to the notion of temporality in self-organizing systems. Can their uses of language mutually illuminate one another?

My goals are the following:

Section 3: to describe the way contemporary scientists of self-organizing systems use time and temporal sequence as a constitutive factor in their analyses;

Section 4: to argue that Christian theology has (at least in the instance of Aquinas) an important resource for thinking about the interaction of God and creation, such that pre-human self-organizing systems can have a role in enacting providence; and

---

[10]Prigogine, *From Being to Becoming: Time and Complexity in the Physical Sciences* (New York: W.H. Freeman, 1980), 206-10; see also Grégoire Nicolis and Ilya Prigogine, *Exploring Complexity: An Introduction* (New York: W.H. Freeman, 1989), 171-78.

[11]See Umberto Eco, "Travels in Hyperreality," in *Travels in Hyperreality*, trans. William Weaver (New York: Harcourt, Brace, Jovanovich, 1986), 9-10. I note the remarks of Prigogine: "What is irreversibility on the cosmic scale? Can we introduce an entropy operator in the framework of a dynamical description in which gravitation plays an essential role? . . . I prefer to confess my ignorance." Prigogine, *From Being to Becoming*, 214.

[12]See Frank Kermode, "What Precisely are the Facts?," in *The Genesis of Secrecy: On the Interpretation of Narrative* (Cambridge, MA: Harvard University Press, 1979), 101-23. Kermode's stoicism is evidenced in the following: "World and book, it may be, are hopelessly plural, endlessly disappointing; we stand alone before them, aware of their arbitrariness and impenetrability, knowing that they may be narratives only because of our impudent intervention, and susceptible of interpretation only by our hermetic tricks" (p. 145).

Section 5: to indicate how and why this language can be transposed into a contemporary understanding for divine-creation interaction.

The title of this article is a rhetorical chiasmus—A:B::B:A. Self-organizing systems are the relatively autonomous instrumental temporality with which divine temporality (i.e., action) cooperates in creation.

3 *Self-Organization, Far-from-Equilibrium Systems, and Chaos*

Physics itself has become interested in the kind of order present in dynamical systems, that is, natural or artificial systems that change over time.[13] Temporal oscillations occur, both periodic and non-periodic, in mechanical vibrations, electronic circuits, chemical reactions, and ecological systems, among others. What can begin as an ordered dynamic motion can, at certain points, reveal no underlying pattern.[14] The system moves to chaos in which no simpler algorithm than a repetition of the points "explains" the data. Systems that dissipate energy are ordered forms of behavior, far from thermodynamic equilibrium, exchanging mass and/or energy with their environments. They have a sensitivity to initial conditions such that the most minute change can drastically affect the later state of the system.[15] Here time itself becomes an *operator* or agent within these systems.[16] Prigogine argues for temporality as an internal function within certain systems.[17] In effect, time becomes a variable in the constitution of the systems themselves.

---

[13]See Arthur R. Peacocke, *An Introduction to the Physical Chemistry of Biological Organization* (Oxford: Clarendon Press, 1983), 148-49; Milos Marek and Igor Schreiber, *Chaotic Behaviour of Deterministic Dissipative Systems* (Cambridge: Cambridge University Press, 1991); Nicolis and Prigogine, *Complexity*; and Prigogine, *From Being to Becoming*. See also Barbour, *Religion in an Age of Science*, 156-65; and James Gleick, *Chaos: Making a New Science* (New York: Viking Books, 1987).

[14]See Nicolis and Prigogine, *Complexity*, 26-28.

[15]See James P. Crutchfield, et al., "Chaos" (in this volume). In fact, Prigogine would go so far as to say that "... [i]rreversibility is the manifestation *on a macroscopic scale* of "randomness" *on the microscopic scale*." See Prigogine, *From Being to Becoming*, 176 and 204-6. For an emphasis upon the mathematical dimensions of chaos, see Denny Gulick, *Encounters with Chaos* (New York: McGraw-Hill, 1992), esp. 84-94 on sensitivity to initial conditions.

[16]Prigogine, *From Being to Becoming*, 206-10; Nicolis and Prigogine, *Complexity*, 171-78.

[17]The phrase "time as operator" is Prigogine's. Though I cannot claim to be thoroughly knowledgeable concerning the physics and mathematics of this assertion, I find Prigogine's position persuasive. He argues for a "microscopic entropy operator *M* and the time operator *T*" as an "internal time quite different from the time that in classical or quantum mechanics simply labels trajectories or wave functions." (Prigogine, *From Being to Becoming*, 209). In this text, "the physical meaning of entropy and time as operators" is discussed in chap. 7, as well as appendices A and C.

An analogy might be drawn to musical forms. The slow progression of a Sarabande or a waltz is quite distinct from the tempo of a Scherzo or a rhumba; part of what makes particular dance movements to be what they are is the timing with its sequences and shifts. This sense of "inner timing" seems true for inanimate systems as well as animate ones in far-from-equilibrium situations. In animate becoming, if the "timing" changes, either the system disintegrates and dies or it evolves into something different. Within a certain range, each system has a temporality appropriate to itself.

As Prigogine maintains, "*three aspects* are always linked in dissipative structures: the *function*, as expressed by the chemical equations; the *space-time structure*, which results from the instabilities; and the *fluctuations*, which trigger the instabilities."[18] One can no longer always deduce or predict what the future of a system will be due to its initial state and a few simple multipliers as though temporal movement were insignificant. As Ford states, "[chaos] is visible proof of existence and uniqueness without predictability."[19] Though it is possible to speak of a system being *attracted* to certain positions in phase space, this attributes to the system a certain internal *teleonomy* that is neither reversible nor always predictable.[20] At certain points in chaotic behavior, a system undergoes a bifurcation sequence in which either/or possibilities are available. A system seems to *choose* a certain course of action. "[T]he bifurcation introduces *history* into physics and chemistry, an element that formerly seemed to be reserved to sciences dealing with biological, social, and cultural phenomena."[21] Nicolis even speaks of self-organizing systems searching for new attractors when driven from equilibrium.[22] As Nicolis says:

> Such ordinary systems as a layer of fluid or a mixture of chemical products can generate, under appropriate conditions, a multitude of *self-organization*

---

[18]Prigogine, *From Being to Becoming*, 100; see Peacocke's discussion of the thermodynamics of dissipative structures in *Biological Organization*, 17-72.

[19]Joseph Ford, "What is Chaos, that we should be mindful of it?," in *The New Physics*, ed. Paul Davies (New York: Cambridge University Press, 1989), 351.

[20] *Teleomatic*, inanimate processes are those whose end states are the simple consequences of their internal, fixed laws—end-directed in a passive, automatic way. *Teleonomic* processes are goal-directed behaviors of living organisms—migration, feeding, reproducing, courtship. It owes its goal-directedness to an internal operator that regulates behavior. *Teleological* processes reflexively intend a goal by virtue of some activity. See Ernst Mayr, *Toward a New Philosophy of Biology: Observations of an Evolutionist* (Cambridge, MA: Belknap Press, 1988); on predictability, see Crutchfield, et al., "Chaos."

[21]Ford, "What is Chaos," 106. See Marek and Schreiber, *Chaotic Behaviour*, 51-102; Gulick, *Encounters with Chaos*, 52-62; and Nicolis and Prigogine, *Complexity*, 71-75. This is *not* to say, as Peacocke points out, that there are absolutely unqualified *un*caused events. Randomness is not non-causal. See Peacocke, *Creation and the World of Science*, 92 and 95.

[22]Grégoire Nicolis, "Physics of Far-from Equilibrium Systems and Self-organization," in *The New Physics*, ed. Paul Davies (New York: Cambridge University Press, 1989), 331.

> *phenomena* on a macroscopic scale—a scale orders of magnitude larger than the range of fundamental interactions—in the form of spatial patterns or temporal rhythms.[23]

The temporal order of the whole as a teleonomy has a governing role in the determination of the integration of the parts.[24]

As a result, causation can operate in a different way than the usual axiomatic unidirectionality. Complex systems respond to the future emerging whole in biological evolution. "When matter is appropriately organized, it becomes sensitive to causes arising from the future instead of just the past."[25] Such hierarchical or top-down causality, a whole "guiding" the parts teleonomically, needs to be explored in this context.[26]

Self-organization is first of all a factor in physical and chemical processes, not simply an element in living systems.[27] Some systems provide the regularities of a chemical clock. This is not just the rhythm of the pendulum, but a temporal

---

[23]Ibid., 316.

[24]"Chemical instabilities involve *long-range order* through which the system acts *as a whole*"(Prigogine, *From Being to Becoming,* 104). Note that here we have the beginning of an argument that the *whole* is not *totally given* in the present. If it only exists as a system in transition, in history, that means that its future state is part of the teleonomy of the present situation, despite the fact that the system in operation does not "know" what its future state is or might be. These positions seem to assume that this is true of pre-human far-from-equilibrium systems. One can argue that this indeterminacy is a condition for the possibility of human freedom.

[25]John H. Campbell, "An Organizational Interpretation of Evolution," in *Evolution at a Crossroads,* 154. The arguments by scientists are both synchronic and diachronic. The part and the whole are synchronically co-implicated, *and* the future states of the whole in relationship to a present state are diachronically co-implicated.

[26]Ibid., 154-61. Note that Küppers maintains that "holistic phenomena are a part of normal physical and chemical experience, and that they therefore do not cast doubt on the validity of the reductionistic program" (Küppers, *Information and the Origin of Life* [Cambridge, MA: MIT Press, 1990], 121).

[27]This is no longer an uncommon assertion. See Hans Meinhardt, *Models of Biological Pattern Formation* (London: Academic Press, 1982), 10, on autocatalysis and his attempt to provide mathematical models for the process of self-organization; and Peacocke, *Biological Organization,* 114-24. David Bohm aims at self-organization and complexity with his notion of "implicate order" opposed to "mechanistic" order; see Bohm, *Wholeness and the Implicate Order* (London: Routledge & Kegan Paul, 1981), 172-86. A more fruitful application of self-organization can be found in Eric Jantsch, *The Self-organizing Universe: Scientific and Human Implications of the Emerging Paradigm of Evolution* (Oxford: Pergamon Press, 1984), though there is probably a not-so-crypto-vitalism in evidence. See similar remarks about the "nesting" of temporalities in J.T. Fraser, *The Genesis and Evolution of Time: A Critique of Interpretation in Physics* (Amherst, MA: University of Massachusetts Press, 1982), 144-75.

stability that realigns itself internally when disturbed.[28] The heartbeat, the circadian rhythms of nature, the cell division cycle—all belong to the same realm of temporal structure and dissipative systems.

The appearance of time as a constitutive function (an "operator") in our understanding of reality can be seen in emerging theories about complex dynamical systems, especially those that are self-organizing, far-from-equilibrium, and chaotic. This is as true in biology as in physics and chemistry.

3.1 *Molecular Biology*

The role of time can be seen in recent theories of molecular biology. Utilizing current information theory, Bernd-Olaf Küppers argues persuasively that there is a continuity between living and non-living matter. Moreover, he argues that the roots of Darwinian natural selection can be explained by the physical properties of matter. Though this may sound "reductionist" to those who hope for some ineluctable, but measurable, principle of life, it is not.[29]

Küppers distinguishes (with others) three phases of evolution: a chemical development in which prebiotic conditions establish nucleic acids and proteins; the self-organization of matter converging upon more complex coupling patterns; and biological evolution with its development of primitive, unicellular elements toward multicellular organisms.[30] In the final phase, the genetic information is optimized and diversified. The transition from the non-living to living matter takes place gradually during the second phase: self-organization.[31]

After arguing against a chance hypothesis or a teleological approach for this evolution toward self-organization, Küppers maintains what he calls a "molecular-Darwinistic" position: that "biological proto-information arose by the selective self-organization and evolution of biological macromolecules."[32] By pointing to selection as "that which produces information," he believes that he has a criterion

---

[28]Meinhardt, *Models*, 322. See also Peacocke, *Biological Organization*, 189-95; and Gleick, *Chaos*, 273-300.

[29]Küppers, *Information*; cf. Küppers, "On the Prior Probability of the Existence of Life," in *The Probablistic Revolution, 1806-1930: Dynamic of Scientific Development*, ed. G. Gigerentzer, L. Krüger, and M.S. Morgan (Cambridge, MA: MIT Press, 1987), 355-69. Even Hans Driesch would argue for the *methods* of physicists in biology, but not their *results*; see Driesch, *The Science & Philosophy of the Organism: The Gifford Lectures, 1907-1908*, 2d. ed. (London: A & C Black, 1929), 32. Cf. Peacocke, *Biological Organization*, 268-72 and 255-63; and Barbour, *Religion in an Age of Science*, 165-68. See also Weinberg, *Final Theory*, 51-64.

[30]Küppers, *Information*, 26. See also Peacocke, *Biological Organization*, 214-41.

[31]Note that a narrative is being constructed about the origins of life.

[32]Küppers, *Information*, 27.

for determining when self-organization makes the transition. The ability to produce information is a "dynamic criterion of value."[33]

Note that value here involves some *normative* (or axiological) dimension to the process; randomness does not exclude teleonomy.[34] Efficiency is determined by internal and external criteria. When this information alters the recipient by making a structural change in it or re-orients the recipient's action, then both novelty and confirmation can occur. Both identity and difference must occur for evolutionary self-organization to be confirmed. "The optimum of production of information may be presumed to lie at the point where there is as much novelty as possible and no more confirmation than is necessary."[35] This biological information "has arisen by the selective self-organization and evolution of biological macromolecules."[36] A *selection value* exists "if the binary-coded mutant sequence corresponds to the reference sequence one 'bit' better than does the master (i.e., parent) sequence[;] then it will be allowed to reproduce more rapidly than the master sequence by a certain factor, which we will term the *differential advantage*."[37]

Temporal sequence is an *intrinsic* dimension of this process. Biological information must include both memory and anticipation of alternative futures. Information must be stored; bifurcating choice must be possible. By this process, "qualitatively new properties emerge in the system that has undergone the transition."[38] This process of temporal succession through which a new whole emerges from individual, but integrative, parts is not a special characteristic of living systems, but one which occurs in self-organization, even in non-living systems.

The continuing success of this temporal sequencing is dependent upon (1) avoidance of equilibrium, that is, a metabolism in which the production of entropy

---

[33]Ibid., 49. See also Nicolis and Prigogine, *Complexity*: ". . . it is selection that enables us to detect, interpret, and transmit the 'message' hidden in the nonlinear, nonequilibrium dynamics of the systems. *Selection decodes information*, and that allows the transfer of complexity from one level to another" (p. 143).

[34]Meinhardt avoids the axiological dimension by defining biological development as "alteration in time." (Meinhardt, *Models*, 13). For an analysis of the reasons for the changes in natural selection, see John Beatty, "Dobzhansky and Drift: Facts, Values, and Chance in Evolutionary Biology," in *The Probablistic Revolution*, 271-311.

[35]Küppers, *Information*, 56.

[36]Ibid., 81. On the development of hierarchical systems see Ernst Mayr, "How Biology Differs from the Natural Sciences," in *Evolution at a Crossroads*, 57-60.

[37]Ibid., 85. Cf. Prigogine, *From Being to Becoming*: ". . . the origin of life may be related to successive instabilities somewhat analogous to the successive bifurcations that have led to a state of matter of increasing coherence" (p. 123).

[38]Küppers, *Information*, 122. Many thinkers point to Driesch's sea-urchin experiments and the teleonomy (what Driesch called entelechy) of an organism without subscribing to his claims for a vitalistic principle distinct from the infrastructural integrations of a biological organism. See Driesch, *Science and Philosophy*, 38-43 and 245.

is compensated for by an infusion of energy; (2) self-reproduction; and (3) mutability, that is, a mutagenicity where self-copying is not error-free.[39]

A fundamental underlying characteristic of dissipative systems is their relationship to the environment with which they exchange matter, energy, and information. The temporality appropriate (of optimal efficiency?) to a particular self-organizing system (i.e., a time pattern that keeps the system operating) negotiates internal teleonomic activity and the constraints of the context without exhausting either the resources of the environment or of the internal program.[40]

One must mark that there continues to emerge a *normative* dimension to the experience of time in scientific (here biological) language. A system has a "better" or "worse" position for maintaining or developing itself. The success or failure of a system's ability to cooperate with the environment is important,[41] but the *game* or *military metaphors* of competition or conflict are a more prominent part of the rhetoric of biological explanation.

---

[39]Küppers, *Information*, 131-32. The Darwinian and neo-Darwinian description of the scale of value for success in this process ordinarily speaks of *competition*. Although it is recognized that Darwin borrowed this image from Adam Smith's economics, who in turn uses Hobbes, the fact that *success in competition* underlies the scale of value is rarely attended to. In Küppers himself, there is a correlative dimension of *cooperation* that is also functional. See the mollifying remarks of G.G. Simpson:

> Advantage in differential reproduction is usually a peaceful process in which the concept of struggle is really irrelevant. It more often involves such things as better integration into the ecological situation, maintenance of a balance of nature, more efficient utilization of available food, better care of the young, elimination of intragroup discords (struggles) that might hamper reproduction, exploitation of environmental possibilities that are not the objects of competition or are less effectively exploited by others. (Simpson, *The Meaning of Evolution* [New Haven, CT: Yale University Press, 1971], 201, quoted in Peacocke, *Creation and the World of Science*, 165-66.)

[40]Küppers, *Information*, 148. See also Prigogine, *Being to Becoming*, 123-26; and Meinhardt, *Models*, 189. On alternative paths for evolution, see John H. Campbell, "An Organizational Interpretation of Evolution," in *Evolution at a Crossroads*, esp. 141-46; and Stuart A. Kauffman, "Self-Organization, Selective Adaptation, and Its Limits," in *Evolution at a Crossroads*, 169-203.

[41]Küppers, *Information*: "The selection value of every single species will as a rule depend on the population variables of the other species taking part in selection, so that every evolutionary "step" changes the structure of the value profile. . . . This means that goal and goal-directedness, even at the level of biological macromolecules, are interdependent" (p. 150). Note that this is *not* to say that some species are better or more important than others; it is simply to explain *why* some systems continue and others do not.

3.2 *Summary*

The language that analysts use about far-from-equilibrium systems maintains that such systems have an inner temporality that governs their organization and operation. Without such a constitutive temporal dimension (pattern and sequence) which affects their very identity as a system, they would either not continue to exist or they would be different entities. Moreover, in such systems, there is an optimal or normative temporality that involves internal and external cooperation with environments. In this sense, their temporality is characteristically open—to a future which is not the same as the past, to an environment that may affect them either adversely or positively. Self-organizing systems, therefore, are an important difference in the structure of our world. They display a self-directing autonomy with an interior teleonomy that entails normative behavior for survival as a system. *That* there is a "before and after" is important in defining a narrative sequence; *how* that narrative sequence is described is also important. Biologists like Küppers use, without attention, metaphors that are primarily competitive and conflictual, rather than cooperative.

4 *The Theorem of the Supernatural*

Bernard Lonergan maintains that the development of the *scientific theorem* of the supernatural in the writings of Albert the Great and Thomas Aquinas provided an explanatory perspective on the interaction between the divine and creation. The supernatural did not add further divine actions to the world. "[T]he idea of the supernatural is a theorem, that [is,] it no more adds to the data of the problem than the Lorentz transformation puts a new constellation in the heavens."[42]

To situate Aquinas's theory of divine transcendence, it will be useful to outline the language about divine action in primordial societies. Such expressions contextualize the theory; they also show something of what Aquinas was criticizing. If we are to understand what the Christian tradition has meant by God's action in and upon the world, we must understand the rhetorical context in which such a thinker was arguing.

4.1 *Divine Action in a Sacral Universe*

In a primordial, often pre-literate, but not pre-symbolic world, religion appears in transformed people and actions. *What* changes things, people, and actions is not within sight or sound or touch, but the *what* exercises force or power and one

---

[42]Bernard Lonergan, *Grace and Freedom: Operative Grace in the Thought of St. Thomas Aquinas*, ed. J. Patout Burns (London: Darton, Longman & Todd, 1971; originally published in *Theological Studies* 2 [1941], 3 [1942]), 16 (page reference is to the 1971 edition).

should respect it.[43] To know why or how, through what mediations and conditions these unseen powers appear, is unnecessary.[44] According to Lévy-Bruhl, the realm of these unseen forces is often vague, including, in an undifferentiated fashion, the dead ancestors (both benign and threatening), the deities, and the spirits of nature.[45] To negotiate the power of these realities, one needs the rituals of one's ancestors, the traditions that protect and heal, the actions that will celebrate the joys without awakening the envy of the gods.[46] It is the shaman or the priest who helps the individual and the community across the thresholds and transitions.

The language used by anthropologists such as Lévy-Bruhl provides categorical reification for the experiences (what he calls the "affective categories")[47] that underlie them. The experiential base, in effect, includes simply the joys and misfortunes of day to day living.[48] The difference, he notes, is that pre-literate cultures assume, believe, and trust that the pleasant and painful patterns of experience are produced by powers that are neither visible nor indifferent to human and cosmic affairs.[49] Because there is a "blurring" of all non-visible causes, because they are not distinguished, one hopes to have all these powers on one's

---

[43]See Lucien Lévy-Bruhl, *Primitives and the Supernatural,* trans. Lilian A. Clare (New York: Dutton, 1935) for a classic, if flawed description. I am here less concerned with whether these analyses fit the self-understanding of primordial societies, than in *how* such societies' religious data have been perceived by analysts.

[44]Ibid., 20; and *idem, Primitive Mentality,* trans. Lilian A. Clare (New York: Macmillan, 1923), 36-37. See also Bruno Snell, *The Discovery of Mind: The Greek Origins of European Thought,* trans. T.G. Rosenmeyer (Cambridge, MA: Harvard University Press, 1953), 191-245; and G.E.R. Lloyd, *Magic, Reason, and Experience: Studies in the Origin and Development of Greek Science* (Cambridge: Cambridge University Press, 1979); esp. 1-8 and 10-58.

[45]Lévy-Bruhl, *Primitives and the Supernatural,* 34.

[46]I would point out that more anthropologists view fear as the origin of religious action than to the celebration of joy. This may have as much to do with Enlightenment prejudices about priestcraft and the normative espousal of human autonomy as with field data. See Lévy-Bruhl, *Primitives and the Supernatural,* 24-27, for an example of the conviction that fear motivates primitive religious activity.

[47]Ibid., 32 and *passim*; see also *idem, How Natives Think,* 22-32.

[48]Lévy-Bruhl focuses upon the negative dimensions of experience (loss, misfortune, disease, death) where the following is representative. "How is [the primitive] to account for these misfortunes which at times seem to have combined to overwhelm him?" He contrasts the response of pre-literate cultures with the contemporary search for answers (Lévy-Bruhl, *Primitives and the Supernatural,* 153-54). See his remarks about the causes of success in *Primitive Mentality,* 307-51. Lévy-Bruhl maintains that "not only the data, but even the limits of experience fail to coincide with our own." But he means by this the fact that data always involve interpretation and that the pre-literate mentality "is accustomed to a type of causality which obscures, as it were, the network of such [secondary] causes." *Primitive Mentality,* 96 and 92.

[49]Lévy-Bruhl, *Primitive Mentality,* 65.

own side (so to speak).[50] The task of religion (if undifferentiated from medicine or agriculture) is to protect people from what appears to be bad and to prolong what is good.

The extended security of the ordinary, the diurnal order, is critical. That is why any intervention in the usual manner of things appears to be dangerous, even malign. "Nature" is not differentiated from the "supernatural" powers; hence, even the most improbable transformations are within the course of things. It is only when their "conception of the inevitable order of nature is *suddenly* disturbed and upset" that they are "troubled" (my italics).[51] *Unusual appearances* are "transgressions" that require negotiation: avoidance, prayer-rituals (in effect, distance without flight), or submission (no distance and absorption). The recognition is that the powerful ancestors and the gods often intervene as unseen forces in experience. "In these frequent occurrences, therefore, the primitive mind pays less attention to the happenings themselves than to the suprasensuous realities whose presence and influence they indicate."[52]

Historical religions, such as Judaism or Christianity, though they have tended to differentiate the realm of the divinity and the causes of nature (at least by placing the divine *over* natural causes—in theories and in some practices), nonetheless use images, stories, and metaphoric symbols that describe the intervention or transgression of God within and among the ordinary course of things and events. So in Joshua's battle with the sheiks at Gibeon, God hurls hailstones killing the enemy soldiers or makes the sun stand still at his prayer (Joshua 10:11-13). God provides an earthquake, rocking Paul's prison and releasing him from jail (Acts 17:25-28). The experiences are an unexpected transgression within the ordinary expectation of events, relayed in images and stories to indicate the activity of God.

These parallels between Judaic, Christian, and non-Christian stories are to be taken seriously, since it is this description of divine transcendence that is "taken for granted" as a prime component in faith or in religious sensibility. It is also the anthropologists' categorization of this experience as "primitive" that marks much theological avoidance of language about divine action in the universe. Does God work in the world "alongside" other causes? Does God only operate in the "surprises," the extraordinary events that overturn or intrude in the ordinary necessities of nature? If one discards these forms of divine interaction with creation, then does one have only subjective, moral suasion on human subjects as the mode of divine action? These questions can only be answered if one understands how metaphors, symbols, and stories (the rhetoric of religion) function in relationship to reflective theology (the dialectics). The same pattern must be studied in the methods of the natural sciences themselves.

---

[50]Ibid., 59-96. Bronislaw Malinowski disagrees with this position, asserting that preliterate cultures distinguish magical from primitive technological devices for agriculture or hunting, though magic continues to be useful for unexpected situations. Cf. Malinowski, *Magic, Science, and Religion and Other Essays* (Boston: Beacon Press, 1948), 12-17.

[51]Lévy-Bruhl, *Primitives and the Supernatural*, 197-98.

[52]Ibid., 57 and 64. See also *idem*, *Primitive Mentality*, 57-58.

4.2 *A Methodological Critique: Aquinas*

Thomas Aquinas, using Aristotle's logic and metaphysics, attempted to understand the relationship between primary and secondary languages in faith, as well as between the two poles of divine-creational interaction. The need to distinguish the "what is divine" from "what is human" within human experience as the operative and cooperative action of God encouraged theoretic discussions. Through Aquinas, Christians have had a theory of the supernatural to account for the specific character of divine action. It is important that it is a *theory*, not an addition of data, "something like the discovery of gravitation and not something like the discovery of America."[53] It is the difference between *going faster* as an experience and *acceleration* as a theory. The latter accords the differing moments of time, distance, and velocity their relative relationship, providing analysis, generalization, and systematic correlation of factors. They apprehend the same data, but acceleration explains both going faster *and* going slower.

The notion of the supernatural offers the same kind of precision in Aquinas's writings. It tries to articulate the various factors at stake (human will, divine love, creation, redemption, evil, etc.) and correlates them much as one would do with distance, time, velocity, and mass. In effect, the Middle Ages always *knew* that the data of their experience were from God; what the theorem of the supernatural offered was a way of *understanding* God's relationship to nature. "Nature" itself was a conceptual abstraction; it was what was distinct from "supernature." Nature was not what was sensed or thought, our common experience; for our common experience included the dimensions of divine activity as well.[54] In this schema, it is not possible to say what is *solely* from human intention and execution. God was the ultimate intelligibility of what is, not an object with multiple attributes (omniscience, omnipotence, etc.) to be explained.[55]

Here I shall outline Aquinas's position on the nature of divine transcendence, the universal instrumentality of creation, and divine action as they might apply to self-organizing systems. Aquinas assumes that he has provided evidence and warrants for God's existence and God's benign intention for creation. Questions of human (as opposed to non-human) interaction with the divine are useful here in so far as they contribute to our understanding of his position on the questions we are posing. To assist in making a useful model of divine-creational interaction,

---

[53]Lonergan, *Grace and Freedom*, 143. See also *idem*, *Insight: A Study of Human Understanding* (New York: Longmans, 1961), 634-86.

[54]See Karl Rahner's interpretation of "pure nature" as a *Restbegriff*, a theological abstraction, that sorts out from our mixed, i.e., graced and sinful, experience what is sinful from what is divine initiative and human cooperation. Rahner, "Concerning the Relationship between Nature and Grace," *Theological Investigations*, vol. 1, trans. Cornelius Ernst (London: Darton, Longman & Todd, 1961), 313-17. See also Piet Fransen, *Divine Grace and Man* (New York: Desclée, 1962), 88-91.

[55]Lonergan, *Grace and Freedom*, 105. See also *idem*, *Insight*, 657-77.

it will be necessary to transpose these classical metaphysical categories into those of an historical ontology (see section 5.3 below).

4.3 *The Nature of Supernature: Divine Transcendence*

The meaning of *supernature* involves several terms: God's transcendence, creation's particular form of identity and self-organization, and their interaction. What links them all is the experience and affirmation of transcendence. "In a more general sense, transcendence means 'going beyond'."[56] Insight and reflection do not simply reproduce images and sense data; they go beyond them. Transcendence is proportionate to the kind of reality about which it is predicated.

Aquinas's understanding of the supernatural was, therefore, a relational, not a reified, term. It was emphatically *not* a term to cover the unusual transgressions of nature in the scriptures! It was an abstraction understood in relationship to nature as a construct. The content of nature and supernature "slides," depending upon the stages of meaning that humans inhabit. In a stage of human development in which nature is the uncontrollable other, God is embodied in theophanies such as the burning bush or Mt. Sinai.[57] In another age, divine transcendence might appear as interpersonal suasion. But a theorem of divine transcendence permits the thinker to understand that God is always *beyond*.

To say that God is transcendent is to maintain that *whatever* creation does (whether human or non-human) is not to be identified with divinity.[58] Divine transcendence cannot be a property attributed to any creature. Every created cause falls within the order of necessity or contingency; but God "produces not only reality," but also their modes of contingent or necessary emergence.[59] God by definition exceeds and determines the modes of both contingency and necessity. The analogy Aquinas uses in *De substantiis separatis*[60] is that of a geometer who not only makes triangles, but makes them isosceles or equilateral at the same time. Were God to produce a contingent effect *directly* in our world with irresistible

---

[56]Lonergan, *Insight*, 635.

[57]See Lonergan, "Religious Experience," in *A Third Collection: Papers by Bernard J.F. Lonergan, S.J.*, ed. Frederick E. Crowe (New York: Paulist Press, 1985), 119-122.

[58]See Gordon D. Kaufman, "Two Models of Transcendence: An Inquiry into the Problem of Theological Meaning," in *The Heritage of Christian Thought: Essays in Honor of Robert Lowry Calhoun*, ed. Robert E. Cushman and Egil Grislis (New York: Harper and Row, 1965), 182-96. In this early essay, he distinguishes scientific and religious language as cognitive, empirical language vs. intersubjective mystery. Images and stories are necessary because of the interpersonal dimension of revelation (p. 193). Again, notice how this is another version of science as "objective" fact and religion as "subjective" value.

[59]Lonergan, *Grace and Freedom*, 108.

[60]*De subt. sep.*, 15 (Works of St. Thomas Aquinas: Marietti editions, para. 137), cited in Lonergan, *Grace and Freedom*, 108.

efficacy, it would have to be God's own self.[61] What the geometer cannot do is to make equilateral triangles with only two sides equal.

What most people think of as divine intervention (the production of a surprising contingent effect with irresistible effectiveness), a *direct* manipulation of "nature" (as an inert object "out there"), a "miracle," is denied by Aquinas. For Aquinas, it is not that some astonishing moments might not appear in nature and history, but that they are not to be identified with divine activity in and of itself.[62] An appropriate theoretic understanding of supernature will provide an explanation for both the usual and unusual course of things. It will also "eliminate intervention's implication of violence."[63]

What occurs in our world (orderly, chaotic, and random) is in need of explanation. 'The supernatural transcendent' is the name given to that realm of ultimate explanation. Nature is the relatively autonomous pole of creation that operates by virtue of its own internal structures and processes, utilizing their own forms of transcendence and self-transcendence. Natural forms of transcendence, however, are dependent upon the supernatural for their relative independence. Nature does not operate on its own *without* divine initiative, support, or direction; theoretically, however, it is possible to distinguish and relate the relative poles of cooperation. Moreover, it becomes clear that the *two orders* (nature and supernature) constitute the definition of grace, God's love poured out in the world.[64]

4.4 *Instrumental Causality: The Intermediary*

The theoretical language that medieval and early modern Catholic theology used to discuss this interaction of the divine and nature was instrumental causality. The notion of instrument was applied to a wide variety of experiences: Christ's humanity, prophecy, the sacraments, but also the peculiar, invisible operations of nature, the influence of magical pictures, and (with Aristotle) the generation of animals.[65] Wherever it might appear that a more-than-finite agent was operative,

---

[61]Lonergan, *Grace and Freedom*, 109. I make a note of the language "direct" and "indirect," since I will return to it section 5.4.

[62]Cf. Thomas F. Tracy, "Particular Providence and the God of the Gaps" (in this volume). Pure chance must be distinguished from causal chance and forms of deterministic unpredictability. These latter two (due to a surprising convergence of local causal chains and our cognitional limits) are often what are called "divine action." Indeed, the images they conjure are the ordinary definition for the "supernatural." Aquinas would deny that they are "solely" divine action.

[63]Lonergan, *Grace and Freedom*, 43.

[64]Lonergan, *Method*, 310, n. 13.

[65]Lonergan, *Grace and Freedom*, 82.

the element itself could be construed as an *instrumentum*.[66] But in medieval life, that was true in all cases. God was never absent, even in death or sin.

Though I will focus upon the theoretical dimensions of the *instrumentum*, particularly as these applied to the Christian sacraments, I must note that the role of magical images, moving statuary that healed, relics that cured diseases and protected in childbirth, the blessing of saints who oversaw a whole host of ordinary tasks from bread-baking and beer-brewing to journeys and coronations cannot be ignored in this context.[67] Anglo-Saxon rituals contained incantations in an amalgam of ancient and current languages to bring about fertile harvests, safe house-raising, and protection from elf-shots. The *Agnus Dei* was a small disk (earlier in wax, eventually in precious metals) that protected against the devil, thunder, lightning, fires, and drowning. All of these were interpreted as "instruments" of the divine power.

Although theologians such as Thomas Aquinas argued strenuously against any magical interpretations of sacramental practice or of these items, still the faithful, including theologians, commonly believed that eucharistic bread could turn into Jesus' physical flesh and blood. Popular stories related the way in which the host was used "to put out fires, to cure swine fever, to fertilize the fields, and to encourage bees to make honey."[68] The relation of the high culture of the theologians and the knowledgeable churchmen and women to the low culture of popular devotion and magic practices is the matrix for the sixteenth-century reformations of Christianity. Is it any wonder that the sciences of the sixteenth through eighteenth centuries saw *their* control over common meanings as more effective than the prior theologies and philosophies?

The general law in the theory we are exploring is that everything is an instrument. *Instrumentum* could be defined as a lesser reality accomplishing the work properly attributed to some proportionately higher reality.[69] In the universe, all realities are moved in relation to some cause higher than themselves. Only God is not an instrument; nor can God be used as an instrument. Hammers do not move of their own accord, except in dreams and fantastic cartoons. To be able to operate

---

[66]See Bernard Leeming, *Principles of Sacramental Theology* (London: Longmans, 1960 ed.), esp. 283-345, for an overview of the various uses of instrumental causality to discuss sacramental efficacy.

[67]See Keith Vivian Thomas, *Religion and the Decline of Magic* (New York: Charles Scribner's Sons, 1971), esp. 25-50; and Valerie Flint, *The Rise of Magic in the Early Medieval Europe* (Princeton, NJ: Princeton University Press, 1991). For an art-historical appreciation of the power of these images, see David Freedberg, *The Power of Images: Studies in the History and Theory of Response* (Chicago: University of Chicago Press, 1989), esp. 82-135.

[68]Thomas, *Decline of Magic*, 35. See also Miri Rubin, *Corpus Christi: The Eucharist in Late Medieval Culture* (Cambridge: Cambridge University Press, 1991), 338-42, and David N. Power, *The Eucharistic Mystery: Revitalizing the Tradition* (New York: Crossroad, 1992), 180-207.

[69]Lonergan, *Grace and Freedom*, 81.

beyond their own proportional ends, instruments must participate in a higher system. For that to take place, some participation is required in the higher cause's "productive capacity."[70] But because Aquinas applies this participation in different ways to differing natural systems, it becomes important to understand what he means in any given instance of instrumentality. It should not be assumed that the primary analogue for instrument is an inert natural object, despite the fact that saws, hammers, and pipes are often his examples.

4.4.1 *Human History as Instrument*

Divine providence (charity, infinite love for the other) in the mind of God is the primary cause of all; but this providence exists in the created universe as governance and fate.[71] Providence is the art in the mind of the divine artisan and fate is the operative instrumentality in history. Providence is the intention of love in divine life for all that is (including the divine self); governance is what takes place in the universe as a result of the natural realm's cooperation or conflict with that intention. Fate is not a cause in addition to God, but the "order of secondary causes; it is their disposition, arrangement, seriation; it is not a quality and much less is it a substance; it is in the category of relation."[72] Fate is the *de facto* pattern of what happens as the divine design unfolds in nature and history by virtue of cooperation or conflict with divine love. Fate is the interaction of divine primary cause and human instrumental causes.[73] There is always a mediated execution of divine providence which in itself is strictly supernatural, an un-owed experience and knowledge of the gift of love. "Extracting" (so to speak) the divine dimension from the human ones is not so simple as pointing to the ordinary and the extraordinary, the usual or the unusual. Hence the language of the supernatural in theology functions as a critical upper blade on the stories, symbols, and metaphors within the primary languages of revelation.

4.4.2 *The Sacraments as Instruments*

Another example of Aquinas's strategy may be seen in his theology of the sacraments, the instruments of divine action *par excellence*. These signs of divine presence and action are invitations and moments of cooperation. They have an instrumental power of their own that lays a permanent existential claim upon

---

[70]Ibid.

[71]Ibid., 82-84. See *STh* 1.116.2, *ad 2um et 3am*; and 1.116.3, in which fate is precisely a "contingent necessity." God's providence orders the universe toward love through both contingent and necessary secondary causes.

[72]Lonergan, *Grace and Freedom*, 84.

[73]Ibid., 89: ". . . [fate] is the dynamic pattern of world events, the totality of relations that constitute the combinations and interferences of created causes. . . ." Cf. Kaufman, "On the Meaning of 'Act of God,'" in *God the Problem* (Cambridge, MA: Harvard University Press, 1972), esp. 136-40. See also Tracy, "Particular Providence."

individuals and objects, but *not without their participation*. The *res et sacramentum* (an intermediate, even symbolic, effect of God's grace) of sacramental life is like fate in the *de facto* patterns of providence in the natural and historical world. It can mean minimal personal cooperation (*non ponitur obex*) or it can be the potential for holiness and communion with the divine. In any case, it is the interaction of objects, persons, and common ritual that provides the *instrumentum*, that is, the mediating reality for the transcendent act.[74]

For Aquinas, it was precisely *through* the internal organizing power of the natural elements themselves that the divine presence and agency is disclosed.[75] The physical signs of nature (sharing bread and drinking a cup) receive their historical (and divine) specification through the power of the words of Jesus whose memory is activated within the present ritual instruments.[76]

In the Eucharist, the material and formal appearances remain, so that "bread" is apprehended, but the "being" or "substance" is changed. This occurs by divine power, the power of the word of Christ, in which what is in common between the being of the natural reality and God is transformed into the being of the divine. It is as though God simply "takes down" the barriers between the divine self and finite creation, so that created reality can be the vehicle for divine love.[77] (Another way of viewing this, from the point of view of the "objects" is to say that, unlike human beings, they have no barriers [no *obex*] other than finitude, to divine presence.)[78] Aquinas compares the operation of words, gestures, and material objects to the mystery of the incarnation of the word in which flesh (a *conjoint instrument*) perceived by the senses discloses the divine.[79] But it must still be noted that the humanity of Jesus is an *instrument*, *not* an unmediated intervention of divine action!

---

[74]See Happel, "The 'Bent World': Sacrament as Orthopraxis," *CTSA Proceedings* 35 (1980): 98-99.

[75]*STh*. 3.60.6, *corp*. and *ad 1um*.

[76]*STh*. 3.60.3, *corp*.; and 3.60.5, *corp*. and *ad 2um*. Divine institution by Christ's words, specifying certain material objects for use, determines the meaning of the natural signs. This is the way *all* natural signs are specified in history; but Christ's words are the evidence of *divine intentionality* for the believer—precisely because the believer has *faith* that in Jesus, God is speaking in human words. This Christological focus is important for Aquinas.

[77]*STh*. 3.75.4, *ad 3am*: ". . . virtute agentis finiti non potest forma in formam mutari, nec materia in materiam. Sed virtute agentis infiniti, quod habet actionem in totum ens, potest talis conversio fieri: quia utrique formae et utrique materiae est communis natura entis; et id quod est entitatis in una, potest auctor entis convertere ad id quod est entitatis in altera, sublato eo per quod ab illa distinguebatur."

[78]Lest this appear as a "romanticism of creation," it is necessary to say that for Aquinas, *all* created reality is in need of redemption. Everything participates in the decline that is initiated by the sin of the first human parents. See Lonergan, *Grace and Freedom*, 43-53.

[79]*STh*. 3.60.6, *corp*.

Not all interpreters agreed with Aquinas.[80] Some (like Henry of Ghent, Okham, Biel, Bonaventure and Scotus) viewed the instrument as an occasion, a condition in which God operates externally.[81] Aquinas held out for the difference of divine action, but not its extrinsic intervention. Others (Cano, Vasquez) argued that the instrument had a moral worth or value in itself that *moved* God to act. Aquinas denied this, since it seemed to make God dependent upon creational conditions. Nonetheless, this position seemed to have great currency through the seventeenth and eighteenth centuries.

Later interpretations (Schmaus, Casel, Vonier) simply argued that there was a "mysterious" power within the instrument that produced a more-than-human effect. Though Aquinas sometimes spoke of the power operative in a word that transmitted divine transformation, he denied an independent sacral vitalism functioning within created reality. The divine and the human remained distinct; the wonder was (and is) that God has chosen to act *through* nature's own self-mediating capacities. There were theological arguments throughout the later middle ages and early modern period about whether it was the material of the instrument itself that was the ancillary cause to divine agency (looking for a causal nexus?). Later thinkers, hoping to avoid a "physicalist" interpretation of the cooperative action in the sacraments, stressed that the instrument provided information to participants about God's intention to make them holy. Or they could be understood as practical signs, like a juridical order or a title-deed, saying something, while simultaneously accomplishing it.

4.5 *Instrumentality and Providence*

Providence is God's characteristic way of being; it is love for the other in and for the sake of the other (*caritas* or *agape*). For Aquinas, it is the prime analogue for knowing and loving. Everything outside of God is an instrument toward or away from that love. So he can quote Augustine approvingly: "God cooperates with good will to give it good performance; but alone [God] operates on bad will to make it good, so that good will itself no less than good performance is to be attributed to the divine gift of grace."[82] But this deals specifically with the nature of human instrument. What about non-human instruments?

If I have interpreted Aquinas correctly on the sacraments, he is arguing that created reality operates at different levels, permitting it to cooperate with divine love each at its own level. Non-human created realities, whether animate or inanimate, interact at their own level of reality as well as between levels precisely through their own internal operation to cooperate with divine providence. So, for example, the ability of bread to bear the historical presence of divinity is due to the

---

[80]Nor do all interpreters of Aquinas agree with my reading of the relationship between signs and causes in the theology of the sacraments. See, for example, Power, *The Eucharistic Mystery*, esp. 219-36 and 269-85.

[81]See Leeming, *Principles*, 287-94.

[82]Lonergan, *Grace and Freedom*, 3.

fact that it is transparent to divine love. Insofar as each element operates at its own level and cooperates with higher schemes of recurrence, it will permit the universal transformation of all things into divine love. *To discover one's own proper instrumentality is to encounter the divine agency in one's being*. In other words, there is no other *causal nexus* than the self-organization of the entity itself.

4.6 *The Language of* Direct *and* Indirect *Divine Action*

To use language about "direct" and "indirect" action seems imprecise in this context. It misrepresents the relationship between the language of the scriptures and it misunderstands the meaning of mediation. The assumption that Christian scriptural and doctrinal languages affirm that there are *special acts of God* or special providence may be true, but intellectual smuggling occurs when the term "direct" is used to describe them.

First, the doctrinal (and Augustinian *theological*) language that is being used is dependent upon metaphoric and narrative primordial language found in the scriptures. What is the relationship between these two or three forms of language? Is the derivation of *no* significance? Why privilege one over the other? Second, within the primary and creedal languages, divine action is often described in anthropomorphic fashion, as extraordinary (by human standards), and as more powerful than nature itself could perform. "Special" seems to discriminate between *unusual* and *usual*. On what basis? Is the criterion smuggled into theology from the extraordinary, superpowerful agency described in scriptural texts?

The problem is in the images of direct and indirect and a misplaced concreteness of their metaphoric intent. Aquinas would have said that God acts "directly" and "indirectly" at all times—namely, by originating, sustaining, and guiding created realities *through* contingencies and necessities. The lack of clarity is caused by misunderstandings about what it means to act through *mediation*. God's transcendent act is providence, love for the other; it is always exercised mediately, in cooperation with the other (first and primordially in triune life; "then" in relationship to created realities). I have no doubt that God acts in and through quantum events; indeed, perhaps the science of quantum events may grant more evidence for the underdetermination of micro- as well as macro-events. But the indeterminism of quantum events is no more opportunity for divine action than the necessities of universal gravity.

For those who use the language of "direct" and "indirect," "indirect" seems to mean "mediated," and "direct" means "without mediation." Sometimes there is the assumption that interpersonal action is unmediated. Most philosophers (whether of language or no) would find this a difficult position to maintain. There occurs "body language," the embodied self-presentation even at the surfaces of one's skin. There is an assumption that God must act without mediation at some point (the "initial" stage) to begin the necessities and contingencies. The sole criterion for determining that it is God who is operating turns out to be a surprising, even uncanny effect from unexplainable causes. It is not simply that the action is accounted an extraordinary event, but that what is extraordinary informs us about a special, particular, and personal action beyond our control. And this becomes the meaning of divine transcendence.

However, for a thinker like Aquinas, divine transcendence is what is always absolutely "beyond." Indeed, it is precisely the divine distinction from creation that permits God to operate immanently and mediately in all created beings. The divine freedom to love involves triune life. God creates only "because" of the divine self-mediations of knowledge and love. This mystery envelops all the mediating contact our world has with God. Any attempt to discover a *causal nexus* is a search for the God of the gaps; it presumes that there is one or another particular negative or unambiguous condition under which one can expect God to act. This solution seems to misunderstand the nature of divine action. God is the necessary, but never the sufficient condition for a created event. God's actions within creation are always mediated by the self-organizing agency of the created reality. Moreover, since all God's actions are efficacious, God does indeed determine effects through the disposition of created causes. That is part of Aquinas's explanation for the effectiveness of petitionary prayer, not an "in principle" ability of God to intervene without intermediary in curing the sick or parting the seas.

What is curious is a continuing search for a *causal joint*. The mediations are what non-human animate and inanimate realities are "about." It is precisely their self-organizing capabilities in relationship to the environment, with their law-governed structures and underdetermination (where that applies), that *is* the *locus* of divine-created interaction. Rocks cooperate as rocks, plants as plants, and dogs as dogs. One does not need to search for a further physical, vital, spiritual, or transcendental "hook" into which God can insert divine operations. This immanence of God to all reality is the place where divine-creational interaction occurs. What the best science tells us about the way the world works *is* what God is telling us about the way divine action operates in our world. If one argues for any moderate realism in knowing, then one must say we *do know* something, however partial and mysterious, about the divine agent in the interaction. This is true not only because God has chosen to speak, but because what we learn about the created world is not deceptive or malicious.

The search for a *causal joint* or *nexus* in physics or biology seems to be *about* something else. (1) Positively, it is an attempt to make more inclusive the criteria or conditions under which a judgment can be made that "God is acting" by extending the conditions from scriptural stories to the natural sciences. In this sense, the indeterminacies of both micro- and macro-level systems appear as part of the contingencies that allow an open-ended universe, a cosmos in which human choice might make a difference. They articulate the conditions for a free universal story. (2) The search for some intimate space in pre-human nature or in the human mind or body where God is "more operative" than in others seeks to preserve a theological and confessional tradition that protects the free sovereignty of God, such as in Biel, Okham, Luther, and Calvin. God could have made another world, but chose to create this one. Nonetheless, according to Okham and the later scholastics, divine power has made no *intrinsic* connections between reason and faith. Divine power is absolute and could be arbitrary. It is this philosophical defense of the arbitrary that may appear useful as a description for divine action. If my arguments are correct, it is not. (3) The term 'direct' often means not just immediate, without mediation, but an action that seems to operate within nature, even along side it, as a form of causation extrinsic to the orderly or random processes of nature. As such, it reduces divine action to created proportions. What *causal nexus* seeks to understand is the moment in which "natural cooperation" occurs. To all these the search for a "divine vitality" within nature misunderstands divine transcendence and nature's relative independence.

5 *A Transposed Melody: Divine-Creational Interaction*

If we are to go beyond the descriptive, but inexact, language of "direct" and "indirect," we must transpose the classical metaphysical categories of Aquinas. We can do so by using the language of self-organization and far-from-equilibrium systems. We do so not simply by seizing upon the instabilities or indeterminacies of such systems, but by noting their fundamental inner temporality. Just as Aquinas used what he believed to be the best available scientific explanations concerning how instruments work, we can integrate such scientific language into an overarching scheme that articulates divine transcendence and God's interaction with animate and inanimate nature.

Much of our interest is with the kinds of interaction between divine and human agents. However, a theory of self-organizing systems provides a way of speaking about non-human agents (and humans) *in their own right*. Indeed, theories of self-organization can specify the internal dynamics of non-human, self-directing realities and specify the *relative independence* of non-human instruments. By developing a hierarchy of such relative independence, from chemical reactions to human teleonomies, it becomes possible to specify the characteristic instrumentality of each, its ability and level of *cooperation* with divine love. For Aquinas, God not only *respects* the proper temporality of creatures; God originates, supports, and encourages them to come to their appropriate completions. For reasons known only to love, God has chosen to mediate divine temporality through the structures of created self-organizing systems.[83]

5.1 *The Temporality of Self-Organizing Systems*

Self-organization theory provides an analysis of how certain kinds of systems work, from inanimate through animate to self-conscious life. The internal clocks of things operate as a self-constituting process by which a reality is what it is and does. In the particular kinds of systems described above, there is the development of a future different from the past, often dependent upon a choice in a bifurcation

[83]Though the language sometimes used by theologians of the world as "God's body" is descriptive, the explanatory value of such a metaphor requires an understanding of the historical, self-organizing structures of the universe's systems to be helpful. See, for example, Sallie McFague, "Models of God for an Ecological, Evolutionary Era: God as Mother of the Universe," in *Physics, Philosophy and Theology: A Common Quest for Understanding*, ed. Robert John Russell, William R. Stoeger, and George V. Coyne (Vatican City State: Vatican Observatory, 1988), 261-63.

process. Such unpredictability offers (perhaps) a condition for the emergence of the new—including the choices of human freedom (at a much higher level of organization).

The "whole" in such a system is relative, a moving viewpoint; it is dependent upon its own evolution. One could speak about it synchronically as well as diachronically. The reality emerges as its future attracts it toward that whole. History in an organism makes a difference to what it is.

The temporality of self-organizing systems articulates a relative independence that is self-directing. This directionality is not totally separated from its environment, but its ability to organize itself can be recognized as encoded within the system itself. Its success is dependent upon its ability to avoid stasis (i.e., to grow and change) and to replicate itself. To be able to do that, there must be some mutability, that is, non-necessity in its replication. (These criteria for completeness are likely true of human actions as well as of lower self-organizing systems.)

5.2 *Providence as Divine Temporality*

God's time is the prime analogue for all temporality. It can be conceived as a "before" and "after" and a simultaneity in which there is only cooperative love. Providence, as the fundamental divine intention, is for "all that is" to be itself in love. For that to take place, there must be an other in whom and with whom a system mediates itself. Reality is intrinsically relational, because God is present as inner relationality. Within divine life, there is the experience of love as identity and difference, the same and other. In a prior article for this research project, I have argued that in God there is an equiprimordiality of mystical simultaneity and narrative sequence.[84]

If we used *our* experience of simultaneity as the prime analogue for God, then we would mix an intense attachment for the other as well as boredom, anger, fear, and rejection that consists of more of the same. If we use our "sequencing" as the norm, we have the experience of our history as a mixture of evil and grace. However, for Aquinas, charity is the ultimate temporality of divine presence. In God, there is only the mutual self-mediation of knowledge and love. Divine love is ever active. In effect, that is to say that God's time is "real" temporality; all other temporalities are analogous and (for Christians) participative. Human beings enter "real" time by more intensely cooperating in and with divine temporality, that is, love and knowledge. As humans become more just, loving, and truth-telling, they share more deeply in divine inner life (temporality). In effect, it means that not only does the finite created reality become more knowledgeable and more loving, but it mediates its own temporality through and with God's intentionality.

An allegory illustrates this theological position. How does paper catch on fire? It must draw closer to a flame already burning. As the paper grows closer to the flame, it reaches combustibility, until finally it too is consumed. God is the

---

[84]See Happel, "Metaphors and Time Asymmetry: Cosmologies in Physics and Christian Meanings," in *Quantum Cosmology*, 103-34.

eternal fire, burning but not consumed; as self-organizing realities draw closer to this burning flame, they participate in the fire itself, eventually becoming that toward which they are drawn. The human fear continues to be (not just since Friedrich Nietzsche, but certainly philosophically thematized by him) that to mediate oneself in and through such an Other will destroy the very self-organizing temporality that human beings struggle to maintain in the context of harsh environments. The Christian claim is that cooperation with an other, while it may be the ultimate risk, is also the primary way to become one's truest self.

5.3 *Temporalities in Cooperation: Mutual Self-Mediation*

The notion of *mediation* is a common one in logic, of course, since it is the term that intervenes between the first principle and the conclusion that follows. In an organic example, one can see the heart as the "source" for the circulation of the blood, but the flow of blood is mediated through the arteries and the veins. In *mutual mediation*, however, the systems operate as a functional whole. The principle of movement from an origin and the principle of control in a system are combined. There are differing centers of immediacy and their mediations overlap. In the same organic example, one can see that the respiratory system supplies fresh oxygen to the whole body, not merely to the lungs; the digestive system provides nutrition, and so forth. The result is a functional whole "that has fresh oxygen, is fed, is under control, and is moving, because there is a number of immediate centers [*sic*] and from each center there flows over the whole the consequences of that center."[85] It is not difficult to see an initial description of a self-organizing system.

If one adds to this notion *self-mediation*, one can see that the different centers of immediacy creating a functional whole are governed by an internal teleonomy[86] that directs the identity of the system itself. Change is not simply a replication of identities, but a genuine integration that produces higher-level functions through the operation of its parts. Lonergan sees a final notion of self-mediation *inward* through which *self-conscious mediation* occurs. Self-consciousness implies a pre-reflexive presence to self that is the condition for all mediations.

> [F]or you to be present to me, I have to be present to myself. This presence of the subject to [the self] is not the result of some act of introspection or reflection. . . . Consciousness is a presence of the subject to [the self] that is distinct from, but concomitant with, the presence of objects to the subject.[87]

God is the perfect self-conscious, self-mediating subjectivity. God is simultaneously present to the divine self as well as present in and through that self-presencing to the other. So God knows and loves all created reality precisely

---

[85]Lonergan, "The Mediation of Christ in Prayer," *Method* 2, no. 1 (March 1984): 5.

[86]Lonergan calls it a finality, but clearly means an internal teleonomy rather than an extrinsic teleology.

[87]Lonergan, "Mediation of Christ in Prayer," 8-9.

through the divine relationality. For Aquinas, this self-presencing activity is completely knowing and loving, but inexhaustibly mysterious. The divine act continues to become "more and more" itself in and through the divine relational identity. Here temporality as sequence can be understood as the deepening of love's mystery and the continual exercise of wisdom; simultaneity can be interpreted as the intense and constant exercise of encountering the other. Instrumentality means that the characteristic temporality of created realities is activated, engaged, and brought to its highest form of complexity by mutual self-mediation with the divine. The gracious action of the divine within our world is not violent confrontation, extrinsic intervention, but a synergistic cooperation. "The integrity of the dialectic of the subject is grounded in the gift of universal willingness or charity."[88] Though this presentation is deliberately brief and descriptive, it does show how Aquinas's understanding of God can be transposed into seeing the divine temporality as the ultimate "self-organizing system."[89]

6 *Final Remarks*

This paper has argued that Christian theology, in the thought of Thomas Aquinas, has had a coherent understanding of divine-created interaction. By developing a clear theorem of transcendence or the supernatural and of universal instrumentality, Aquinas was able to articulate the basic ways in which inanimate, animate, and human secondary causes cooperate or conflict with the divine act of love for the universe (providence). These terms can be transposed into an historical ontology and a language of mutual mediation such that all levels of reality have their relative autonomy. Contemporary science, with its analysis of self-organizing systems, provides an understanding of the regularities and contingencies of inanimate and animate created realities. Their language permits us to understand how an open, flexible universe can provide the conditions for cooperation among these systems and with divine action without conflict or violence to the integrity of creation.

There is an optimism inherent in this analysis; it is born of a religious conviction that though the cosmos (whether human or non-human) is flawed and finite, its internal logic is not vitiated, malicious, or deceptive. Images, the body, and the non-verbal are no more (and no less) prone to sin than reason. Within the temporal being of "nature," self-organizing, living, self-conscious beings can engage with their environment in a cooperative way. Ultimately, it argues that self-conscious creatures may learn that cooperating with the ultimate environment, an unfathomable Other, will not do violence to their own complex teleonomies.

---

[88]Robert M. Doran, *Theology and the Dialectics of History* (Toronto: University of Toronto Press, 1990), 502.

[89]I am indebted for my current reformulation of the argument to criticisms and questions raised especially by Nancey Murphy, Robert John Russell, Thomas Tracy, Ian Barbour, and to a former colleague at Catholic University, James R. Price III, now at the University of Maryland, Baltimore.

The Christian claim, however, goes further. It maintains that this mysterious enveloping environment is involved in a *mutual self-mediation* with creation.[90] When one is in love, one mediates oneself in and through an other who is discovering, planning, negotiating his or her personal identity in and through oneself. This is mutual self-mediation. Christians claim that they are not merely projecting themselves abstractly into an alien environment to mediate themselves, but that the Other has chosen out of love to mediate the divine subjectivity in and through natural self-organization (because God is ultimately a community of mutual self-mediation!). The story of the Christ could have been quite different than it was. Jesus could have mediated himself in some other fashion; but he did not. He chose to offer his life for others in self-sacrificing generosity. In this action, he operated as though neither the natural nor human environment nor God were an enemy. In loving creation, entrusting his own life to others—even in death—faith claims that there is here a divine love. This is what I have called elsewhere the "double dative of presence." We are present *to* the divine who in that same movement is present *to* us. What we discover in this fragile and stumbling process of mediating ourselves and our world is an antecedent lover and friend.

[90]See Lonergan, "Mediation of Christ," 17-19.

REFLECTIONS ON CHAOS AND GOD'S INTERACTION WITH THE WORLD FROM A TRINITARIAN PERSPECTIVE

Jürgen Moltmann

In this brief paper I want to describe five models about the God-world relation that are found frequently in traditional and contemporary systematic theology. Then I want to comment on how these models function in the current discussions about the theological significance and implications of chaotic, complex, and evolutionary systems in nature. In doing so I will adopt an explicitly trinitarian conception of God and suggest its importance to these discussions. I take this paper to be something of a thought experiment, but that is, after all, entirely fitting to the theological program as a whole.

1 *Five Models of the God-World Relation*

Contemporary theologians often read the "book of nature" symbolically and sacramentally as a sign and manifestation of a greater, more encompassing divine system. In doing so, they are asking not only "How does it work?" but also "What does it mean?" and "What is its significance?" The response to these questions leads directly into a topic of fundamental importance in systematic theology, namely the relation of God and the world. A variety of models have been developed to respond to this broad topic. Here I would like to comment briefly on five of the most significant ones.

1.1 *The Thomistic Model of Double Causality*

The Thomistic model draws on a rich tradition and flourishes in contemporary theology. According to this model, God, as *causa prima*, acts through the *causae secundae* which God has created. God's action in the world is therefore an indirect, mediated action. It allows for the relative independence of secondary causes, which serve as the "instruments" through which God acts. Conversely, the secondary causes point to God as primary cause and allow us to reason by inference from the facts of nature back to the One who causes them. We can say here that the instrumentalization of the world is at the same time its transcendental transparency in the face of the brilliance of its beauty and order.

1.2 *The Interaction Model*

Though found in a variety of theological contexts today, the interaction model is most frequently associated with process philosophy. Compared with the Thomistic model, where God determines everything but is determined by nothing, the process model insists on a reciprocal influence between God and the world. Thus, God acts in the world according to the divine primordial nature but God also experiences the world according to the divine consequent nature. To use a human analogy, God speaks and God hears. This model is broad enough to contain the essential features of the Thomistic model, but not vice versa: the interaction model cannot be integrated as a whole into the Thomistic model.

1.3 *The Whole-Part Model*

A third model widely used today makes use of the concept of "whole-part" derived, most often, from biological systems theory. The essential idea is that of the "whole" as made up by its parts and yet as qualitatively more than, and different from, the mere sum of its parts. The idea is particularly relevant in considering complex and chaotic systems, such as those studied in the physical and biological sciences, where we find a form of top-down causality in which the whole determines the coordination of its parts. In this model, causal relations provide only one possibility within a multitude of interconnected relations, making this model clearly superior to the previous models we have considered. Using this model we can form important conclusions about God's indirect effect upon the world. We do so on the basis of top-down causality by applying the analogy of "similarity in the face of ever greater dissimilarity." Thus, just as the whole is in relation to its parts, so God acts in relation to the world as a whole.

1.4 *The Model of Open Life Processes*

I want to consider another important model derived from biological systems theory. This one specifically emphasizes the open character of biological processes. Every life system has a fixed past which corresponds to its present reality and a partially open future which corresponds to its possibilities. Drawing upon a theological insight first expressed by Augustine, we may say that its present is thus constituted by the convergence of its remembered past and its anticipated future. In short, the present is both present past and present future. The process of evolution and the progressive differentiation of increasingly complex life systems in nature corresponds to the hermeneutic circle of present past and anticipated future in human history, or, in general, to tradition and innovation. According to this model, the openness to God to the world process can be deduced from the openness of life systems to their future. One can say that open life systems are read as real symbols of their own future even as that future is transcendentally being made possible.

1.5 *Models of Creation and Incarnation*

Finally I shall discuss two specifically theological models: those of creation and of incarnation, that is, of inhabitation by virtue of God's self-limitation (what the Kabbala calls *tzimtzum*). According to these models, God as eternal and omnipresent restrains Godself to allow creation to be, thereby giving it time and providing it with a habitat of its own. God's omniscience is limited by God to such an extent that the future is open and experimental even to God. At the same time, the eternal and omnipresent God is self-limited such that God can inhabit the temporal, finite creation and impel it from within without destroying it (*schechina*), guiding it to its completion, that is, to its eternalization and divinization.

The challenge for this model is whether it can explain the previous, simpler models and integrate them. That it should be capable of doing so follows from the fact, mentioned above, that complex systems can comprehend and explain simpler ones, inasmuch as they contain the latter. On the other hand, a complex system cannot explain an even more complex one because it cannot know the whole of which it is a part.

2 *Three Comments on How These Models Function in the Discussion of Divine Action and Chaos*

First of all I would like to comment on the model of "God's action/interaction with the world" as it is understood in several of the essays in this volume.

2.1 *Theism, Deism, Pantheism, and the Theistic Model*

"God's action/interaction with the world" is perhaps best understood as a *theistic model* of God, one derived from the paradigm of "reality as history." God is understood as an absolute subject, different from the world, and one who either intervenes at will or preserves the created order. In the same way as God saved Israel with a "mighty arm" through the miracle at the Sea of Reeds, God called creation into being out of nothing and then brought forth order out of chaos. God creates, maintains, saves, and perfects the world, for the world is "God's handiwork."

It is not necessary for me to describe this theistic model any further. The modern reaction to it was essentially the emancipation of modern culture and science from God. The more it was recognized that the laws in society and nature work independently, the more the theistic model declined and in its place a *deistic model* was established. Here God's action is only needed to explain the world's contingent beginning. Today deism evokes the image of God going into retirement, but, ironically, in the Baroque period it was actually a "theology of glory" which conceived of God as resting on the Sabbath after creating the "best of all possible worlds." This is abundantly illustrated by the hymns of praise dating from that time.

The contemporary reaction to the theistic model is not so much the atheism of the nineteenth or the agnosticism of the twentieth centuries. It is, instead, the elaboration of the *pantheistic model*. According to this model, self-organizing matter and self-developing life are divine. For an interesting current example of pantheism, see the writings of Erich Jantsch.[1]

---

[1]Erich Jantsch, *The Self-Organizing Universe: Scientific and Human Implications of the Emerging Paradigm of Evolution* (New York: Pergamon, 1980).

If one is to do justice to the phenomena embraced by the pantheistic model without relinquishing biblical-Christian theism, then what commends itself to me is the *trinitarian model*. It allows one to speak of "God and the world" in a threefold fashion: God the father creates through the Logos/Wisdom in the power of the Holy Spirit. All things exist "out of" God, "through" God, and "in" God, and God not only transcends the world but is also immanent in the world. Through God's creative spirit (*ruah*), God is present and active in all creatures (Wisd. of Sol. 12:1), weaving the patterns of relationship within the community of creation. "In transfusing into all things his energy, and breathing in them essence, life and movement, he is indeed plainly divine."[2] "The Holy Spirit is life-giving Life, the Mover of the universe and the Root of all created being" (Hildegard von Bingen).

By "cosmic Spirit" we mean that power and space in which fields of energy and material conglomerations form themselves, the symmetries of matter and the rhythms of life, the open systems in their growing complexity. According to this model we can say that God acts upon the world not so much through interventions or interactions, but rather through God's presence in all things and God's *perichoresis* with all things. What emerges is the image of the mutual indwelling of God in the world *and* the world in God through the Spirit. This mutual interpenetration fashions the community and the sympathy of all things. From this perspective, "God's actions in the world" and "God's interactions with the world" are only a part of God's comprehensive perichoresis with all things and with their relationships. Causality is also only one aspect of this network of relationships. The self-organization of matter and the self-transcendence of all that lives are the divine Spirit's signs of life in the world and demand therefore neither an atheistic nor a pantheistic interpretation.

2.2 *Creation and the Future of the World*

My second comment touches upon the theological framework for dialogue with science. I believe that Nancey Murphy and John Polkinghorne are correct to identify God's continuing action in the created world in terms of sustenance, providence, guidance. This avoids the creationist identification of *creatio originalis* and *creatio continua*. On the other hand, the model of "God's continuing action in the world" often takes on theistic overtones which limit this action one-sidedly to the conservation and preservation of the world that has already been created (i.e., the original, temporal creation). When this happens we fail to give equal attention to the future of creation, that is, toward the new, eternal creation.

In her paper, Nancey Murphy poses the following question: "Are there states in between this final state, when God will be all in all, and the present state of God's hiddenness in natural processes?"[3] My response is that *creatio continua* is God's influence on what has already been created and should therefore be

---

[2] Calvin, *Inst*. I, 13, 14.

[3] Nancey Murphy, "Divine Action in the Natural Order: Buridan's Ass and Schrödinger's Cat" (in this volume).

distinguished from the creation of the world. It is God's influence on time, for time was established along with creation. This influence both preserves and renews the world, as the Psalmist writes: "And you renew the face of the earth" (Ps. 103:30). God's preserving activity in keeping creatures from self-destruction shows God's faithfulness, while God's renewing activity with creatures reveals God's hope. The preservation of the world occurs through God's patience, God's "long suffering." God's "omnipotence" is God's forbearance: God gives to the world time. The renewal of the world takes place through anticipations of the new creation of all things through rebirth to life. Thus, if we speak of "God's continuing action in the world," we must also speak of God's intention and goal. The eschatological horizon of the new creation of all things, the divinization of the cosmos and of the eternal creation, has been lacking in the discussion up to this point or, at least for me, it has not been given adequate consideration.

What is the future of creation and the goal of providence? If one assumes that the world is in the process of "becoming," then one implicitly expects the final future of the world to be a state of "being." In other words, if one supposes that the world is "unfinished," then one expects that some day it will be finished and completed. If one understands the systems of nature with their openness to time merely as "systems that are not yet closed," then one counts on their final culmination and closure. In that case the world's future would consist of its "being" taking the place of its "becoming," its time being replaced by its eternity, and its potentiality becoming actuality.

But biblical Jewish and Christian eschatology hopes that God will make creation into God's dwelling place and temple, and that the glory of God, having renewed and completed creation, will "indwell" it. This implies the participation of finite creatures in the infinite existence of God as well as the participation of the eternal God in the creatures' temporal existence. Therefore the indwelling of the living God's unbounded fullness does not mean the end, but rather the openness par excellence, of all life systems. They participate in God's creative abundance of possibilities and so gain an eternal future and an eternal history.

These theological insights seem increasingly consistent with the discoveries of modern science. As we now know, chaotic, complex and evolutionary systems of matter and life are built up in such a way as to display a growing openness to time and to an abundance of possibilities. These possibilities increase as their complexity increases, thus expanding the scope of their open, indeterministic behavior. It would be difficult to imagine their future in a world system which has been brought to completion, one in which all possibilities have been realized and the future has become wholly a part of the past. Indeed, that would mean the "eternal death" of the world, which is only thinkable if the world comes to its end without God. If, however, our starting point is the expectation of God's "indwelling" of creation, then the future of the world can only be imagined as the openness of all finite life systems to the abundance of eternal life. In this way they can participate in the inexhaustible sources of life and in the divine creative ground of being.

This means, however, that we can no longer think of God's being as the highest reality (*ens realissimum*) for all realized potentialities, but rather as the highest possibility and as the enabling of all potential realities: "All things are possible for God." As that which enables potentiality, God's future is in fact the source of time.

2.3 *Open System vs. Closed System*

My last comment relates to systems theory. If the systems of matter and life known to us are best described as "open complex systems," then should the universe as a whole be conceived of as a "closed system" or as an "open system"? Open systems are participatory and anticipatory, capable of communication and able to "experiment" with the differences between tenses or times. Their reality is fixed in a relatively stable fashion; their possibilities grow proportionately to their degree of complexity. Their behavior is only partially determined by their reality; it is also, though only partially, undetermined. In "closed systems" the amount of matter and energy is constant, whereas in "open systems" it is not, for they are by definition matter and/or energy receiving systems.

If we conceive of the universe as a system closed in itself, then the open systems of matter and life are only subsystems and movements within a stable whole. If we conceive of the universe as an open system, then all individual systems participate in this openness. In this case we do not need to posit a constant amount of energy, but rather we might expect to receive a constant inflow of energy. But how should we conceive of the source of this energy? Perhaps it would be possible to extend the idea of the self-transcendence of life, of which biologists speak, to the self-transcendence of the universe, with God's transcendence as the "outer cause" and the "inner motive" for this self-transcendence. In this case the world would be a "system open to God" and God a "Being open to the world." We might then look for "traces of God" (*vestigia Dei*) not so much in the realm of regular or irregular realities, but rather in the realm of potentiality. By potentiality I mean first of all that which is really possible, but even more so I mean the Foundation who enables all potentiality.

I realize that this is an experiment in thought. But experimental theology also belongs to "systematic theology," because in the conception of God it deals with an "open" system, not with a "closed" one. As with all theology, it is itself an experiment with God.

THE GOD OF NATURE

Langdon Gilkey

In this paper I want to focus on two inter-related questions: first, whether nature's processes imply a god at all, and second, if they do, what sort of god they might imply. These moves are, of course, a part of an old tradition: that of natural theology, in which an overview of nature, of history or of human existence itself, that is, a *philosophical* overview, is taken to be able to lead us to a rationally secure, if religiously incomplete, "knowledge" of God. They also presuppose the claim that science gives us a picture of nature in which there are, arguably, signs of the sacred, "traces" of the divine presence and activity.[1]

Although what follows can fairly be called a modest and somewhat hesitant member of the family of natural theologies—perhaps a distant cousin—nonetheless I wish to emphasize two points about it which diminish, if they do not extinguish entirely, the confident claims of most natural theology. The first is that natural theology represents philosophical reflection, or, better, logical reflection of a philosophical sort. It is generated by the urge to understand with as much coherence, clarity, and accuracy as possible the entire welter of experience, to articulate the structure and the meaning of the whole as we experience it in all its variety and ambiguity. Religious existence is generated—usually and for most of us—otherwise. It arises out of sharp awareness of the dilemmas of life, when life is experienced as estranged, at sea, and some disclosure of rescue, of reconciliation, of illumination is received.

Such experiences, of course, have for humans arisen continually in relation to nature. When they did—as throughout archaic life—then nature was experienced as laced with the sacred. But such a religious relation to nature is vastly different from the relation implied in a scientific inquiry into nature, an engineering use of nature—or even a vacation enjoyment of nature! With nature turned over largely to science, technology, and leisure, our modern religious discourse, our myths, rites, and religious confidence, have mostly concerned the human self, its community, and the wider history in which it lives—as the great modern "myths" of Progress, of the Material Dialectic, and of Humanism illustrate. Religions arise out of disclosures to selves in community, in relation to the crucial realities in which those communities exist. If that latter reality be nature, then religion centers

---

[1]As I have written elsewhere, these were only, at best, hints; actually they were questions raised by special inquiries that sprang up unexpectedly there, but received no answer from these special inquiries. Nonetheless, they remained puzzles persistently there: questions about power, about order, and especially puzzles about the strange "story" or meaning, despite randomness, disorder, death, and negation that seemed implicit in the process—and mocked us, like elusive ghosts, each time we looked carefully at nature.

on nature in relation to the community. With us, however, our crucial environment has been social and historical, and hence it has been there that fundamental religious disclosures have arisen. The common thread, often remarked, between modern historical and humanistic ideologies on the one hand and the religious traditions of Judaism and Christianity on the other is that in all of these it is persons in communities and communities in history that represent the loci of disclosure (the latter are the so-called "historical religions"), whatever other differences they may exhibit.

This effort of natural theology never represents, so it seems to me, the initiating source of religious knowledge, and hence it cannot function as the final criterion of this knowledge. The center of religious apprehension lies elsewhere: first, it does not appear out of inquiry itself, scientific or philosophical; and, second, it does not arise for us in relation to nature, but rather in relation to the existential dilemmas of existence, of self, of community, of history. In our tradition it is among the "People of God" that disclosure takes place, not in scientific or philosophical reflection—though much in both may open one's eyes and ears to the disclosures within the religious community. Correspondingly, Buddhists' "knowledge" appears in the higher consciousness through meditation and in participation in the tradition, not in the sort of reflection even their philosophical "theology" represents. The reflections of the religious consciousness on the structures of nature and history—natural theology in this modest sense—do not, therefore, either initiate and certainly they do not exhaust the knowledge of God. The important knowledge of God is not philosophical; it is "religious," that is, it is on the one hand existential, and on the other hand it is communicated through symbols to the community that acknowledges the most fundamental disclosure to its history and witnesses to that knowledge.

In fact reflective knowledge of God of a philosophical sort does not tell us very much. Like the traces that point our minds beyond the processes of nature, such knowledge raises questions about that process—and, as we shall see, it begins to suggest an apprehension of the divine presence there, the dim outlines of the mystery of our origins. What it does do, and does significantly, is to represent the first step in the articulated relation between what we know *religiously* or existentially of God in relation to our community, our selves, and our history, and what we know *elsewhere* and by other sorts of inquiry into community, self, history, and nature. In each case, a "doctrine" is a correlation of what is known of God in revelation with what we know by other means of self, of community, of history, and of nature. In this essay we are interested in what nature can tell us of God—and this represents the smallest enlightenment on the divine mystery, but it adds an important illumination to all the rest.

Second, and as an implication of the above, what is known of God in nature represents by no means the center of the knowledge of the divine for most religious traditions, and certainly not of God for the Christian. The mystery of being and so of our being is at best only partially unveiled or clarified here; far more important for our tradition, as for its biblical roots, is the presence and activity of God within history, and especially within the communities of the covenant, the Hebrew community and the subsequent community established around the person and work of Jesus the Christ. If in nature the divine power, life, and order (law) are disclosed through dim traces, and if the divine redemptive love is revealed only in ambiguous hints in and through the tragedy of suffering and death, it is in the life of Israel, in the life and death of Jesus, in the promises to the people of God, and in the pilgrimage of that people, that all of this is disclosed in much greater certainty, clarity, and power. Moreover, when once God is known in this historical disclosure to the community, then these traces in nature themselves relinquish much of their dimness and become genuine signs of the power and the order of the will of the God of faith. As Calvin put it, "when one puts on the spectacles of scripture, then the signs of God in creation are seen truly for what they are."

The consequence of these points is that what I am trying to do here, while important, is not the vital center of Christian theology. It is important because the deeper understanding of nature's mystery and value in and for itself is very important, and because the relation of God to natural processes is an essential part of our understanding, not only of nature, but of God and of ourselves. How else are the immensity, the transcendence, the wonder, power, order, and glory of God to be experienced, known, and articulated? How else are we to know ourselves as truly children of nature yet as ones called by our common creator to responsibility for nature's integrity, nature's value for itself, and nature's preservation? And finally, it is, I suggest, only through a knowledge of nature inclusive of the signs in nature of transcendence, of ultimate power, order, and value that the mystery, the depth, and the richness of nature can be experienced, articulated, and valued by us. An important part of our knowledge of nature's reality is a religious apprehension of nature as well as a scientific, a technological, an aesthetic, and a responsible (a moral) knowing of nature. And basic for all of these is care in treasuring and encouraging a bodily as well as a spiritual joy in nature, in nature's beauty, variety, and richness, experiences which we have almost lost. What we are doing, therefore, has its point as a theological/philosophical articulation of the presence and activity of the divine in nature.

Let us turn to our picture of nature as seen through reflection on the contemporary sciences and their apprehension of nature. This scientific knowledge of nature represents an extraordinarily reliable (i.e., valid) picture of the way nature's forces work; it is also a limited picture, an abstraction from the richness and depth of nature, an abstraction that omits not only much of the mystery within nature's processes but even more omits the subject, the scientist—also part of nature—who conducts the inquiry and paints our present picture. World and self, objective reality and interpreting subjects, arise together, constitute necessary conditions for any knowing of "nature," and so alone each represents a relative abstraction. Any full account, therefore, of the whole must include both of these. Interestingly enough, evolutionary science gives us new possibilities for this inclusion; for nature also produces the subjects that know nature.

What sort of world, then, appears from the perspective of science as a valid yet partial understanding? (1) It is a dynamic process, a stream of energy coalescing into events, into matter, and into the larger societies of events we call entities. This stream represents, therefore, the fundamental power to be of natural existence; power is the first characteristic of the universe science discloses to us. (2) This process is strangely defined and definite as it reaches actuality; here lie the unavoidable, even the necessary, conditions for each present. Nonetheless, as science, from contemporary physics (including—perhaps—chaos theory), through evolutionary biology, to psychology and social science reiterates, each present also faces indeterminate possibilities. It is apparently open to alternatives, to unpredictable "jumps," "changes," "mutations"; what consistency it represents at this point seems hardly necessitated or determined but only persistently predictable. Destiny and spontaneity represent the general categories of process as science describes it.

(3) Nonetheless, this passage from actuality to possibility is itself characterized by order, an order, therefore, spanning the entire stream of events. Thus, it is an order reaching into the defined past and yet pervading the movement on into the open future—a movement characterized by indeterminate microcosmic events, by indeterminate mutations of organisms, and by the more radical indeterminateness in the continuity of historical process. It is, moreover, an order running through each dimension of this hierarchy, as does the stream of power. Causality and order pervade natural existence as its constituent aspects, but they appear "analogously," in different dimensions, on different levels of nature. (4) There is continuing novelty, a novelty that represents a "break" in the conditioned past but also a process in continuity with that past. The genuinely new appears in process; not only change within given forms, but even more a change of forms. Thus, new possibilities, still "not-yet" in the immediate past, become actuality in the present; open alternatives become definite choices chosen; openness is transformed into what appears (afterward) to be necessity—or to have been necessity, as we seem to see clearly in the field of dynamic chaos. In a deeper way, evolutionary theory has underlined this coming to be of mutations, as does reflection on history; cosmology as a whole assumes it and describes its scope and importance. Time and its passage effect a "modal" change from possibility to actuality, from the openness of an infinity of options to the definiteness, even the necessity, of actual finitude.

This sequence of novelties, of new not-yet possibilities becoming actual, seems itself to embody an ascending order, an order of increasing value. Value is, as Alfred North Whitehead suggested, intensity of experiencing, of feeling, of self-reality combined with creative interrelations beyond itself; value is at once value for itself and for others, richness of experience united with richness of relatedness, relations in the present and for the future—these in combination and in contrast. There seems little question that if this be a useful description of value—and another set of standards of value would also serve—then there has in fact been an increase in value as there has in fact been a fundamental morphology of forms. As our discussion of hierarchy indicates, there are now beings who in their self-awareness and their interrelating with others experience value for themselves, value in others, and value for others to an extent almost infinitely richer than was the case in the early organic world and certainly in the inorganic one. The biologists seem inescapably to recognize this "ascent" of forms and so of value whenever they write about organisms—and yet as soon as they do this, they deny it officially. Of course, whether that increase of value is a result of "chance," that is, of random mutations, or of some kind of inner teleology, or even of some form of divine "providence," remains an issue which the mere increase of value, even if admitted, does not settle. Certainly, if it be admitted, it creates a reflective problem for a purely naturalistic view of things.

These five principles seem to be aspects of nature and of nature's processes entailed directly by what science knows of nature—though they represent entailments and inferences not always drawn by scientific accounts. Many such scientific accounts are intent on an analysis that reduces their explanatory principles to the more universal and abstract terms of physics and chemistry. The "quest for certainty" can lead reflection in science, as it often did in religion, to ignore many facts and their implications. Actually, these aspects of nature are on the way to becoming principles of ontology or of metaphysics. Since these aspects are drawn from the inquiries of the physical sciences, they represent abstractions from the whole of experience and so, by implication, from the whole of experienced actuality. As we noted, they omit the subject of inquiry and of reflection on inquiry—and thus the entire "inside" of being as we experience our own being as knowers and doers. They also omit the experience of the *other*, and so the ontological character of organic and human relations, of community, and so of history. Finally, they omit the order and thus the possible teleology of process both outside and inside.

There are, therefore, left over the fascinating problems of how a self-maintaining system of things comes to be and how it can be understood to embody both an order (there are no police), and an order moving into levels or dimensions of increasing value. When all these omitted but crucial factors (ranges of data, one might say) are taken into account, the resulting reflection about the structure of actuality—now nature *and* history, scientific objects *and* scientific subjects (scientists), scientific theory *and* science—becomes ontology or metaphysics, the effort to understand the structure of being *qua* being, of what it means for any entity at all to be. Insofar as any theologian asks what the major symbols of his or her tradition mean or imply with regard to the structure of reality, and seeks to explicate as far as he or she can that meaning (of creation, providence, human being—and so of nature as well), that theology must borrow an ontology or forge one for itself. Thus arises from theology's side the interesting problem of the relation of theology to ontology or metaphysics, a problem the theologian cannot possibly escape.

Reflection, we have said, must reach this philosophical level for two reasons relevant to our present enterprise (of natural theology): First, if the implications of science for our understanding of nature are to be articulated—and not left to wither in a reductionist side street—our reflection must recognize that science implies and even necessitates a wider metaphysical base or framework if it is to be rationally and not irrationally grounded. Second, if the implications for theological construction—for our understanding of our traditional religious symbols in the light of our contemporary world—are to be articulated, theological reflection must include an ontological explication of its symbols if those are not to be conceived literalistically and "ontically" on the one hand or empty of content on the other.

There is, however, a further reason in religious reflection for this philosophical level of discourse. If any attempt at all is made to show the reasonableness, the credibility, and the persuasiveness of a religious symbol in relation to ordinary experience and its data, then such a reflective structuring of all of experience—a metaphysics or ontology—is essential. Proofs of all sorts, strict, "hard" proofs and loose, "soft" ones, those that demonstrate, those that suggest, and those in between—all are "if-then" propositions, whatever otherwise they may claim to be. That is, they presuppose an already articulated (or assumed) reflective structuring of all of experience: be it Platonic, Aristotelian, Newtonian, Hegelian, evolutionary -progressive, mechanistic, or "process"; and then on the basis of that universe *so understood*, they conclude that "God" (or whatever is here established) is necessary to that universe. If one accepts that metaphysical vision of things, then the proof of God probably holds; if one does not accept that vision—or if one concludes that metaphysics itself is impossible—then the proof of God as essential to the universe so understood never gets started. Such proofs are persuasive as well as viable when a common structuring of the world can be widely assumed; proofs accomplish little of what they intend (they may still instruct us about what is here believed) if there is a wide plurality of competing visions abroad in the land. The same may result if—as in our contemporary case—many intellectuals suffer under the illusion that they share no such metaphysical vision but that through the special sciences they can "see things as they are." In any case, if we are to develop further an answer to our question of what nature, as understood in modern science, might imply about God, then we must perforce move from the ontic (special sciences) to the ontological; that is to say, we must re-present these aspects of nature just rehearsed in more inclusive, universal form, namely, as principles of all experience, categories applicable to all entities whatsoever, symbols that represent "the universal traits of being" —"what it means for anything to be," as Aristotle put it, or in Whitehead's words, "those principles from which actuality never takes a holiday, that are there and never not there."

With this in mind, let us re-present or re-phrase these aspects of nature as suggestions for a possible metaphysics or ontology. Were this the beginning of an ontology rather than a short treatise on religion and science, we would have perforce to enlarge these suggested principles into a coherent and adequate articulation of all of experience. And, if we were seeking to provide a full philosophical theology, we would also have to justify such ontological principles not only in relation to the width of experience but even more in the light of the implications of relevant Christian symbols about God and God's relations to nature, history, and human being.

There are, we suggest, five fundamental ontological characteristics of all that there is in our experience—all of finite actuality or, in theological language, all of "created reality": (1) Temporality or passage—the appearance of what is, the vanishing of what is, and the further appearance of what is new. (2) The definiteness or determinedness of actuality as the given; what is achieved in process is definite, a new unity in and for itself, and so effective beyond itself—and (as 1 makes clear) then it vanishes. (3) The role of possibility, of genuine possibilities that have not been actual before, of novelty—these are somehow "there" as relevant possibilities, and then they become actualities. One cannot understand nature as process, or human existence in history as contingent, open and "intentional" without the category of possibility as paradoxically an aspect of "reality." (4) The role of order as self-maintaining—that is, as an aspect of the nature and behavior of actualities in passage, a role that spans and so transcends the dichotomies of past and present, of achieved actuality and possibility, and of the vanishing present and the impinging future.

(5) There is apparently self-determination in that actuality, a self-constitution at each level of being from the spontaneity evident in inorganic existence, through self-direction in organic and social life, to the self-choosing and autonomy of human existence. On each level given conditions (destiny) inexorably set the terms and limits for spontaneity or, at higher levels, for freedom. We never experience any freedom greater than the capacity to choose those particular possibilities inherent in the given, more than the possibilities provided by our "destinies"—we can, as Søren Kierkegaard said, only "choose ourselves." This ontological characteristic is known more vividly from the inside, in our "existential" experience and awareness of ourselves as subjects, in our experience of personal relations, of choices, of inquiry and reflection, and of political and historical activities. There nature—for we *are* nature—is known from the inside, and on that basis nature can be reinterpreted throughout its extent both in terms of self-awareness and in terms of the awareness of the other as person. For these same categories of definiteness and spontaneity, of the given, and of open alternatives, of destiny and freedom, make their appearance analogously and at different levels all around us in nature. If "spirit" thus be the result of nature and hence an illustration of nature's creativity and mystery (as evolutionary science surely implies), then these analogical signs of spirit, in spontaneity and self-direction, may be taken as "signs" of the ontological or metaphysical relevance of spirit as it is known in personal awareness and in personal relations.

These hints and traces of reality as characterized by spirit and meaning, as well as by power and by causality, are universal throughout nature. They become clearer in personal and historical experience, where the level of moral ultimacy—as reality and as demand—is ever present. Our present question, however, is whether what we have said about nature has any persuasive implications for the reality and the nature of God: granted this is only the dimmest of glimpses of the mystery of God, what might that mystery begin to look like?

What, then, of God in our age? Are there grounds in the preceding for speech about God or the search for God, for confirmed commitment to God? Beyond—and also through—the abstractions of the scientific understanding of nature, nature's reality has manifested itself as power, as spontaneity or life, as order, and as implying a redemptive principle, a strange dialectic of sacrifice, purgation, redemption, and rebirth. In nature each of these appear in creaturely guise, as vulnerable and ambiguous as well as creative, as non-being as well as being. Each, therefore, represents a trace of God, a limit question, to which, I believe, God remains the sole reasonable answer. Each leads to an ontological principle or structure which appears to characterize all that is; let us look at them again.

(1) The power of nature is the power of existence, the power to be. Yet existence and its power are radically both contingent and temporal. Nature apparently has a beginning; its origin is, therefore, a puzzling and important question. Even more important, existence comes, and then as quickly it goes; it is energy-matter in process, through and through temporal. It is here in the present, and then it vanishes into the past and so into non-being. It is, and then as quickly it is not. How then does the past, which immediately has vanished, at the same time so thoroughly ground the existence and shape the character of the present, which

it surely does? How is it that the conditioning, the causing, the effecting and affecting of the present by the vanishing past happen? How is it that we can understand our common experiences of continuity and continuing presence, of causality and influence—of our being and our action from the past into the future?

Some deeper, more permanent power must be there, must continue from moment to moment, if this "temporal power," this power to be over time of finite existence, known to science, to all our common experience, and in our inner sense of ourselves, is to be possible. For what is now present vanishes away, and yet its effects help to form the now on-coming present. What power is it that does not vanish but spans the gulf of past and present—if temporality be so real and so fundamental? 'God' is, therefore, first of all the name for that omnipresent, ever-continuing and so eternal power, this necessary and hence non-contingent source of contingent, temporal being. This unconditional power has many names in different religions, different cultures, and different philosophies; for example, in "process thought" this is the function, not of God, but of Creativity. In all of them, reality as the temporal and radically passing, must also be Reality as permanent if our experience is to be made intelligible; but the finite, vanishing realities around us (and in us) have no such capacities. Therefore, as Thomas Aquinas said, some necessary being is itself necessary to explain this experienced contingency—"and this we call God." 'Continuing and necessary Power' is the first name of God.

(2) This power manifests itself originally as spontaneity and later as life. It is determined, caused and "frozen" once it appears and becomes definite; but it is undetermined, open, as it now becomes what it is to be. Certainly in chaotic systems, nature exhibits "jumps" and unexpected turns; it faces options and alternatives that cannot be predicted beforehand. Whether this is a sign of genuine openness in the future is still a debatable question, as the papers in this volume suggest. We, however, experience real openness from the inside as the inescapable pressure in our present on each of us to decide, to choose among the options facing us in each impinging moment coming to us from the future. In us this power, latent in nature's spontaneity and in its capacity for the new, becomes what we call "freedom": reflection, deliberation, and decision among real alternatives. Process becomes out of the past; but in the present it constitutes itself out of possibilities before it in its future, bringing together in the present actuality alternatives which once were only possibilities. Events around us, events as objects, all appear, when we look at them, to be determined, caused—even if, as in chaotic systems, in unpredictable ways. They are now, of course, but when we look at them they are already in the past. Events happening inside us on the other hand, events as subjects and so in our present, are open; nature in us facing its future is open and not determined. Even science as a creative human project requires in the subject doing science deliberation and decision, reflection and assent, as in analogous ways do all human and hence all cultural and historical experience. Thus, for science as a human enterprise the future is open, even if it seems "closed" for the inquiries of science. Hence it is that in both nature and history genuine novelty is possible, the new appears and becomes actual. Novelty and openness lie in the spontaneous power of the present to determine itself, to be self-creative, to be a relative, dependent, yet creative source of what is original and unique.

Now neither the achieved past nor a set future can account for this spontaneous power to become, the power of each present to constitute itself—lest we return again to determinism and lose novelty. Science is always tempted to speak only of determinism—as it looks carefully at its objects of investigation (which, as we noted, lie in the immediate past). But science as a human project, as the creative work of human subjects, is laced with openness, with deliberation and with centered actions, with self-constituting freedom, directed at the future and its projects for the future.

'God', therefore, is the name for this continuing ground of freedom, of spontaneity in nature, and of self-constitution in humans and so in history. We do not create our freedom through our freedom, our power to choose ourselves and shape our future; such power of self-constitution is given to us as an essential aspect of ourselves, of being selves at all—and we cannot, try as we may, avoid it. This human experience of dependence in our very freedom represents a deeply religious experience of the sacred gift of freedom. It is, incidentally, essential for all else theology says about our spiritual dependence on God (the grounding of our freedom in God) and our estrangement when our freedom seeks to ground itself in itself alone. We are as dependent on God in our spirit as we are in our existence or our life; hence God is the source of our freedom as an essential aspect of our being. This divine grounding of freedom is analogously disclosed in all the evidences of spontaneity and the origination of the new in nature's life. This freedom or self-constitution is, therefore, given to us from beyond ourselves, from a source beyond ourselves, beyond our world and our past, as is our existence itself. This source lies beyond both the world which our choices affect and beyond our freedom which does not create itself. Again, this source must include both world and freedom, both determined past and open future, as the creative ground of each and the principle of unity of both. As Friedrich Schleiermacher argued, this is another sign or pointer to God, now as the continuing ground of our freedom.

(3) This spontaneity or freedom—this openness leading to novelty—yet remains mysteriously within an order, an order that pervades and unites the objective cosmos, our inner deliberating consciousness, and external history. The presence of real novelty within a dependable and calculable order, and of intelligible order characterizing the unexpected yet pervasive appearance of novelty, are genuine "wonders," even miracles. Perhaps the most intriguing traces of the divine lie here. How random spontaneity and genuine freedom can unite with continuing order is a puzzle baffling alike to scientific inquiry and philosophical reflection as it is to observers of history. 'God' is, therefore, the name for that unlimited reality spanning the entire ordered past and the entire open future, uniting into an ongoing order achieved actuality on the one hand with the open possibilities of the novel future on the other, uniting destiny from the past with freedom in and for the future.

God is thus the unconditioned power to be—yet present in each of the tiniest particles of existence, the Infinite Power of Being in all that is. God is the transcendent ground of freedom; the source of all that we know of freedom in ourselves and others—and of the analogous levels of spirit in nature around us, creative in each quantum jump as in each human decision. God is the eternal source of order amid novelty, uniting the determined past with the possibilities latent in the open future—and thus God is as essential for the unity and development of nature as for its being.

Without this transcendent yet immanent principle active in all events we face the very nest of contradictions we have outlined: a world outside of us both determined at some levels and open at others, impossibly coupled with the creative mind and shaping, self-constituting spirit of those who inquire into and manipulate that determined/open world. Neither one—the determined/open world of material causality nor the ungrounded minds also implied by scientific inquiry—is true to our experience. Such an unreflected view, in which stark opposites are posed in uncoordinated juxtaposition, represents on both sides the most extreme sort of abstraction and thus an incoherent and perilous dualism. On the grounds, then, of the coherence of central ideas and adequacy to experience, theism seems more reasonable, philosophically, than its far more popular alternatives.

God is only dimly known here, barely perceived and stumblingly described—as is the wonder and mystery of nature through which God is thus dimly known. As we have seen, nature is an image of God, a creaturely reality of immense creative power, order, and value, an image of the sacred, and hence a finite reality or value for itself. Nature is also the instrument through which God created us and now sustains us. Nature is for itself and for us the medium through which God's power, life, and order are communicated to us.

To the religious consciousness, however, there is more of nature, more of us, and especially more of God. God is more deeply, clearly, and truly known, as is appropriate, in our personal and communal existence, and there we also are more deeply and clearly known. There God is to be known as the creative power of our own existence in time, the ground of our being; the source of the freedom that makes us ourselves, the ground, therefore, of the responsibility that makes us persons; and hence God is disclosed as the ordering or moral principle of our temporal and yet free existence. Finally, central to all we know of God is redemptive love bringing us back to unity again. Here lies the center of our Christian message: in conscience, in law and promise, in prophetic judgment, in incarnation, teachings, atonement, and resurrection. But it is the same God who can, by eyes that search there, be seen dimly but pervasively in the wonders of nature. To know God truly is to also know God's presence in the power, the life, the order, and the redemptive unity of nature. Correspondingly, to know nature truly is to know its mystery, its depth, and its ultimate value—it is to know nature as an image of the sacred, a visible sign of an invisible grace.

IV

ALTERNATIVE APPROACHES TO DIVINE ACTION

GAPS FOR GOD?

Willem B. Drees

Chaotic and complex processes are as much a part of natural reality as linear and simple processes. Theories of chaotic and complex systems have made clear, even more than before, that a naturalistic explanation may be available even in the absence of predictability. In this sense, they result in a shift in our understanding of 'understanding.' Theories of complex systems have, by the explanations they offer and the shift in the concept of explanation, closed gaps in our understanding of nature. Thereby they have enhanced the strength of a naturalistic view of reality.

The first section will develop this in a discussion of John Polkinghorne's understanding of the unpredictability of natural processes as a locus for divine action in the world. The second section will give a more general discussion of 'explanation' as understood in contemporary philosophy of science. One might discern two conceptions of explanation in contemporary philosophy of science. Ontic views of explanation consider an event explained if it is understood as a possible consequence of a causal mechanism. Epistemic views of explanation consider phenomena and laws explained if they are seen as part of a wider framework. Chaotic processes can be considered explained, or explainable, within each of these conceptions of explanation. Thus, this more general argument supports the conclusion that a quest for gaps in chaotic or complex processes is misguided.

Denying such gaps within natural processes does not foreclose all options for a religious view of reality, as the framework does not explain the framework itself (upon an epistemic view of explanation) and the mechanisms do not explain the mechanisms (upon an ontic view of explanation). Thus, questions about the whole and about the most fundamental structures of reality are not excluded. One way of developing such an argument of a more general nature, focusing on the world as a whole, is Arthur Peacocke's model for particular divine action via "top-down causation." Some weaknesses in this position will be pointed out in the third section. Hence, rather than seek to understand God's action *in* the world, we might attempt to envisage the world *as* God's action. Limit questions about reality are persistent. Discussions about religion in relation to the whole of reality (cosmology) and to the most fundamental structures of reality (e.g., quantum physics), as well as discussions about the nature of scientific and religious knowledge (which, in different ways, seek to acquire a view of the world from a point within the world), are much more credible than a quest for gaps in the chaotic and complex. Even if complex phenomena within reality are understood naturalistically, the world as such is not thereby explained. Hence, there remains room for a sense of wonder and gratitude. The world may still be seen as dependent upon some source which transcends the world.

Before embarking on the critical discussion of the proposals by Polkinghorne (section 1) and Peacocke (section 3), I wish to say that I focus on these authors because I think their approaches are good examples of their kind and deserve a significant role in subsequent discussions. I will emphasize those aspects where I perceive problems rather than the many aspects which I admire and appreciate.

A second preliminary remark: I will not distinguish in detail between chaotic processes and processes of self-organization, though in the first and second sections the focus will be on limited predictability (which some take as a sign of openness), whereas in the third section the focus will be on the appearance of order (which some take as a sign of a "self," or even of "intentionality").

A third, more fundamental preliminary remark: Arguments are shaped by assumptions and by the audience one has in mind. This essay is primarily written with an audience of people like myself in mind: those who take science seriously and who hold that there is power (for better and for worse) in religious traditions and symbols, but who are not easily persuaded by traditional doctrine. On my view, intellectual investigation in our time has to take the sciences very seriously, and thus has to favor them over other alleged sources of knowledge, whether astrology, common sense folk wisdom, or religious traditions, including the Christian tradition. Otherwise, such investigation runs the risk of demanding "less than it could of theologians and more than it should of scientists."[1] The burden is on theology rather than on science. As Peacocke has put it, the retreat to conservative positions is "a sign not so much of a recovery of faith as of a loss of nerve before the onslaught of new perceptions of the world."[2] With Gerd Theissen, I am convinced that only by deeply immersing ourselves in science, rather than stopping short of the innermost sanctuaries, the tradition may show up in a new light.[3]

---

[1]L.B. Eaves, "Adequacy or Orthodoxy? Choosing Sides at the Frontier," *Zygon* 26 (1991): 496. Thus I deviate from William Stoeger (see his article, "Describing God's Action in the World in the Light of Scientific Knowledge of Reality" [in this volume, p. 241]). At the conference he took the position that we should start from presuppositions which favor neither the sciences nor religion. This seems to me to assume more equality and neutrality than is warranted. Unlike Stoeger, I do not consider as neutral presuppositions the existence of an active God and the reliability of knowledge provided by the sources of religious knowledge, such as scripture, tradition, and experience. In giving primacy to the sciences in intellectual matters, I also deviate from contributions which employ highly theological language.

[2]Arthur R. Peacocke, *Theology for a Scientific Age: Being and Becoming—Natural and Divine* (Oxford: Basil Blackwell, 1990), ix.

[3]Gerd Theissen, *Biblical Faith: An Evolutionary Approach* (London: SCM Press, 1984), xi.

1 *Polkinghorne's Defense of Divine Action in the Openness of Processes*

Theories regarding chaotic behavior introduce an openness into our description of the natural world which has been missing from classical Newtonian physics so far. This openness would be a kind of "local contingency"[4] which might allow for human or divine free will and human or divine agency. This view has been eloquently defended by Anglican priest and theoretical physicist John Polkinghorne.

In a bottom-up description of the physical world, the onset of flexible openness is signaled by the myriad possibilities of future development which present themselves to a complex dynamical system. In a quasi-deterministic account they arise from the greatly differing trajectories which would result from initial conditions differing only infinitesimally from one another. Because of their undifferentiable proximity of circumstance, there is no energetic discrimination between these possibilities. The "choice" of the path actually followed corresponds not to the result of some physically causal act (in the sense of an energy input) but rather to a "selection" from options (in the sense of an information input).

> It is by no means clear that information input of the kind described originates solely from animals, humankind, and whatever similar agents there might be. I do not believe that God is contained within the mind/matter confines of the world, but it is entirely conceivable that he might interact with it (both in relation to humanity and in relation to all other open process) in the form of information input. . . . God is not pictured as an interfering agent among other agencies. (That would correspond to energy input.) Instead, form is given to the possibility that he influences his creation in a non-energetic way.[5]

The laws of nature allow for gaps where one might envisage divine and/or human action. Central to this argument is the possibility of information input without energy input, thus without interfering with physical laws regarding energy.

I question this line of argument for several reasons, especially insofar as it appeals to chaos theories and the like. Even if there is no difference in energy between two states, energy might still be needed to change the system from one definite state to another. This energy is taken from the background (with its

---

[4]See Robert John Russell, "Contingency in Physics and Cosmology: A Critique of the Theology of Wolfhart Pannenberg," *Zygon* 23, no. 1 (March 1988) 23-43, for a definition of "local contingency."

[5]John C. Polkinghorne, *Reason and Reality: The Relationship between Science and Theology* (London: SPCK Press, 1991), 45. See also *idem*, "The Laws of Nature and the Laws of Physics" in *Quantum Cosmology and the Laws of Nature: Scientific Perspectives on Divine Action*, ed. Robert John Russell, Nancey Murphy, and C.J. Isham (Vatican City State: Vatican Observatory, 1993; and Berkeley, CA: Center for Theology and the Natural Sciences, 1993), 437-48.

non-zero temperature acting as a source of energy fluctuations), but it is not a choice made by the system itself. Self-organization should perhaps be more properly named "hetero-organization," organization triggered by the immediate environment. In a sense Polkinghorne grants this when he speaks about information input, which sounds like external determination rather than self-organization, though he also uses the argument to argue for human self-determination.

Polkinghorne's argument is based upon the possibility that there is a significant difference in output, even though there is no energy difference in the input. Take, for instance, two different mental acts corresponding to a choice between two options. We experience them as different in information content, not (*qua* mental acts) as physically different in energy or labor involved. However, that may well be an illusion, due to the enormous amplificatory powers of the central nervous system.[6] Theories on chaos and self-organization show just this, that amplificatory powers of physical systems with respect to small initial differences are much more impressive than was previously thought. One should avoid confusing zero and close to zero in this context; it is essential to Polkinghorne's position that the energy input is absolutely zero rather than almost infinitesimally close to zero, as in the case of low energy events acting as switches modulating processes which expend larger, observable amounts of energy.

In *Science and Providence* Polkinghorne takes the example of a bead at the top of an inverted U-shaped wire. In this case, he argues, there would be no energy barrier between the options of moving the bead to the left or to the right; God could act without input of energy.[7] An objection to this claim is that if God were to act without input of energy, God's action would have to be infinitely slow. Technically speaking, a basic rule of quantum mechanics is $\Delta E \cdot \Delta t \lfloor /2$ (just like the better known uncertainty relation for position and momentum). Hence, if the energy is to become zero, the time will have to extend to infinity. However, infinitely slow action is ruled out, as the decisive input of information should take place before energetic disturbances have changed the situation.[8]

The relation between providing information and spending energy can also be argued more positively. Information which is embodied physically through two (or more) distinct states, representing 0 and 1, requires a minimal energy-

---

[6]Daniel C. Dennett, *Elbow Room: The Varieties of Free Will Worth Wanting* (Cambridge, MA: MIT Press, 1984), 77-78.

[7]Polkinghorne, *Science and Providence: God's Interaction with the World* (Boston: Shambhala, 1989), 32.

[8]George F.R. Ellis suggests in his essay "Ordinary and Extraordinary Divine Action: The Nexus of Intervention," (in this volume, p. 377) that intervention without energy can take place by controlling the timing of quantum events, say the decay of an excited atom. This seems to violate the same Heisenberg uncertainty relation between energy and time at the other end: controlling the timing precisely would imply a major indeterminacy in energy.

difference to write or read. This is a consequence of the second law of thermodynamics. The relevant inequality[9] is $\Delta I \quad \Delta E/k_B T \ln 2$, which is equivalent to $\Delta E \lfloor \Delta I \epsilon k_B T \ln 2$.

There seems to be no basis in physics for the claim that there is transfer of information without transfer of energy. Invoking quantum physics to provide the missing premise is problematic. Working with a mixture of classical and quantum physics, eclectically invoking the one which fits best at any specific stage of the argument, may be acceptable in a pragmatic context; however, it is methodologically unsatisfactorily once one aims at more fundamental, metaphysical claims. Once one understands the world as a quantum world throughout, one has to deal with the question whether there is a similar divergence of trajectories in the quantum world as there is in the classical world of chaos. I am not capable of judging that issue but, as far as I am aware of the technical literature on "quantum chaos," the question has been answered in the negative.

Polkinghorne is cautious about quantum chaos. He offers another argument for ontological openness:

> [I]f apparently open behavior is associated with underlying apparently deterministic equations, which is to be taken to have greater ontological seriousness—the behavior or the equations?[10]

This preference for the phenomena (unpredictability) rather than the current explanation (*deterministic* chaos) is problematic, at least given Polkinghorne's defense of critical realism. Defenses of critical realism argue from explanatory power to ontology, that is, to the reality of the theoretical entities postulated in the explanation. In this sense, "epistemology models ontology," as Polkinghorne affirms and which he interprets as "acquired knowledge is a guide to the way things are."[11] The disagreement is as to what constitutes the "epistemology," or the "acquired knowledge," which the convictions about the underlying entities and processes have to follow. Polkinghorne seems to assume that the "epistemology" which is to be followed is the limited predictability (which he sees as an indication of ontological openness). However, the epistemology is much richer than the observation of limited predictability. The epistemology includes the theory which explains that unpredictability and the processes by means of non-linear, deterministic equations. In that sense, a comparison with the analysis of quantum uncertainty is mistaken, as there the theory allows, at least on some major interpretations, the conclusion of "genuine indeterminacy."[12] Polkinghorne prefers to interpret unpredictability as a sign of ontological openness, bypassing

---

[9]Leon Brillouin, *Science and Information Theory*, 2d. ed. (New York: Academy Press, 1962), 681; see also John D. Barrow and Frank J. Tipler, *The Anthropic Cosmological Principle* (Oxford: Clarendon Press, 1986), 661.

[10]Polkinghorne, *Reason and Reality*, 41.

[11]Polkinghorne, "Laws of Nature," 440.

[12]Polkinghorne, *Reason and Reality*, 42; and *idem*,"Laws of Nature," 440.

the (deterministic) explanatory theory available. Sticking to the phenomena and discarding the available explanation is not a critical realist strategy, but an empiricist one.

The move from a deterministic theory to ontological openness is also problematic in relation to the science at hand: there is no new principle involved in chaotic systems. There are, of course, new discoveries of order in and through chaos. And the iteration characteristic of self-organizing systems—that the outcome of one stage, say a living organism, is itself the starting point of the next stage—enriches our understanding of the historicity of nature.[13] Still, the scientific study of self-organizing, complex, and chaotic systems has not revealed new gaps to be filled by some external actor. Although complex systems exhibit behavior as if they were guided by an external organizing principle or an intentional self, the theories show that such behavior is explainable without invoking any such actor—whether a self, a life force, or a divine Informator. As such, chaos theory is the extension of the bottom-up program to complex systems rather than a suggestion of the existence of some "top" from which "intentional causation" as "information" proceeds downwards.

Polkinghorne acknowledges that the use of openness as the causal joint between God and the world seems like a "God-of-the-gaps" strategy, even though God is not competing as "an alternative source of energetic causation." However, Polkinghorne argues that there is a fundamental difference between these gaps and earlier gaps, which "were epistemic, and thus extrinsic to nature, mere patches of current scientific ignorance."[14] I do agree that there is a fundamental difference, but it works in the other direction. Whereas in the case of epistemic gaps reflecting ignorance one might maintain an agnostic stance with respect to the possibility of a regular scientific explanation, with the advent of chaos theories there is no reason for such an open attitude. Though there is unpredictability, there is also an underlying theory. To claim gaps is not merely to remain agnostic where we do not know, but to go against what is currently taken as knowledge—the unpredictability of systems which are described by deterministic equations. At this level, we are not confronted with any indications of "gaps" in the processes, unlike the situation at the quantum level and at the cosmological level.

The newly won scientific insights regarding complexity change our view of the world. Unpredictability is, of course, very relevant beyond the strictly scientific context, especially in the context of ethics. To what extent are we responsible if we have only a limited view of the consequences of our actions? For instance, limited predictability and the instability of systems is very

---

[13]See Bernd-Olaf Küppers, "Understanding Complexity" (in this volume); and *idem*, "On a Fundamental Paradigm Shift in the Natural Sciences," in *Selforganization: Portrait of a Scientific Revolution*, ed. Wolfgang Krohn, Gunter Kuppes, and Helga Nowotny (Boston: Kluwer, 1990), 51-63.

[14]Polkinghorne, "Laws of Nature," 446.

relevant in assessing the risks of a *Jurassic Park*. But it does not offer or undergird a specific view of divine action in individual events or of a causal joint between God and the world. Unpredictability is metaphysically uninteresting, at most a necessary but insufficient condition for metaphysical openness. Peacocke's alternative intends to avoid the interventionistic approach of seeking gaps as specific causal joints. He envisages God's interaction with the world as a whole. It is to such ideas that we will turn below. But before turning to alternatives, I would like to consider briefly the concept of scientific explanation in relation to arguments about gaps, as it allows for an additional argument against the attempt considered here to find gaps for divine action within unpredictable and "self-organizing" processes.

2 *Explanations and Gaps*

If there is in reality an openness which allows for divine action (see Polkinghorne's position in the preceding section), there should be elements in the processes which are not explained sufficiently otherwise. Thus, it seems relevant to spend some time on the notion of *explanation* as it functions in the context of the sciences.

Explanation is one of several notoriously difficult concepts.[15] The "classic" view is the *covering-law model* of explanation, which explains an event on the basis of one or more general laws and one or more conditions.[16] On this view, an explanation is similar in structure to a prediction from initial conditions and a law. The covering-law model is of limited value, as the notion of laws is more adequate to the physical sciences than to the other natural sciences. Besides, the connection between explanation and causation is absent in many cases. For instance, the height of a flagpole is calculated on the basis of the laws of optics, the position of the Sun and the length of the shadow—but it is not caused by these.

Given the problems with the covering-law model, contemporary philosophers have offered other views of explanation. They seek to give accounts of explanation which incorporate not only predictive power but also some other features which make for successful explanations and justify the move from explanatory power to approximate truth. There seem to be two

---

[15]See Philip Gasper, "Causation and Explanation," in *The Philosophy of Science*, ed. Richard Boyd, Philip Gasper, and J.D. Trout (Cambridge, MA: MIT Press, 1991), 289-98; and Philip Kitcher and Wesley C. Salmon, eds., *Scientific Explanation* (Minneapolis, MN: University of Minnesota Press, 1989), of which one essay has been republished independently as Salmon, *Four Decades of Scientific Explanation* (Minneapolis, MN: University of Minnesota Press, 1989).

[16]C. Hempel and P. Oppenheim, "Studies in the Logic of Explanation," *Philosophy of Science* 15 (1948): 135-75.

kinds of conceptions of explanation, an epistemic one and an ontic one. For instance, Philip Kitcher stresses unifying power[17] while Richard Boyd emphasizes a realist view of causes.[18] Kitcher's epistemic approach sets the phenomena in a wider theoretical framework. Boyd seeks an ontic approach, a quest for the causes or mechanisms involved. Such ontic and epistemic approaches "are not mutually exclusive, but, rather, complementary."[19]

With the distinction between epistemic and ontic views of explanation, we can now return to our reflections on chaotic processes. These processes existed, of course, before they were labeled as such. However, with the discovery by E. Lorenz of a simple set of three equations which showed erratic behavior in 1963, we enter the era of chaos theory. This theory deals with deterministic equations governing processes which are extremely sensitive to minute differences in initial conditions, and which therefore allow only limited predictability.

On an epistemic view of explanation, chaotic processes are explained since they fit into a wider theoretical framework, parts of which have been around for centuries (e.g., differential equations), other elements of which were developed more or less around the same time as chaotic processes were recognized (e.g., fractals). Chaos theory has not diminished the unity of explanatory accounts, but rather has increased it as more phenomena are now treated within the framework of mathematical physics.

An ontic view of explanation is not so much oriented to the structure of our knowledge as to the availability of a mechanism which would explain the phenomena under consideration. With respect to chaotic processes of limited predictability, such a causal account is readily available. Even though we could not have predicted a specific storm two weeks in advance, since we were unable to observe in sufficient detail all the conditions at that moment (e.g., the butterfly effect), we have no problem envisaging a possible causal mechanism which resulted in that storm. We cannot predict the numbers that will come up when we throw a pair of dice. We are unable to predict which way the bead will fall along an inverse U-shaped wire. But in either case, we can envisage how it may have come about the way it actually came about (e.g., due to minute influences from the air, the surface, etc.). Predictability is not a necessary condition for explainability.

---

[17]Kitcher, "Explanatory Unification," *Philosophy of Science* 48 (1981): 507-31; reprinted in *The Philosophy of Science*, ed. Boyd, et al.

[18]Boyd, "Observations, Explanatory Power, and Simplicity: Toward a Non-Humean Account," in *Observation, Experiment, and Hypothesis in Modern Physical Science*, ed. Peter Achinstein and Owen Hannaway (Cambridge, MA.: MIT Press, 1985); reprinted in *Philosophy of Science*, ed. Boyd, et al. For other views of explanation besides the epistemic and ontic views considered here see Bas C. van Fraassen's "pragmatic view" (*The Scientific Image* [Oxford: Clarendon Press, 1980], 97-157); and Salmon's "modal" conception of explanation (*Four Decades*).

[19]Salmon, *Four Decades*, x.

We prefer deductive explanations, which tell us that given the conditions and the laws (or the mechanism, or the framework, or whatever understanding of explanation is involved), the event was certain to happen. However, we also have situations where we can say that given the conditions and the laws, it was likely (say with 0.95 probability) or less likely (say with 0.3 probability) to happen. If a theory predicts that an event might happen, say with a probability of 0.3, and it happens, would one say that the theory has explained the event? It certainly explains the possibility of the event, even though it does not explain its occurrence on this occasion. This is the situation which we face, for different reasons, in the case of chaotic processes such as the weather, and in the context of quantum theories. For the moment, I will restrict the discussion to chaotic processes. There the problem is not so much due to an intrinsic openness in reality as described by the laws (as might be the case in quantum physics) but to our limited knowledge of the actual situation.

With respect to a probabilistic explanation necessitated by such insufficient knowledge of the conditions I agree with the following summary of an argument by Richard Jeffrey:

> [W]hen a stochastic mechanism—e.g., the tossing of coins or genetic determination of inherited characteristics—produces a variety of outcomes, some more probable and others less probable, we understand those with small probabilities exactly as well as we do those that are highly probable. Our understanding results from an understanding of that mechanism and the fact that it is stochastic.[20]

Jeffrey's position is expressed within the context of an ontic view of explanation. On this view, we may well have unpredictability without inexplicability, and without an opportunity to postulate openness in the processes involved.

There is a wide range between the explained and an inexplicable event which would be linked with a genuine "gap" in nature, or in our understanding of nature. In between are phenomena which could be explainable but are currently inexplicable as we do not have the correct theory yet—such was the situation with the discovery of "high temperature" superconductivity in ceramic materials. Even without an explanation we assume the phenomenon to be explainable in terms of physics, probably known physics, but otherwise with a modification of known physics. There may also be phenomena which are explainable but will never be predictable—as is the case with chaotic processes. As the events will never be fully predictable, one can never exclude particular divine action hidden in the unpredictability. However, as I see it, if there is no indication of or need for such an assumption of openness and divine action, the assumption is not justified. Quantum uncertainty, such as in the decay of a nucleus, may be of a different kind. Here we have good grounds to

---

[20]Salmon, *Four Decades*, 62, referring to R.C. Jeffrey, "Statistical Explanation vs. Statistical Inference," in *Essays in Honor of Carl G. Hempel*, ed. Nicholas Rescher (Dordrecht: Reidel, 1970).

exclude an ordinary cause or "hidden variable," and thus an explanation of the limited predictability as a consequences of an unobserved but real physical process. However, even with quantum physics we need to be cautious, as quantum physics will be modified or replaced, and is open to various interpretations.

So far, we have not made an explicit distinction between explanations of particular facts or events and the explanation of laws. However, this distinction is relevant in this context. Most accounts of explanation, including the traditional covering-law model, are primarily concerned with the explanation of facts, assuming a framework (laws, mechanisms, or the like). To some extent, the framework assumed can be considered as a fact to be explained in a wider framework—as Ohm's law for electrical currents can be explained in the context of a more general theory of electromagnetism in combination with some solid state physics. There are sequences of explanations. Chemists refer to astrophysicists for the explanation of the existence of elements and to quantum physicists for the explanation of the bonds between atoms. Somehow, these sequences converge: various questions about the structure of reality are passed on until they end up on the desk of fundamental physicists (dealing with quantum field theory, superstrings, etc.) and questions about the history of reality end up on the desk of the cosmologist. As an American president is said to have had written on a sign on his desk: "The buck stops here." Thus, the physicist and the cosmologist may well say "Only God knows."[21] This particular position of physicists and cosmologists in the quest for explanation may make it clearer why they get drawn easily into philosophical and theological disputes in a way foreign to geologists, biologists, or chemists. It is not the claim that it is an effective, fruitful, or feasible heuristic strategy to explain all phenomena from "first principles"; calculations and derivations may be beyond our capacities. The argument is that there are limit questions concerning the scientific enterprise. These limit questions show up most clearly in physics and cosmology, and—I would like to add to the example from Misner and Weinberg—in philosophy of science, since on the desk of the philosopher of science rest questions about the nature of the explanations and arguments offered, and the role of human subjects therein.[22]

---

[21]The image of handing questions from one desk to another is taken from C.W. Misner, "Cosmology and Theology," in *Cosmology, History, and Theology*, ed. Wolfgang Yourgrau and Allen D. Breck (New York: Plenum, 1977). See also Steven Weinberg, *Dreams of a Final Theory* (New York: Pantheon Books, 1992), 242. It may be that the distinction between structural and historical questions breaks down in quantum cosmology, but that makes no difference for the argument. On quantum cosmology see C.J. Isham, "Quantum Theories of the Creation of the Universe," in *Quantum Cosmology*.

[22]I only realized after completing this paper that this argument about cosmology, physics, and philosophy of science is parallel to the conclusion I reached in an earlier contribution on time in cosmology, where I argued for two options, a Platonic cosmological one and a constructivist one. See Willem B. Drees, "A Case Against

The position that all phenomena can be explained in a framework which would be "incomplete" only with respect to questions about the basic structure and the whole, is naturalistic. It is well captured in a phrase from the philosophy of John Dewey: "Mountain peaks do not flow unsupported; they do not even rest upon the earth. They are the earth in one of its manifest operations."[23] Such a naturalistic view of reality fits well with contemporary science and contemporary philosophical reflections on the concept of explanation. It is at odds with the quest for an openness for divine action in complex or chaotic processes. However, such a naturalistic view does not exclude theological options at other levels. There still may be speculative theological answers to questions about the framework, the laws and initial conditions, or whatever is assumed.

Quantum physics is one of the options; this is emphasized by George Ellis, Nancey Murphy, and Thomas Tracy in this volume. I do agree that this is a more appropriate level for envisaging divine action than any process at a higher level of reality. However, I am nonetheless skeptical about the use of quantum physics to envisage divine action. One reason is that quantum indeterminacy might be resolved either via a modification of quantum physics or via a different interpretation. Another reason is that indeterminacy may be an opportunity for a metaphysical supplement to physical causes, but, in my opinion, it does not require such a move. There is no need to adhere to a metaphysical principle of sufficient reason, even though the principle of sufficient reason is a good heuristic notion within any naturalistic approach. I will leave off with these brief remarks, as quantum physics will be the topic of a future conference. The next section will consider an attempt to articulate a notion of divine action at the level of the world as a whole, drawing on our understanding of processes of self-organization in macroscopic systems.

3 *Top-Down Causation as Divine Causation*

An alternative to an interventionistic view of God's action within processes in the world has been presented by Arthur Peacocke in his *Theology for a Scientific Age*. It relies on the notion of top-down causation. I will briefly present the idea of top-down causation and its application in the context of theology, before making some critical comments.

Temporal Critical Realism? Consequences of Quantum Cosmology for Theology," in *Quantum Cosmology*.

[23]John Dewey, *Art as Experience* (New York: Minton, Balch & Co., 1934); also quoted by H. Fink in *Free Will and Determinism: Papers from an Interdisciplinary Research Conference, 1986*, ed. Veggo Mortensen and Robert C. Sorensen (Aarhus, Denmark: Aarhus University Press, 1987), 51.

3.1 *Some Examples of Top-Down Causation*

There are physical and chemical systems in which we find coordinated behavior of billions of individual molecules. Chemical clocks and the Bénard reaction are examples of this. The system exhibits a global pattern as long as certain conditions at the spatial boundary are maintained. Individual molecules behave according to this global pattern, rather than in the manifold of possible ways described in the statistics of an ideal gas. As Peacocke formulates it, after describing spatial or temporal (rhythmic) patterns:

> In both these instances, the changes at the micro-level, that of the constituent units, are what they are because of their incorporation into the system as a whole, which is exerting specific constraints on its units, making them behave otherwise than they would do in isolation.[24]

Bernd-Olaf Küppers has on various occasions presented theories of self-organizing systems as theories regarding "boundary conditions."[25] There are spatial boundary conditions, such as the two plates which set the non-equilibrium which gives rise to Bénard convection. More relevant to our understanding of reality are DNA molecules, which shape the development of each organism and may be seen as a kind of initial condition. Boundary conditions are, of course, a traditional feature in physical descriptions—corresponding to the freedom of the experimenter to choose a certain experimental set-up. However, in the case of the DNA of organisms, we do not deal with such almost totally contingent boundary conditions. The boundary conditions which are initial to one stage are the outcome of the preceding step, and so on. They are thus the product of a long iterated sequence that gave rise to organisms with biological complexity.

The relation between mental phenomena and brains is sometimes referred to as another example of top-down causation. According to Peacocke, top-down causation would provide a middle ground between an unacceptable Cartesian dualism of two entities and a physicalist reductionism of mental states to brain states. As far as I understand the discussion, the notion of top-down causation is invoked in an attempt to clarify and illuminate the relation between mind and brain, rather than our understanding of the brain and mind being invoked in order to explain top-down causation. That is the major reason for caution in appealing to this example. Another reason may be that a reductionistic approach, if it includes the environment-organism interaction and the difference between a first-person and a third-person account, has a stronger case than is granted by authors who see the need for top-down

---

[24]Peacocke, *Theology*, 53-54.

[25]See Küppers, "Paradigm Shift."

causality to explain the mind/brain problem.[26] However, that would lead us into a discussion which far exceeds the scope of the present paper.

3.2 *Top-Down Causation as a Model for the God-World Interaction*

Peacocke exercises welcome caution in pointing to the inadequacy of all human models and metaphors regarding God.[27] That, however, does not keep him from an attempt to think through the model of top-down causation.

> In the light of these features of the natural world, might we not properly regard the world-as-a-whole as a total system so that its general state can be a "top-down" causative factor in, or constraint upon, what goes on at the myriad levels that comprise it?[28]

On this view, divine action could make a difference without violating in any way the regularities and laws. Besides, the model also envisages how natural events, including human decisions and actions such as prayer, could contribute to the state of the whole. This truly would be a model of dialogue between humans and the divine. There is a further gain in this model:

> For these ideas of 'top-down' causation by God cannot be expounded without relating them to the concept of God as, in some sense, an agent, least misleadingly described as personal. . . .
>
> My suggestion is that a combination of the notion of top-down causation from the integrated unitive mind/brain state to human bodily action . . . with the recognition of the unity of the human mind/brain/-body event . . . *together* provide a fruitful clue or model for illuminating how we might think of God's interaction with the world. . . . In this model, God would be regarded as exerting continuously top-down causative influences on the world-as-a-whole in a way analogous to that whereby we in our thinking can exert effects on our bodies in a 'top-down' manner.[29]

This is better conceived of in terms of transfer of information than of energy. The result is more than a general influence on the world: "Initiating divine action on the state of the world-as-a-whole can on this top-down causative model thereby influence particular events in the world," without ever being observed as a divine "intervention."[30]

Peacocke acknowledges the problem that a transfer of information requires a transfer of energy at the levels with which we are familiar. However, he locates it at a peculiar place, at the interface between the world-as-a-whole

---

[26]Accounts of the brain and mind without an appeal to "top-down" causation are given by Dennett, *Consciousness Explained* (Boston: Little, Brown & Co., 1991); and John R. Searle, *The Rediscovery of the Mind* (Cambridge, MA: MIT Press, 1992). See also *Scientific American* (September 1992), which is dedicated to the mind/brain question.

[27]See Peacocke, *Theology*, 90, 167 and 188.

[28]Ibid., 158 and 159.

[29]Ibid., 161.

[30]Ibid., 163.

and God, rather than within the natural order. "This seems to me to be the ultimate level of the "causal joint" conundrum, for it involves the very nature of the divine being in relation to that of matter/energy and seems to me to be the right place to locate the problem."[31]

3.3 *God as the Top in Top-Down Causation?*

The issue of energy and information has already been considered above. Here, I will consider the application of the notion of top-down causation to the world-as-a-whole.

The example of the Bénard cell is a clear instance where the conditions at the boundary determine the behavior of billions of individual molecules. However, this is also an example where one could replace the term "top-down causation" by "environment-system interaction." That environment which sets the temperature at the boundary plates is a physical system, just as is the system in which the Bénard cells occur. There is nothing peculiarly global about the experiment; all influences can be traced to local phenomena within the space-time framework. For instance, setting the boundary conditions has no immediate impact on the behavior of molecules at some distance from the boundary; it takes some time to settle into the coordinated state. In the case of DNA the relations are also traceable as local relations within the spacetime framework. The DNA shapes the development of an organism. The environment has an impact on the survival of the organism, the mutations in its DNA, the shuffle of DNA in sexual reproduction, and so on.

In both instances, there is some sense in which a whole (such as the state at the boundary-plates, or the DNA) serves as the boundary for the system, while the next stage of the whole (for instance the DNA of the next generation) is shaped by the development of the system (the organism) in its environment. However, there is no sense in which the system-as-a-whole has any mysterious causal influence. All causal influences can be traced to local physical influences within the system or between the system and its immediate environment. Boundaries are local phenomena, rather than global states of the system-as-a-whole.

In taking top-down causation as the point of departure for describing the relation between God and particular events, there is a significant extrapolation from particular environments to the encompassing notion of "the world-as-a-whole." God is introduced as the one who sets the boundary conditions for the world-as-a-whole at the global level. This seems to me to be problematic, if not to say unwarranted, with respect to the sciences at hand. In the examples which led to the notion of top-down causation, there is always an important role for the physical environment. One could say that in the example of the Bénard cells it is the environment which acts as the "top," setting the temperature at the plates and thereby the state of the system. And in the DNA example, it is the

---

[31]Ibid., 164.

preceding history which has resulted in the DNA that serves as the boundary condition for the organism that is to develop. When we start talking about "the world-as-a-whole" then the notion of a global context, of an environment, becomes a metaphor. In science, we always deal with a context which is also captured in terms of the same laws of physics. This is, it seems to me, an instance of the distinction between relative information, as it arises in the scientific context, and absolute notions with roots in an idealist philosophy and which keep cropping up in theological use of the science at hand.[32]

The problem is not only the presence of absolute notions, but the idea that in the natural realm there can be activity proceeding from such absolutes. With Peacocke I agree that this may be a more appropriate location for "the causal nexus" than any place within the world of natural processes. However, as a quest for an understanding of a *causal* nexus between the divine and the world it still interferes with any (assumed) completeness of the natural account. A promising alternative which avoids such interference is the reflection upon the natural account itself, and especially the themes of the existence, order, and intelligibility of the world.[33] Rather than seeking an understanding of divine action in the world, the world itself is understood as God's action. Whatever strength scientific explanations have, there always remain limit questions about reality and about understanding. These may evoke an attitude of wonder and gratitude. Even when phenomena within the world are understandable in a naturalistic way, the world as thus understood may be interpreted from a religious perspective as dependent upon, or created by, a transcendent source.[34]

---

[32]I owe this distinction to Küppers, during a preparatory meeting in Castel Gandolfo in December 1992. The same problematic move from a relative to an absolute lies behind the distinction between "the future as present," as belonging to the realm of physics, and "the future as future," as the domain of theology, a move which is important for some German and Swiss Protestant authors on theology and science. See A.M. Klaus Müller, *Die präparierte Zeit* (Stuttgart, Germany: Radius Verlag, 1972); and C. Link, *Schöpfung: Schöpfungstheologie angesichts der Herausforderungen des 20. Jahrhunderts*, Handbuch systematischer Theologie, Bd. 7/2 (Gütersloh: Gerd Mohn, 1991), 444.

[33]See Michael Heller, "Chaos, Probability, and the Comprehensibility of the World" (in this volume).

[34]This is developed further in W.B. Drees, *Religion, Science and Naturalism* (Cambridge: Cambridge University Press, 1996). One of the elements in the articulation of a combination of a naturalist view of religion and a religious view of naturalism is the need to differentiate between scientific realism and theological realism; neither does the one build upon the other in the way models in the sciences build upon each other, nor is theological realism defensible along the same lines as defenses of scientific realism.

DESCRIBING GOD'S ACTION IN THE WORLD IN LIGHT OF SCIENTIFIC KNOWLEDGE OF REALITY

William R. Stoeger, S.J.

1 *Introduction*

I intend this brief essay as a "trial balloon." I shall sketch how we can describe God's action in the world, accepting with critical seriousness both our present and projected knowledge of reality as we have it from the sciences, philosophy and other non-theological disciplines, and our present knowledge of God, his/her relationship with us and our world, and his/her activity within it.

By saying that we shall accept the knowledge we have from both ranges of experience with "critical seriousness," I mean accepting it as indicating something about the realities it claims to talk about, after carefully applying the critical evaluations of such claims which are available within the disciplines themselves, and within philosophy and the other human sciences. This obviously involves beginning with a number of definite presuppositions, some of which favor neither the sciences nor religion and spirituality, and some of which do. But it also involves the presupposition that the claims of each have been carefully examined in the light of the different ranges of experience and certain principles of interpretation and validation. I shall not spend time here going through that process step by step, but instead shall simply assert some general results in each area which derive from such a distillation. It will be somewhat obvious to those in the respective fields what critiques I have applied to reach the results I shall assert. Then I shall attempt to marshall these results into a roughly- sketched, integrated theory of God's action in the world.

The input into this integrated, coherent theory of God's action will not consist of highly technical assertions—either from science or from philosophy and theology—but rather assertions which more or less describe the general character of the world as we know it from the contemporary sciences and the limits of our knowledge of it, and the general character of God's action in the world as we know it from contemporary Christian belief and theology. The latter has already developed a great deal in response to the input and challenges mediated to our culture by the sciences. In other terms, we wish to attempt to describe more adequately God's action in the world, given that we know that the world, its structures, and processes, are presently best described in such and such a way (from the sciences and philosophy) and that God and his/her relationships to the world are presently best described in such and such a way (from theology and philosophy). What we know from each set of disciplines must critically interact with what we know from the other set according to certain principles (which we shall later outline). This interaction should modify each set of disciplines—particularly in our interpretation of the conclusions each one reaches at a philosophical level—and allow us to describe God's action in the world in an integrated way.

Implicit here, as Stephen Happel has pointed out to me,[1] is the methodological problem of how these two languages are to be integrated. This is an issue which is important, but one which is best treated after allowing the interaction to occur via the critical apparatuses which are already available and functioning. The two languages of science and religion/theology, though different, are not isolated from or out of contact with one another. They continue to be in dynamic interaction in our common cultural and academic fields.

In describing what we know about the world and about God, and his/her relationship to us and to physical reality, I need to employ a language, a set of categories, and certain philosophical presuppositions. In particular, I assume a weakly critical-realist stance and use some of the language, categories, and metaphysical presuppositions of Aristotelianism and Thomism, most notably the notions of primary and secondary causality. Other categories might have been chosen and other assumptions might have been made instead. I have chosen these because, in my opinion, they are more adequate to both the scientific and the theological data, and lead to fewer difficulties in explicating the essential differences between God and his/her creation, the relationships between them, and the ideas of divine immanence and transcendence. It is important to note also that I use the term 'law' in the context of both physical processes involving inanimate entities—"the laws of nature"—and free human actions. 'Law' is any pattern, regularity, process, or relationship, and by extension that which describes or explains a pattern, regularity, process, or relationship. Thus it applies, in the range of the ways I use the word, not only to the inanimate and non-human, but also to the human and the divine. 'Law' is a word used to specify, describe, or explain order. It does not necessarily imply determinism. As I use it throughout the paper, modified by various adjectives and adjectival phrases, its meaning should be clear.

1.1 *Presuppositions*

An obvious presupposition we make in pursuing this discussion is that the sciences give us some knowledge of reality. We are not able to specify that correspondence precisely, because we do not have an independent handle on reality as it is in itself. Furthermore, our knowledge of it is always only provisional and corrigible, and its certainty is only relative, not absolute.[2] But we are still reasonably persuaded to

---

[1]Personal communication. Here and elsewhere in this paper I am indebted to Happel's very helpful comments.

[2]See Frederick Suppe, *The Semantic Conception of Theories and Scientific Realism* (Chicago: University of Illinois Press, 1989), 475; and William Stoeger, "Contemporary Physics and the Ontological Status of the Laws of Nature," in *Quantum Cosmology and the Laws of Nature: Scientific Perspectives on Divine Action*, ed. Robert John Russell, Nancey Murphy, and C.J. Isham (Vatican City State: Vatican Observatory, 1993; Berkeley, CA: Center for Theology and the Natural Sciences, 1993), 209-34.

maintain that there is correspondence, however precarious and uncertain it may be. The care we exercise in validating and confirming scientific knowledge indicates that this is what we as scientists are intending to do. And unless reality is extraordinarily malevolent and contrary, the intersubjectively applied criteria used in scientific observation, theory, and experiment assure us that the sciences give us *some* purchase on the structures and the dynamics of the physical, chemical, and biological world of which we are a part. We presuppose in doing this that in its interaction with us, reality reveals something of what it is. It could be very devious, it is true, but we presume it is not so devious as to reveal nothing of itself in the phenomena we observe.

The other key presuppositions we make here may not be so obvious or common. They relate to God and to divine action, and to our knowledge of it through Christian belief and theology, according to the critical principles of discernment, validation, confirmation, and interpretation which are applied in these areas. First of all, we presuppose that God exists and is and has been actively present and involved in our lives and in our world. How this action, presence, and involvement are to be described and understood will be modified—even significantly modified—in the conversation with the natural sciences. We are not attempting to prove Christian doctrines by appealing to scientific evidence, but rather attempting to re-articulate and understand theological truths in a more satisfactory way by looking at the relevant knowledge available to us in the sciences and other disciplines. As Happel has said, "religion and theology are put in conversation with the data, concepts and language of scientific performance and theory."[3]

Secondly, we presuppose that the sources of revelation, the scriptures, tradition, and our living experience as believers who are individually and communally open—more or less—to God and to God's action, do give us some reliable knowledge about God and about his/her action in our world. As in the sciences, this is very limited and corrigible knowledge, subject to error and modification, particularly with regard to interpretation and understanding of that revelation, and of our overall response to it. And, as in the sciences, it too is dependent on the careful application of critical principles of interpretation, discernment, and confirmation suitable to the experiences being examined. We might also mention that the limits and uncertainties of this knowledge derive both from the extraordinary but limited character of the revelation we have available, and perhaps most of all from our own limitations and lack of openness to receiving, interpreting, and living out that revelation.

1.2 *The Aim of Our Discussion*

The aim of our discussion is simply to describe God's action in the world in terms which are faithful to Christian sources of revelation and consistent with what we know from the sciences about reality, its structure, evolution, and processes,

---

[3]Private communication.

especially in view of the self-organizing capabilities of matter, from the chaotic and dissipative structures evident even in inanimate systems to the complex systems of living organisms themselves. One of the key issues here is causality. How can we speak of divine causality within the world as we know it, without compromising scientific and philosophical principles—without using an interventionist model, for instance?

But how can this aim be pursued? Where is the common mode of inquiry to be found? How do we distill relevant information concerning God's action in the world from the sciences and from the sources of revelation? Those questions are very difficult foundational ones. But I do not think they can be answered from an a priori perspective. As I have mentioned already, I am assuming that we have used and are using the relevant tools of philosophy, philosophy of science, the critical methods proper to scripture studies, historical and systematic theology, and hermeneutics to do this. I am also assuming that we can begin to integrate these results through the common ground of understanding and language which our various specialized languages share with one another. They are not, as I have stressed above, completely isolated from one another, nor are the experiences to which they appeal.

2 *What the Sciences Tell Us about Ourselves and Our Universe*

If we generalize from the vast knowledge of the universe and all that makes it up, including living and conscious beings like ourselves, we can say that at every level there are self-ordering and self-organizing principles and processes within nature itself, which can adequately describe and account for (at the level of science) its detailed evolution and behavior, the emergence of novelty, possibly even of consciousness, the inter-relationships between systems and levels, and even the various laws of nature themselves—and the unfolding of all this, its diversification and complexification, from an epoch very close to the "initial singularity" or Big Bang. Some of these principles and processes are well known and understood, and others are at present only conjectured or suspected. No outside intervention is necessary to interrupt or complement these regularities and principles at this level. Nor is an *élan vital* called for to explain living things—nor an *élan spirituel* at the next level of development. At the level of the sciences there are no "gaps," except the ontological gap between absolutely nothing and something.[4]

---

[4]Some—for instance Ellis, Murphy, and Tracy—consider the indeterminacy at the quantum level to be an essential gap which requires filling (see their papers in this volume). Though this view needs much more careful discussion than is possible here, my assessment is that indeterminacy is not a gap in this sense, but rather an expression of the fundamentally different physical character of reality at the quantum level. It does not need to be filled! To do so, particularly with divine intervention, would lead in my view to unresolvable scientific and theological problems. The demand for a cause to determine the exact position and time of an event misconstrues the nature of the reality being revealed. Quantum events need a

This general conclusion is strongly supported by detailed conclusions from physics, chemistry, biology, molecular biology—particularly from those emanating from the studies of complex systems, information theory, molecular biophysics, and by the promised or envisioned advances in these fields. The gaps in scientific knowledge have not all been filled, but they are gradually being filled by new discoveries. And it has become clear that appealing to divine intervention is not an acceptable means for doing so. Nature itself is open and capable of realizing new possibilities in a whole variety of ways. Even in the surprising transitions from inanimate beings to living ones, from living ones to conscious ones, from conscious ones to human ones, it seems very unlikely that any intervention from outside natural processes was involved. Material, physical reality is much richer in its possibilities, particularly when it is in a highly organized form, than we usually think. At the same time, an analysis of the sciences, the theories and the laws of nature which derive from them, makes us very aware of their limitations. The knowledge given us by the sciences—like all human knowledge—is imperfect, provisional, corrigible. In particular, it only very imperfectly describes the regularities and underlying inter-relationships, necessities and possibilities, and structures which constitute reality.[5] Through the sciences we do not know reality as it is itself; we do not know it directly, interiorly, comprehensively, exhaustively, as we would like to know it—as God must know it. So, although we have through tremendous sustained effort and genius come to unravel a great deal about reality, we are far from comprehending it at its ultimate depths.

In particular, from the sciences we still are unable to answer the questions, why there is something rather than nothing, why there is order rather than disorder, and why there is openness to novelty—to new and more complex entities—rather than just sterile uniformity. That is, why are there "laws of nature" in the first place? And why these "laws of nature" and not some others? In fact, not even philosophy can adequately answer these questions.

A third conclusion stemming from the sciences is one which is not usually mentioned but one which I believe is quite important—but not for the first reason that will probably occur to us: The laws of nature and nature itself constrain but under determine what develops or occurs. Great possibilities are left open in nature. It is very pliable. This does not mean that nothing happens, obviously, but it does mean that uncorrelated coincidences often end up "filling in" what is needed to complete determination. It is this pervasive feature of reality—along with others, such as its knowability and its localizability—which enables human beings and other animals to manipulate and harness reality, and even to know it. We can fly in airplanes, build bridges, and heal the sick, precisely because the laws of nature as we know them, and perhaps even as they are in themselves, under determine events. In fact we are who we are as human beings because of this important feature—we can decide to do things which otherwise would not happen

---

cause and have a cause, but not a cause determining their exact time and position of occurrence, beyond what is specified by quantum probability (the wave function).

[5]See Stoeger, "Contemporary Physics."

within the constraints imposed by physics, chemistry, and biology. Some of this underdetermination is due to the indeterminism and unpredictability of physical systems at the quantum level and to the unpredictability of both simple and complex systems on the macroscopic level. As we have seen in studying the behavior of chaotic, nonlinear, or nonequilibrium systems, very slight changes in the initial conditions or the boundary conditions can severely alter how they will behave, and what sort of self-organizing behavior they will manifest. However, the underdetermination of phenomena by the laws of nature is due to much more than these important sources of indeterminism and unpredictability. It is due primarily to the freedom that exists in establishing initial conditions and boundary conditions throughout nature. An agent can, with some expenditure of energy, change initial conditions and/or boundary conditions of a system or, even more importantly, construct new systems, thus determining outcomes much different from those which would otherwise occur.

"Aha! You have pointed this out in order to leave room for divine intervention!" someone might say. In fact I have not, because, as we shall see, this underdetermination of reality by the laws of nature does not easily allow for divine intervention—at least not direct divine intervention—because that would involve an immaterial agent acting on or within a material context as a cause or a relationship like other material causes and relationships. This is not possible; if it were, either energy and information would be added to a system spontaneously and mysteriously, contravening the conservation of energy (and we just do not have substantiated cases of that happening) or God would somehow be acting deterministically within quantum indeterminacy, which presents a number of serious scientific and theological difficulties.[6] No, I have pointed out this feature of reality in order to emphasize the potentiality, flexibility, and scope for newness that is within nature, as well as the many different levels of agency which operate within it, including the types of agency we exert as human beings.

Before going on to summarize what revelation tells us about God and divine action, we should point out that the sciences themselves are limited in dealing with personal agency and personal relationships. In some ways psychology and sociology deal with the phenomena related to these, but I think we are all aware of the limitations under which they labor in their quest for knowledge in these profound and mysterious areas.[7]

3 *What We Know from Revelation and Our Reflection upon It*

From revelation, and partially from reason, we know that God exists, created the universe and all that is in it, reveals him/herself to people, loves and cares for us, continually acts within material creation, particularly now through Jesus and the

---

[6]See n. 1 above.

[7]Arthur Peacocke, *Theology for a Scientific Age: Being and Becoming—Natural and Divine* (Oxford: Blackwell, 1990).

presence of his spirit among us, and calls us to share his life and mission forever—a promise which will be fulfilled only after our deaths.[8]

Here a couple of conclusions stand out in reference to the issue we are probing. Though it is not the primary revelation of God, the first is that he/she is somehow the answer to the question, "Why is there something rather than nothing?"—and to the other similar fundamental questions we posed above. He/she created what is not God from nothing. But how that was done is still very much a mystery, as well as whether or not creation is eternal—does God create from all eternity? How that was done is understandable only to God, at the very depths of the divine being. We know in a very limited way how it was done by looking at nature as revealed to us by the sciences—or let us say, we know how it was not done!

A second conclusion from revelation is God's motivation for creation and for his/her interaction with the world—it is God's goodness, God's innate drive as God to share that goodness, and God's love both for him/herself and for all that he/she creates and holds in existence. So, interpersonal relationships are of paramount importance to God—as are the values of goodness and truth. This is true of God in him/herself— God as Trinity. But it is also true of God's relationships *ad extra*. This divine priority is most fully expressed in the Incarnation of the Son of God in Jesus, and in the sending of the Spirit. But it is manifest throughout creation at every level.

A third conclusion is that creation itself is good, and an expression of God's goodness and love. Therefore, it makes perfect sense that it should reflect to some extent who God is and what his/her characteristics are. Also, the more complex and capable beings are, the more they reflect who God is—including humanity, which is made in the image and likeness of God. This perspective—the priority of the values of goodness and truth, along with reverence and respect for all that is—is consistent with the importance and value God gives to personal relationships.

A fourth conclusion is that, although God reveals him/herself through everything in creation, God's most particular revelation is in terms of persons and personal relationships involving generous, self-sacrificing love and forgiveness. And our principal way of responding to God's revelation is in those same terms. So we experience revelation as personal and social, God among us—as creator and source of life, yes, but also as a personal presence and force who loves, invites love, gives and invites giving, forgives and reconciles, and invites forgiveness and reconciliation. The created, inanimate, and non-personal levels of reality, though they exist in their own right and reveal God and God's goodness, power, and love in their own way, and give glory to God in their own way (they cannot do otherwise!), exist also to enable the development and maintenance of persons to whom God can reveal him/herself and with whom God can maintain a personal

---

[8]As Happel points out (private communication), this creedal summary is deceptive. The meaning of the language used is neither static nor agreed on by all who accept it. It will change, even radically so, as we live out of and reflect upon our individual and common experience of God's presence and action among us.

relationship leading to the full and harmonious union of the divine with created reality. The degree to which this is desired by God is expressed in creation itself, in the Incarnation and all that follows from it, and in the sending of the Spirit.[9]

These are the principal conclusions flowing from Christian revelation which I wish to highlight. Our endeavor now will be to bring these conclusions into critical interaction with what we know about reality from the sciences, as outlined in the preceding section. As I have already emphasized, these conclusions will have to be re-articulated and modified as a result of this interaction. For instance, the strong anthropocentrism of this particular articulation would have to be significantly mitigated. And the radical non- objectifiability of God would have to be factored in, on other, more theological and religious grounds.[10]

4 *God's Creative Action and Science—Primary Causality*

I have already emphasized that the sciences—physics in particular—do not explain or account for existence or for the general order of the universe. They presuppose it. They do not answer the question, Why is there something rather than nothing? They can deal very well with questions of origin in which—as is usually the case—the origin of a structure or an entity derives from something else which already exists, for example, the origin of children from their parents. But the sciences do not deal with ultimate origins. They cannot bridge the gap between nothing (which includes no potentialities and no physical laws—*absolutely* nothing) and something—or even between God and nothing else and God and something else not God; and it is not clear that any branch of human knowledge can adequately address this fundamental issue.

The God of Christian revelation, belief, and spirituality, however, is an adequate answer to this question—though this answer, adequate as it may be, is somewhat impervious to adequate understanding on our part. It does not adequately tell us *how* God bridged that gap. God is the one who in some way has brought something out of nothing; God is the agent of *creatio ex nihilo*. In one way this is not accurate, for, as I have already implied above, God has always existed as a "necessary" being. He/she *is*, as the uncaused cause, or primary cause, as Christian theology has traditionally described him/her. So something (i.e., God) has always existed. There was never "absolutely nothing," if something exists. What we really want to say is that the only explanation for something created to emerge from the absence of anything created is God. This affirmation, as I have just said, does not particularly deepen our understanding—*how* this happened, the

---

[9]My emphasis here on the priority of persons does not deny the wider role the Spirit has throughout the created order, and the impact of the Incarnation on the cosmos. Nor does my formulation properly describe the relationship of non-conscious entities to the divine presence and their essential mystery.

[10]José Porfirio Miranda, *Marx and the Bible* (Maryknoll, NY: Orbis Books, 1974), 40ff.

details—but it is, strictly speaking, an adequate answer to the fundamental question we are considering.

It should be clear, furthermore, that this is not basically a temporally weighted answer to the question of existence. It does not necessarily imply that there was a state or situation when there was nothing besides God, and then at some juncture God created entities other than him/herself, and with them time, space, etc. As Thomas Aquinas[11] realized, it could be that God has created from all eternity—that created reality is eternal in the sense that it has no temporal beginning (there was never a state in which God existed and created reality did not), but it is still radically contingent on God.[12] There may have been a beginning of time, but that is by no means essential. Ultimate origins are essentially ontological, not temporal. In fact, I believe a good argument can be made for eternal creation on the basis of who God must be as God. If God is of his/her very nature *bonum diffusivum sui*, infinite love, and therefore creator, then he/she was always and eternally such. Therefore, in order to fully realize who he/she is, creation must in some sense, at least in intention, be an eternal process. This may at first seem to infringe on God's freedom to create. But it really does not do that at all. His/her creating is perfectly free, but is also a natural consequence of God's very nature. Nor does this mean that God or God's love is dependent on creation for self-origination. God and God's love must be sovereign. But God's love must also be fruitful, and that one principal manifestation of its fruitfulness be an eternal created order is not surprising.

This primary divine, existence-endowing causality is always operative, holding things in existence, charging them with realization. It is essential to conceive primary causality very differently from the causes—secondary causes—we discuss and deal with each day. The primary cause is not just another one of these—it completely transcends them and provides their ultimate basis in reality. There are no gaps is the secondary causal chain, but the whole chain demands a primary cause to support and sustain it. Without the primary cause there is no explanation for its existence or for its efficacy.[13]

But it is not just a question of existence. It is also a question of order. What accounts for the order which exists in nature—in the universe? Why is there order rather than complete disorder? Again this is not a question which can be answered by the sciences. In the same bald and impoverished way as before,[14] however, the existence of God does provide an adequate answer: God is the ultimate source of order in nature and in the universe, and of both necessity and contingency—and therefore of any possibilities which might emerge from their interaction. A

---

[11]Thomas Aquinas, *Summa Contra Gentiles*, 1.44.

[12]See Ernan McMullin, "How Should Cosmology Relate to Theology?" in *The Sciences and Theology in the Twentieth Century*, ed. A.R. Peacocke (Notre Dame, IN: University of Notre Dame Press, 1981), 39ff; and Stoeger, "Contemporary Physics."

[13]Stoeger, "The Origin of the Universe in Science and Religion," in *Cosmos, Bios, Theos*, ed. Henry Margenau and Roy A. Varghese (LaSalle, IL: Open Court, 1992), 254-69.

[14]Ibid.

consequence of this, of course, is that God is ultimately the source of the underlying regularities, constraints, and behavioral relationships and patterns which are imperfectly described by the laws of nature we formulate.[15] The question why the world behaves this way rather than some other logically possible way can only have an ultimate answer in God as creator. He/she is the well-spring of both necessity and possibility in nature.

5 *God's Creative Action*—Creatio Continua *and Secondary Causality*

There is an important corollary to the foregoing discussion, which takes us into a brief consideration of God's continuing creative action in the universe, conceived now more richly than simply as *just* divine existential conservancy. It is that a principal mode of God's activity in the world at the level of inanimate and non-personal beings is precisely through the underlying regularities, constraints, and relationships he/she has established in nature, and which we sometimes refer to as "the laws of nature."[16] This is a very rich way of looking at nature—as the expression of God him/herself and as one of the fundamental ways in which God acts within the world. The regularities, constraints, and relationships are as they are by God's allowance or choice—he/she works through the secondary causes of our world. They give God's presence and action concrete form. As new possibilities are realized God becomes present and active in new ways.[17] They express how God desires the world to be—the necessities that are imposed along with the contingencies, the possibilities and the openness to development and to novelty.

If we put this into an evolutionary context, then, and consider what we know of the complexification of structure and the diversification of physical, chemical,

---

[15]See ibid.; and *idem*, "Contemporary Cosmology and Its Implications for the Science-Religion Dialogue," in *Physics, Philosophy, and Theology: A Common Quest for Understanding*, ed. Robert John Russell, William R. Stoeger, and George V. Coyne (Vatican City State: Vatican Observatory, 1988), 219-44.

[16]I prefer to reserve this term for our imperfect formulation of the underlying regularities, constraints, and relationships we discover, or our models for those. However, we must distinguish between "the laws of nature" as God knows them, and the "laws of nature" as we have imperfectly and provisionally formulated them.

[17]Though the general primary-cause-secondary-cause approach to the problem of God's action in the world is very traditional, I believe that it is the only one that holds much promise. Owen Thomas ("Recent Thoughts on Divine Agency," in *Divine Action*, ed. Brian Hebblethwaite and Edward Henderson [Edinburgh: T & T Clark, 1991], 35-50.) arrives at a similar conclusion, that in the current state of discussion only the theories involving either primary and secondary causality or process theology even approach adequacy. I am thankful to Russell ("Introduction," in *Quantum Cosmology*; and in this volume) for this reference. In my view, the approach of process theology, though attractive in some ways, has unresolved philosophical and theological problems, particularly with regard to the doctrines of God, creation, and Christology.

and biological processes from a time shortly after the Big Bang, we see that we can conceive of God's continuing creative action as being realized through the natural unfolding of nature's potentialities and the continuing emergence of novelty, of self-organization, of life, of mind, and spirit, as the universe expanded and cooled. Within this perspective, God's direct intervention—in the sense of operating outside of the regularities, constraints, and relationships he/she has established, or abrogating or mitigating them in any way, either *ad hoc* or regular, to fulfil some higher purpose—fails to make much sense if God is really God, though it cannot be ruled out. Even if intervention in the underlying principles, relationships, and regularities as they are in themselves sometimes occurs, it is still clear, from critical reflection upon both scientific knowledge and the knowledge we have from faith, that the operation of the laws of nature, from the divine perspective, is a principal channel of God's active presence in our world, and as such is an expression—inadequate and imperfect though it may be—of who he/she is. Thus, our investigation of these regularities, constraints, and relationships, and our imperfect formulation of them in scientific theories and in our "laws of nature," articulates an important mode of divine activity in created reality.

I shall have more to say about this later when we discuss God's action within personal and social contexts. Looking forward briefly to the issues which will emerge there, we see that it is crucial to distinguish carefully between the "laws of nature," the regularities, constraints, and relationships realized in nature, as we have conceptualized and formulated them, and the "laws of nature" as they in fact function in created reality—from God's full and complete point of view, so to speak—which somehow includes the internal or interior relationship he/she has with nature, with us, and with other created entities.[18] We immediately see the importance of this distinction—since our very limited account and formulation of these "laws" may leave out crucial relationships (even constitutive relationships) which organize the inanimate and unconscious world at a very profound level, which function to subtlety link the personal and the non-personal, or which subordinate the non-personal to the personal. We are not fully able to see how this might happen, but we begin to see something of it in the underdetermination of physical reality and its vulnerability to human agency, which can mold it within its constraints to our intended use, for better or for worse.[19]

From our point of view, manifestations of this may be interpreted by us as contravening the "laws of nature" simply because we have not fully understood them, whereas in fact they are in perfect accord with the "laws of nature" as they are in reality. In other words, God may act in a purely "natural" way within the relationships and regularities he/she has established and maintained, but in a way which we see as supernatural intervention simply because we have not yet come

---

[18]Stoeger, "Contemporary Physics."

[19]Cf. Peacocke, *Theology for a Scientific Age*.

to comprehend fully the relationships and regularities (the "higher laws") which obtain.[20]

In light of this, it is clear that the distinction we often make between the natural and the supernatural really derives from our limited perspective on reality, and our imperfect knowledge of it. We simply do not know enough to put everything together. Where there are gaps in our knowledge we always seem to insert God's direct intervention, with the implication that there is a concomitant abrogation, mitigation, or suspension of the "laws of nature." Again, the distinction between the two rather different meanings "the laws of nature" may have is some help to avoiding this confusion.

6 *Problems of Conceiving Direct Divine Action and the Need for It*

God's action in the world through the regularities, constraints, and relationships he/she enforces, as we have sketched it in the previous section—through the "laws of nature" as they are in themselves—is indirect.[21] God establishes an order within which processes occur and constraints are imposed. These processes and constraints lead to the evolution of structures and even of other, higher-level processes which govern their behavior, and to the emergence of new and more complex entities which are able to reproduce and evolve further. The whole process culminates in entities which are conscious, able to know, free and capable of making decisions, and able to harness and control reality within certain limits. All this has been orchestrated by God—so to speak—through the divine establishment and maintenance of the "laws of nature."

We can easily understand God's indirect action, because we are familiar with analogous instances of indirect action in our human experience—using an instrument, making a machine, or constructing a program which will perform some function for us, setting an organization or a group into action to carry out some series of commands directed toward fulfilling some desired end we have conceived. God does something analogous in establishing and maintaining the "laws of nature."

But conceiving or modeling God's direct action is a very different kettle of fish. We have the experience of what "direct action" means within human experience. It means active involvement without an intermediary—the agent does what he or she intends personally, without asking someone else or triggering something else to do it. Any action will always have a direct component and indirect components. No action by an agent can be completely indirect. When I

---

[20]Will all events be 'lawful' in his extended sense? Referring to how I characterized 'law' in the introduction and in this section, that may very well be the case. However, it needs more careful consideration than I can give it here. Certainly, relative to a more restricted notion of law—as what is generalizable—some events will fall outside its comprehension, e.g., what is important and significant in its radical particularity.

[21]By 'direct' I mean unmediated; by 'indirect' I mean mediated.

contract a firm to repair my roof, I indirectly repair the roof by doing so, and the supervisor indirectly repairs the roof by directing his subordinates to do so and telling them how to go about it, but I *directly* act to initiate the contract with the roofing firm (picking up the phone to call them, showing them what needs to be done, making and communicating the decision to accept the estimate on the proposed work, signing the contract, etc.) and the supervisor *directly* acts to put his roofers "into motion." It will be the same in God's indirect action in the world. We see the results of the indirect components, and even have access to the agents through whom God is acting indirectly. But we know or conclude that there must be a component of God's perceived indirect action which is direct. At some stage—some "initial" stage—he/she acts without intermediary to initiate the intended action or create a range of necessities and possibilities, for instance, by directly establishing fundamental laws of nature and the fundamental constants or their primordial antecedents. At some level we know that God's direct action was and is necessary to ground and maintain existence of everything that is not God, and to enforce the regularities, constraints, and interrelationships which we refer to as the "laws of nature" and which endow reality with its interlocking levels of order, necessity, and possibility.

But our ability to model God's direct action seems to encounter an insuperable barrier at this point. Our experiences of acting directly no longer provide a helpful analogy or model for what divine direct action must be. Essentially, even though we know that at some fundamental level God is and must be acting directly, we never have direct experience of his/her doing so! We always experience divine action as indirect—even though the action may sometimes seem to operate outside of the "laws of nature" as we understand them. And we never have experience of God acting directly—even though we have assurances from revelation that he has and does, in creation, in the Incarnation, within the realm of the personal. We would apparently not be able to determine if a particular consequence were the result of God's direct action, instead of God's indirect action through a channel or instrument we are not aware of or do not understand. Thus, an apparent divine intervention on our behalf—a miracle—in answer to our prayers, for instance, a healing of a disease of paralysis which cannot be explained by contemporary medical science, does not of itself manifest the direct action of God, though it does manifest God's personal loving and life-giving action towards us. We always experience it through some intermediary datum or agent—through some sacrament. Even when there is no obvious cause—we just find ourselves well whereas before we were ill and dying—*our experience* of this is mediated by what has occurred unexplainably in our bodies. Our experience is not of any direct encounter with God, however mystical (in the extraordinary sense) that may have been. Furthermore, there is no assurance that the proximate cause of the healing, miraculous as it is, was not effected by God operating through a "regularity or law of nature" which is beyond our present knowledge or understanding or through an intermediary agent, that is, a prophet or an angel.

My point is that, though the extraordinary character of the event, which is outside what we normally expect in similar situations, leads us to believe that God is personally responding to our needs and prayers, this does not of itself indicate that the divine action is direct. It may indicate, however, that it is special, particular, and personal; I shall have more to say about this later. Even in terms of the Incarnation, no one, not even Mary, had an unambiguous *experience* of direct divine action, however personal and gratuitous it was.

Another possibility for divine action, however, is what St. Ignatius of Loyola refers to as "consolation without previous cause," as being an unequivocal sign of God's active graced presence in a religious experience.[22] This may be, but it still is not at all clear that it is an experience of direct divine action! It may be an unequivocal sign of God's presence and action, but it is very difficult to assess critically as an experience of God as a direct, unmediated cause. Perhaps the only place where we shall experience that is in the beatific vision.[23]

The key point to this discussion is simply that we have no experiential basis upon which to model God's direct action with regard to created reality.[24] Thus, although we know it occurs, it is apparently inaccessible to our experience and therefore to our detailed understanding. The case of God's action in creating from nothing and maintaining in existence is essentially direct divine action—perhaps the clearest case of it. But here again, that extraordinary and pervasive relationship of creatures with the divine, in which we ourselves participate, occurs at the very core of our beings and is hidden from our eyes.

7 The Problems of the Primary-Secondary Causal Nexus, of Double Causality, and of Top-Down Causality

There are a series of unsolved problems related to divine action, which flow from this discussion of the impossibility of adequately articulating or modeling God's direct action towards a creature. From what I have said above, it is clear that God's direct creative action *ex nihilo* is not susceptible to experiential "detection" or probing. In a sense, in order to answer how it happens, we would have to be God! We have some access to the "why" because of revelation—in terms of God's

---

[22]St. Ignatius of Loyola, "Rule for the Discernment of Spirits," *The Spiritual Exercises*. See Karl Rahner, *The Dynamic Element in the Church*, trans. W.J. O'Hara, vol. 12, *Quaestiones Disputatae* (New York: Herder & Herder, 1964), for a theological account of this.

[23]Cf. F. Suppe, "The Scientific Vision and the Beatific Vision," paper presented at the "Notre Dame Symposium on Knowing God, Christ, and Nature in the Post-Positivisitic Era," University of Notre Dame, April 14-17, 1993.

[24]Russell (private communication) insists that we distinguish three different ideas which I have tended to conflate here: (1) knowing where God acts directly (such as at the quantum level or in the free moral agent); (2) having an immediate experience of such a direct act; and (3) being able to model the act itself. My point here is that, though we may know or suspect that God acts directly in a given place or situation, we are never in the position to model it, simply because we do not have access to it in its immediacy. We have mediated experience of it, but no experience of the direct action itself, which is precisely what is in question.

goodness and love. When we turn to other categories of direct divine action, the same obstruction is found.[25]

7.1 *The Primary-Secondary Causal Nexus*

A key issue is the direct action of God with regard to secondary causes, through which he/she acts indirectly. How does God operate on a secondary cause, other than by bringing it into existence and conserving it in existence, so that it is the instrument for carrying out his/her intentions? In some cases this is just by maintaining it in existence and continuing to endow it with the nature or properties it has, and we do not know how that is done. But in some other examples, there is more going on—that is, in the sacraments, in the prophets, who are inspired to speak for God, in individuals who are in personal relationship with him/her. In other words, the causal nexus between God and any other cause or entity—incontrovertible as it is as a necessary condition for what we experience—is shielded from and inaccessible to our probing. Does God simply inject information or intention into secondary causes, inducing them to act on his/her behalf? Does this happen within the framework of the physical and other laws of nature, as we imperfectly know them? Or does it instead at least sometimes involve an abrogation or a fulfillment of those laws in terms of higher laws operating in the realm of the conscious and the personal and transcending those of physics, chemistry, and biology? We do not know for sure. I would strongly suspect that the last is often, though not exclusively, at work, simply on the basis of the priority the personal seems to have for God, as is clear to us from revelation.[26] But it seems extraordinarily difficult to substantiate that suspicion independently and to model such a causal nexus in concrete terms.

One of the difficulties here is simply that, in speaking of God's causal activity, we are trying to speak about a cause which is radically different from any other cause we experience—God is the primary cause. And we have no direct experience of this sort of causality. He/she is never one cause among many others, and cannot be conceived in his/her activity on the pattern of the created causes which we are and which we experience.[27] God's causal activity completely transcends secondary causality, and at the same time is perfectly immanent in secondary causality, supporting it and giving it efficacy. To use metaphors, God as primary cause is much more interior and present to creatures than they are to one another as secondary causes. But at the same time, on the basis of our *lack of direct experience of it*, God's causality is extremely subtle and hidden, and does not

---

[25]Arthur Peacocke discusses this problem at length—the problem of the "how," or what he refers to as the "causal joint" (Peacocke, *Theology for a Scientific Age*). He suggests that the resolution of it can be approached by locating creation "in God" and applying "top-down" causality, God acting on created reality.

[26]At the same time, however, we must find a way of avoiding an overly anthropocentric theology.

[27]Stoeger, "Origin of the Universe."

interfere with necessities, regularities, and freedoms with which secondary causes are endowed, except in response to a higher or more personal law.

7.2 *Double Causality*

Another problem with God's direct action in the world which is connected with this issue of the causal nexus is what we might call double causality. This is not so difficult in light of the conclusions we have already reached, but it bears mentioning. It is essentially: How can we have two adequate agents causing the same effect—God as primary cause and the secondary causes through which God is working?

There are several rather different issues which must be distinguished here: (1) God as primary cause acting to maintain secondary causes in existence, with their own particular capabilities, tendencies, and limitations, without further determining how they act to produce their effects (the underdetermination we were speaking of earlier); (2) God not only acting as primary cause to maintain secondary causes in existence, but possibly working through secondary causes to produce an effect God desires, a special or particular effect, outside of the ordinary pattern of what we would expect; (3) God inviting secondary causes to act in a certain way, but not determining or forcing them to do so; and (4) God apparently being a sufficient cause for an effect—directly or indirectly—and some created cause apparently being a sufficient cause for the same effect.

Regarding this last issue, I believe that the only problem here may be our confusion concerning what constitutes a sufficient cause—or reason—in a concrete case, along with whether the sufficient cause is acting directly or indirectly. For example, if one cuts the stem of an apple hanging on the tree and the apple falls to the ground, we might at first think that the person cutting the stem is the sufficient cause for the apple to fall, but that sufficiency presupposes a context in which other causes are acting, namely gravity. Without the action of gravity the apple would not fall. Nor is gravity a sufficient cause for the apple to fall; nor is God, who at some level instantiated the "laws of nature,"—they are necessary conditions, but not sufficient ones. The apple must be free to fall before gravity can cause it to fall. Applying this example to divine action, we see that God is never the sufficient condition for an effect occurring—though he/she is always a necessary condition for what occurs and sometimes contributes (in situations involving free moral agents) to the further conditions needed to constitute sufficiency. Correlatively, a secondary cause is never an absolutely sufficient condition for an effect, only what might be called a relatively sufficient condition—given that other normal conditions are fulfilled, that is, that gravity is acting.

Again God as a necessary condition for the existence of something, or of anything, is not in doubt, but God as sufficient condition is always in question. This is undoubtedly an aspect of divine kenosis (or self-emptying) and hiddenness in created reality—that God withholds his/her capability of being the sufficient condition of particular effects. For instance, God is not the sufficient condition of my existence—by relying on secondary causes (my parents and the processes of reproductive biology) divine causal sufficiency is surrendered. This is true even with respect to an event like the Incarnation. God invites it, but does not force it. The *fiat* of Mary was essential to the concrete realization of the Incarnation.

Now that they have been distinguished, the other issues, (1) through (3), lead to fairly straightforward resolutions. I shall not discuss (1) and (3) further, as the only one which may cause a problem is (2), that of God possibly determining a special or particular effect through secondary causes. The situations where this occurs are in God's personal action towards a person open to his/her presence and activity, in God's activity through impersonal, animate, or inanimate beings or causal chains, and more clearly in the cases where God apparently has directly or indirectly rigidly fixed general patterns of physical behavior, relationships, and structures in the "laws of nature."

In the first situation, God somehow communicates love and mercy—God's life-giving presence—in a particular experience or concrete event to a person or group. This rarely involves even the appearance of abrogation of the laws of nature, but instead a certain coming together of events which seem purely coincidental but which speak strongly of God's care and love to the person concerned. Does God really marshall such natural occurrences in these ways? Or is it rather that God sensitizes or inspires the person to whom he/she wishes to communicate the divine active presence in whatever naturally occurs, by means of the laws of nature we normally experience, and those higher laws of which we have no adequate understanding? In either case we are dealing with God's intended action toward a particular individual or group as a perceived response to faithfulness, openness, prayer, petition. And in either case, we must deal at some juncture with God's direct action on secondary causes and how that direct action is effected.

In these personal secondary causal situations, there is always some form of personal relationship between God and the created person—an openness and free initiative of God towards the person, and of the person towards God. And this mutual relationship is expressed in a whole network of manifestations. Furthermore, the person's cooperation is not forced, but free. How does this personal communication take place? Something analogous occurs in human relationships and in human agency involving other persons—acting to have someone else do something, either by command, by suggestion, by request, usually based on a previously established relationship of some sort between the persons. The key problem, as always, in this relationship between God and the created personal agent is that of the causal nexus—how does God influence or inspire someone? Oftentimes it is indirect—through an event, another person, reading and reflecting upon scripture, an idea or an emotion. But at some point there must be—at least according to our analysis so far—some direct connection, communication, or component of divine action with respect to the created agent. There must always be, it seems, some direct divine communication involved at some stage in the designation of a prophet, in the issuance of a call or vocation, and certainly in very special events like the Incarnation and the Resurrection. How is this direct link realized?

And then there is the second situation, God's action through impersonal secondary causes, in which the agents or instruments are not free to act or not act. Despite this difference from the previous case, the same issue arises—the way in which God directly causes or constrains some created beings to act as secondary causes.

In either case how does God do this? What is the nexus between God and the secondary causal instruments? We do not know. But perhaps we can begin to understand in terms of human agency and action. We do something very similar, do we not? We act through secondary causes. We decide to do something—to build a bookcase, to type a letter, to make a pot of coffee. And working through our bodies—directing our eyes, our hands and our fingers to perform very complex, goal-directed series of operations using tools and instruments, we bring all sorts of secondary causes together to aid in finishing our bookcase, completing the letter, and brewing the coffee. Undoubtedly, God is able to do the same thing, but with great reverence for both his/her creation and for the freedom and independence of the persons with whom God is communicating, for the character and the individuality of the beings, whether they be personal, animate, or inanimate, and their interrelationships, through which he/she is working. Although this is a description of top-down causality, which will be briefly discussed later, my point here is not its character but rather whether and how we can describe or model the causal nexus between God and secondary causal instruments. We understand something about our interaction with the material world around us, because we are material (but we certainly do not yet understand the relationship between our minds and our bodies!). It is considerably more difficult to understand the direct causal nexus between God, who is immaterial and uncreated, and the material secondary causes.

And yet a profound nexus there must be—whether it is more "interior" to the created causes, or more top-down and "exterior." I shall very shortly suggest how the immanence and transcendence of God may provide the key to understanding this problem of the "causal joint."[28] Before doing so, it is worth pointing out that, as we come to understand that the material and the immaterial are not essentially different, but intimately united at every level, and how this sameness in difference functions, we will perhaps come to some better appreciation of God's direct interaction with secondary causes. This will be paralleled, I hope, by progress in understanding how mind-body issues can be resolved. Both advances will help our analogy between human agency and divine agency to yield more fruit.

Answering this question "how?" concerning the direct causal connection between God and secondary causes requires a detailed knowledge and understanding of God and of God's causal and personal relationships with persons and with other creatures. Our inability to answer that question reflects the profound inadequacy of our knowledge of the divine. Still, to the extent that we know something about God—thanks to revelation and our reflection upon it—we can move in a promising direction.

As I have mentioned, this is in the direction of God's immanence and transcendence, particularly as they are realized in God's transcendent primary causality as a cause unlike any other. The key point is that God acts immanently

---

[28]Peacocke, *Theology for a Scientific Age*.

in nature—in every "nook and cranny" of nature, at the core of every being and at the heart of every relationship—to constitute and maintain it just as it is and just as it evolves. God constitutes things as they are and as they act—with freedom or without freedom, personal or impersonal—and maintains this constitutive relationship with them, with great efficacy, but also with great reverence and respect for the individuality and character of each and of the network of relationships they have with one another. This constitutive presence of God at "the heart of things" is so pervasive that from a strictly scientific perspective we do not notice it. There is nothing we experience or encounter, either exterior or interior, that is without it. God is fully and actively immanent precisely because God is fully transcendent.

Transcendence implies complete availability, accessibility, and active presence at every level—that is, immanence. What is transcendent is not trapped or constrained by a given level of being, a given relationship, or a given perspective, and so is available to all. There are no principles or regularities or relationships needed other than the secondary causes, regularities, and relationships which are vulnerable to scientific and philosophical investigation. But God's transcendent/immanent primary causality is always immediately and immanently endowing them with existence and with the intricate and dynamic order and interrelationships they enjoy.

Creation is a limited expression of the divine being. The direct causal nexus *is* the active, richly differentiated, profoundly immanent presence of God in created beings and in their interrelationships. It *is* at the same time their limited and specific participation—inclusion—in God's own existence and interrelationships as Trinity, which is utterly transcendent and immaterial but also radically open to and available for the realization of finite possibilities. The presence of God in each entity constitutes the direct, the immediate, relationship of that entity with God, and therefore is the channel of divine influence in secondary causes. This approach by no means resolves the mystery or answers the question, but it serves to locate it where the answer almost certainly lies. I shall discuss some of these issues again when I deal with God's personal action. Here I have briefly looked at the causal problems associated with this approach. There I shall focus more on the experience and intention involved in such modes of divine action.

The final situation in which God has determined an effect through secondary causes is where either directly or indirectly he/she has rigidly fixed (determined) general patterns of behavior, structures, relationships, constraints—the structure of atoms, for instance, and the periodic table with all the chemical laws embodied in it, the operation of the "laws of physics." Among all the possible and apparently internally consistent ways in which physical reality could behave, only this one is realized. And, if God exists and is the primary cause, he/she must have either chosen this realization, or allowed it to develop from some other more primordial laws.

In either case, God at some point or in some way acts directly to effect them, and continues to act directly and immanently to conserve them. Again, we have the "nexus problem," for which we have no real solution—other than the observations made above concerning the immanently and transcendently interior active presence of God in all that is. God chose to make the world the way it is, however much he/she allowed it to develop on its own. God implements that choice by initiating and maintaining an existence-endowing (constitutive) relationship with the possibilities he/she wishes to realize. The choice of a particular instantiation and its *direct* implementation—whatever the number of allowed outcomes—was necessary at some level. From revelation, we appreciate some of the motivations directing that choice, in terms of freedom and the primacy of love, dictating a world in which God remains involved and caring, but in which we remain free and able to freely give or refuse love and service to God and to one another.[29]

7.3 *Top-Down Causality*

In this discussion we are already aware of the final problem we shall briefly discuss, that of top-down causality. The brief discussion of human agency above provided examples of top-down causality—a human being building a bookcase, typing a letter, brewing a pot of coffee—in which an entity of higher complexity or possessing greater versatility determines or causes entities at lower, more fundamental levels to behave in a certain way—in a more organized and coherent way than they would do otherwise. In the hierarchical layers of organization and complexity which characterize our universe, top-down causality is pervasive. Although some causal influences operate from lower levels of organization to higher levels, constraining and also enabling what more complex entities do, other causal influences act from the top down to marshall and coordinate less organized constituents into coherent, cooperative action in service of the more complex organism or system. A precondition for this being possible is the radical underdetermination of effects by the "laws of nature" at lower levels (the freedom and the need to establish initial conditions, or boundary conditions)—rendering nature very pliable within certain limits. There is really no problem here—just a characteristic of reality which requires proper recognition and careful analysis. Obviously in the case of divine action, we have the ultimate case of top-down causality, in which the essential issues challenging our understanding are those which we have already discussed.

8 *Divine Action within the Context of the Personal*

The realm of divine action which is especially important for the meaning, orientation, and direction of our lives is that of the personal. In fact, within the context of Christian revelation at least, the focus of divine action is on the personal and the communal—God's continual active presence with and on behalf of his/her people, drawing them closer to God, and sharing the divine life ever more fully with them as individuals and as groups. God takes the initiative in our regard, invites us and enables us to establish a relationship with him/her, gives life, reveals

---

[29]Cf. George F.R. Ellis, "The Theology of the Anthropic Principle," in *Quantum Cosmology*, 367-405.

Godself, heals, punishes, reconciles, forgives, transforms, renews, saves—out of love and care for persons. The ultimate manifestation of this is in the Incarnation, and in the life, death, resurrection of Jesus, and sending of the Spirit of the Incarnate One, who is Wisdom, Word, Child of God. It is only as an afterthought, so to speak, but a very important afterthought, that revelation and our response to it in faith speaks of God's creative action with regard to the whole context within which God personally directed saving and transforming activity takes place. It is obviously important from many points of view, but falls outside the primary focus of attention in much of revelation.[30]

Within the context of our present interest in articulating more adequately divine action in light of the self-organizing capabilities of material systems, we may think that God's personal action falls outside our primary focus. There is a sense in which this is true. But there is also a sense in which it is completely false. If our understanding of God is primarily as a Trinity of persons—with all that that implies within the Christian tradition—then all divine action, however impersonal it may seem, in its consequences or manifestations, must be seen in terms of the personal, of personal relationships, and of the preconditions for the emergence of the personal within the universe. This is certainly true from the standpoint of our faith and the knowledge which we have based on divine revelation. However, it is far from clear simply from the standpoint of the physical and the other natural sciences, even though there are indications that point in that direction (e.g., the coincidences which point towards an "anthropic principle," however vacuous the actual logic of those arguments may be without the presupposition of God's existence). Our procedure is really to take both areas of knowledge seriously and let them critically interact with one another, as we have already done in dealing with other issues. What are the consequences of doing so on this subject of the priority of the personal in divine action and on manifestations of divine action at the level of the impersonal and inanimate through the underlying physical constraints and regularities and the self-organizing capabilities we see in reality?

The clearest answer would be that all of what we see manifested in the natural world has been established for the purpose of securing the priority and dominance of the personal and of personal relationships within creation, and to enable created persons to relate freely and lovingly with one another, with the rest of creation, and with God. Profound as this is, there is nothing new here which we would not have

---

[30]See Richard J. Clifford, "Creation in the Hebrew Bible," in *Physics, Philosophy, and Theology*, 151-70. In saying this, however, we must not separate what is personal and self-conscious from God's action in its deepest form in inanimate creation. The focus of much of revelation on the personal should not insulate us from attending to and celebrating God's active presence in all creation. In fact, in light of both what we know from revelation and from contemporary sciences, part of our commitment must be to emphasize our profound unity with the rest of creation, to learn from it by contemplating it, and to take a more enlightened responsibility in caring for it and fostering reverence for it. Though we must be faithful to revelation in terms of the priority of the personal, we must be faithful to all that it offers us, and we cannot continue to indulge in an overweening anthropocentrism.

known before delving into the self-organizing behavior of matter. But is there anything else?

Yes, I believe that in George Ellis's "Christian Anthropic Principle,"[31] we see a deep compatibility among the autonomous ways in which physical, chemical, and biological laws operate at every level of nature—particularly in the self-organizing capabilities of matter and systems composed of matter at every level. The core of this compatibility is the relative independence and freedom of created reality to evolve and organize at every level without direct divine intervention or interference (except at the most radical ontological level) with a richness of inherent potentiality and possibility. Correlative with this, as I have already mentioned, is God's relative hiddenness in creation; God has created and is creating it, but at the same time is radically setting it free to become itself, to discover itself, to become conscious of itself, to become free and to become independently personal and social, to discover its roots and its ultimate origins, to respond freely to the invitation to enter into relationship with the community and society of persons which is God, its source and origin.

In a sense the fact that we are made "in the image and likeness of God" necessitates an infrastructure like we have. One which needed the constant intervention of God—divine direct action to fill gaps and to negotiate the difficult transitions between nonliving and living, living and conscious, conscious and knowing—would be a creation which would be very unfree and incapable of becoming itself, discovering God as a person (and not just as a demiurge and problem-solver), and entering into a loving relationship with that God. Nor would such a creation be very compatible with God's self-communication to it. In short it would be a creation unworthy of God, and one which did not adequately reflect who God is.

If we take this point of view, then there is one other point that falls into place. If the personal has priority, then relationships are of the utmost importance. And what we see throughout creation is a reflection of this—the central role that constitutive relationships play at every level. Entities are as they are at every level not just because of the parts that constitute them but because of the relationships which exist among the components. The whole is always greater than the sum of the parts because of these relationships. And the different interactions which obtain—for instance those of gravitation, electromagnetism, the weak and strong nuclear forces in physics—and the behaviors they allow and forbid help to determine these interrelationships. We are able to some extent to describe these regularities, patterns, constraints, and relationships through the "laws of nature" we formulate.

But, as I have insisted before, these are only imperfect descriptions of the intricate network of regularities, constraints, and relationships which actually operate, linking everything with everything else, but also constituting each entity's individuality and relative independence. If there are phenomena which seem to fall outside of the regularities we are able to describe securely, or situations in which

---

[31]Ellis, "Theology of the Anthropic Principle."

they do not seem to hold, then undoubtedly there are "higher" laws at work. These laws somehow reflect more fully the dominance of the personal, or the essential role of relationships. These in turn are more intricate, complex, or subtle, than we are yet capable of understanding and modeling, but they would be thoroughly compatible with the nature of all the entities involved and of their relationships with one another, with the personal, and with God, if we were to completely understand those relationships.

9 *Conclusions*

This has been a sketch of my synthesis of a model of God's action in the world, taking seriously both revelation and the knowledge of reality we have from the sciences. There are aspects of divine action which we are able to understand somewhat better by letting these two areas of our knowledge critically interact and dialogue with each other. There are other aspects which seem to be thoroughly resistant to our understanding, particularly that of the nexus between God and the secondary causes through which God acts or between God and the direct effects of divine action, as in *creatio ex nihilo*. The analogue of human agency is of some limited help here. However, the principal barrier seems to be that we can only know that critical nexus—an adequate answer to "how" divine causality operates in this circumstances—if we are divine, or if God reveals such knowledge to us. Otherwise we do not have enough knowledge of the key term in the nexus—God.[32]

---

[32]My special thanks to all those who have given me comments on a previous draft of this paper or who have discussed aspects of it with me, especially Stephen Happel, Ian Barbour, Tom Tracy, Nancey Murphy, George Ellis, Arthur Peacocke, Denis Edwards, Bob Russell, Wim Drees, and John Polkinghorne.

GOD'S INTERACTION WITH THE WORLD:
The Implications of Deterministic "Chaos" and of Interconnected and Interdependent Complexity[1]

Arthur Peacocke

1 *Introduction*

At the beginning of the seventeenth century, John Donne lamented the collapse of the medieval synthesis but,[2] after that century, nothing could stem the rising tide of an individualism in which the self surveyed the world as subject over against object. This way of viewing the world involved a process of abstraction in which the entities and processes of the world were broken down into their constituent units. These parts were conceived as wholes in themselves, whose lawlike relations it was the task of the "new philosophy" to discover. It may be depicted, somewhat over-succinctly, as asking, first, "What's there?"; then, "What are the relations

---

[1]This paper is a summary and (I hope) a clarification and development of some ideas elaborated more fully in my *Theology for a Scientific Age: Being and Becoming—Natural and Divine* (Oxford: Blackwell, 1990; 2d ed., enlarged, London: SCM Press, 1993 and Minneapolis, MN: Fortress Press, 1993; references are to the 1990 edition unless otherwise indicated). It is also developed from a paper entitled, "God's Interaction with a 'Chaotic' World: Some Scientific Reflections" (Durham, England: meeting of the Science and Religion Forum, March 22, 1991). An earlier version and a fuller account of "chaos" theory in relation to the issues discussed here, and its historical setting, is to be found in a lecture, "Natural Being and Becoming—The Chrysalis of the Human," given by the author at the Georgetown University Bicentennial, April 4-5, 1989, and published in *Individuality and Cooperative Action*, ed. Joseph Earley (Washington, D.C.: Georgetown University Press, 1991), 91-108. Its present form is the result of further reflection on the issues as a result of comments made on a draft paper by participants in the Vatican Observatory/Center for Theology and the Natural Sciences Conference at Berkeley, CA, August 1-7, 1993 on "Chaos, Complexity, and Self-Organization: Scientific Perspectives on Divine Action"; and by Brian Austin, Philip Clayton, Steve Knapp, Nancey Murphy, and Iain Paul. I am grateful to all of these for their helpful critiques.

That a "whole-part" (or "downward/top-down") constraining relation (such as that proposed as illuminating various systems-parts relations, including the brain-body relation—section 3 below) could provide a clue to how God might be conceived of as interacting with the world (sections 4.2 and 4.3) was originally expounded by the present author in lectures given on May 9, 1987, at a Rewley House course of the Department of External Studies of the University of Oxford.

[2]"'Tis all in pieces, all coherence gone," John Donne in 1611 (in his *Anatomie of the World, The First Anniversary*).

between what is there?"; and finally, "What are the laws describing these relations?" To implement this aim a *methodologically* reductionist approach was essential, especially when studying the complexities of matter and of living organisms, and the natural world came to be described as a world of entities involved in lawlike relations which determined the course of events in time.

The success of these procedures has continued to the present day, in spite of the revolution in our epistemology of the physics of the sub-atomic world necessitated by the advent of quantum theory. For at the macroscopic level that is the focus of most of the sciences from chemistry to population genetics, the unpredictabilities of quantum events at the sub-atomic level are usually either ironed out in the statistical certainties of the behavior of large populations of small entities or can be neglected because of the size of the entities involved.[3] Predictability was expected in such macroscopic systems and, by and large, it became possible after due scientific investigation. However, it has turned out that science, being the art of the soluble, has concentrated on those phenomena most amenable to such interpretations. What I intend to point to are some developments from within the natural sciences themselves that change this perspective on the natural world in a number of ways which might bear significantly on how we can conceive of God's interaction with the world. After an initial setting of the scene and exposition of the part played until recently by the notions of predictability and causality in the scientific world-view, I shall focus on recent ideas of deterministic "chaos" (including "order out of chaos") and on whole-part constraint (or "downward/top-down" causation) in interconnected and interdependent complex systems. Whether or not these new perceptions are of assistance in conceptualizing more coherently and intelligibly than hitherto how God interacts with the world is then considered.

2 *Predictability and Causality*

What do the sciences tell us is there in the world? By far the greater proportion of the sciences are concerned with the region which lies between the atomic and the cosmological and which manifests a hierarchy of complexities, while at the same time being a diversity-in-unity. Each "level" of natural complexity usually has a corresponding science that focuses upon it and one "above" focusing on the next, more complex level, which thereby represents a new emerging feature of reality, descriptions of which are often not wholly reducible to those applicable to the lower levels out of which it arose.[4]

---

[3]There is also the possibility that some quantum events at the micro-level can be amplified so as to have macroscopic consequences, e.g., in the case of non-linear dynamic systems (see sections 2.4 and 4.1.2).

[4]For references on the literature on reductionism, see *inter alia* the diagram of the "hierarchy" of scientific disciplines in Peacocke, *Theology for Scientific Age*, fig. 3, p. 217. See also, *idem*, *Creation and the World of Science* (Oxford: Clarendon Press, 1979), chap.

The world is also—almost notoriously—in a state of continuous flux, and it has, not surprisingly, been one of the major preoccupations of the sciences to understand the changes that occur at all levels of the natural world. Science has asked "What is going on?" and "How did these entities and structures we now observe get here and come to be the way they are?" The object of our curiosity is both causal explanation of past changes in order to understand the present and also prediction of the future course of events, of changes in the entities and structures with which we are concerned.

The notions of explanation of the past and present and predictability of the future are closely interlocked with the concept of causality. For detection of a causal sequence, in which, say, *A* causes *B* which causes *C* and so on, is frequently taken to be an explanation of the present in terms of the past and also predictive of the future, insofar as observation of *A* gives one grounds for inferring that *B* and *C* will follow as time elapses, as in the original *A-B-C*. . . succession in time.

It has been widely recognized that causality in scientific accounts of natural sequences of events is only reliably attributable when some underlying relationships of an intelligible kind between the successive forms of the entities have been discovered, over and beyond mere conjunction as such. The fundamental concern of the sciences is with the explanation of change and so with predictability, and often therefore with causality. It transpires that various degrees of predictability pertain to different kinds of natural systems and that the accounts of causality that are appropriate are correspondingly different.

2.1 *Relatively Simple, Dynamic, Law-Obeying Systems*

Science began to gain its great ascendancy in Western culture through the succession of intellectual pioneers in mathematics, mechanics, and astronomy which led to the triumph of the Newtonian system, with its explanation not only of many of the relationships in many terrestrial systems but, more particularly, of planetary motions in the solar system. Knowledge of both the governing laws and of the values of the variables describing initial conditions apparently allowed complete predictability with respect to these particular variables. This led, not surprisingly considering the sheer intellectual power and beauty of the Newtonian scheme, to the domination of this criterion of predictability in the perception of what science should, at its best, always aim to provide—even though single-level systems such as those studied in mechanics are comparatively rare. It also reinforced the notion that science proceeded, indeed *should* proceed, by breaking down the world in general, and any investigated system in particular, into its constituent entities. So it led to a view of the world as mechanical, deterministic, and predictable. The concept of causality in such systems can be broadly subsumed

---

4.1; *idem, God and the New Biology* (London: Dent, 1986; reprint, Gloucester, MA: Peter Smith, Magnolia, 1994), chaps. 1 and 2 (references are to 1994 edition); and Ian Barbour, *Religion in an Age of Science,* The Gifford Lectures, vol. 1 (San Francisco: Harper-SanFrancisco, 1990), 165-72.

into that of intelligible and mathematical relations with their implication of the existence of something analogous to an underlying mechanism that generates these relationships (the generative mechanism).[5]

2.2 *Statistical Properties of Assemblies*

Certain properties of a total assembly can sometimes be predicted in more complex systems. For example, the relationships expressed in the gas laws (excluding phase changes) are still determinable in spite of our lack of knowledge of the direction and velocities of individual molecules. It is well known that the predictability of events at the atomic and sub-atomic level has been radically undermined by the realization that accurate determinations of the values of certain pairs of conjugate quantities (e.g., momentum/position and energy/time) are mutually exclusive. This introduces a fundamental uncertainty (given by the Heisenberg uncertainty principle [HUP]) into the quantitative description of events at this micro-level and reinforces the unpredictability in many respects, though not in some significant ones (e.g., the classical gas laws) of such macro-systems.

Other related kinds of unpredictability are now also accepted with reference to other systems at the sub-atomic level, for example in radioactive decay (see section 4.1.2 [ii], below). All that is known is the *probability* of a given atom breaking up in a particular time interval. This exemplifies the current state of quantum theory, which allows only for the dependence on each other of the probable values of certain variables and so for a looser form of causal coupling at this micro-level than had been taken for granted in classical physics. But note that causality, as such, is *not* eliminated, only classical, deterministic (Laplacian) causality.

2.3 *Newtonian Systems Deterministic yet Unpredictable at the Micro-Level of Description*

That there are Newtonian systems which are deterministic yet unpredictable at the micro-level of description has been a time-bomb ticking away under the edifice of the deterministic/predictable paradigm of what constitutes the world-view of science since at least the turn of the century. For example, the French mathematician Henri Poincaré pointed out that, since the ability of the (essentially Newtonian) theory of dynamical systems to make predictions depended on knowing not only the rules for describing how the system will change with time, but also on knowing the initial conditions of the system, such predictability was extremely sensitive to the accuracy of our knowledge of the variables characterizing those

---

[5]Rom Harré, *The Principles of Scientific Thinking* (London: Macmillan, 1970), chap. 4.

initial conditions.[6] Thus, it can be shown that even in assemblies of bodies obeying Newtonian mechanics there is a real limit to the period during which the micro-level description of the system can continue to be specified, that is, there is a limit to micro-level predictability, the horizon of what Wildman and Russell call "eventual unpredictability."[7] There is certainly not predictability by us over an indefinite time.[8] This is so, in spite of the deterministic character of Newton's laws, on account of our inability ever to determine sufficiently precisely the initial conditions.

For example, in a game of billiards suppose that, after the first shot, the balls are sent in a continuous series of collisions, that there are a very large number of balls and that collisions occur with a negligible loss of energy. If the average distance between the balls is ten times their radius, then it can be shown that an error of one in the 1000th decimal place in the angle of the first impact means that all predictability is lost after 1000 collisions.[9] Clearly, *infinite* initial accuracy is needed for the total predictability through infinite time that Laplace assured us was possible. The uncertainty of movement grows with each impact as the originally minute uncertainty becomes amplified. So, although the system is deterministic in principle—the constituent entities obey Newtonian mechanics—it is never totally predictable in practice.

Moreover, it is not predictable for another reason, for even if *per impossible* the error in our knowledge of the angle of the first impact were zero, unpredictability still enters because no such system can ever be isolated completely from the effects of everything else in the universe—such as gravity and, of course, the mechanical and thermal interactions with its immediate surroundings.

Furthermore, attempts to specify more and more finely the initial conditions will eventually come up against the barrier of the HUP in our knowledge of key

---

[6]See, e.g., Henri Poincaré, *Science and Method*, trans. F. Maitland (London: Thomas Nelson, 1914), 68. Earlier considerations of this kind also appear in the works of J. Hadamard (1898) and P. Duhem (1906). For details, see David Ruelle, *Chance and Chaos* (Harmondsworth: Penguin Books, 1993), chap. 8, nn. 1 and 3, pp. 175-6.

[7]See Wesley Wildman and Robert John Russell, "Chaos: A Mathematical Introduction with Philosophical Reflections" (in this volume).

[8]John T. Houghton, having defined "chaotic behavior" as "the unpredictable situations . . . in which given infinitesimally different starting points, systems can realize very different outcomes,"(p. 42) later makes the same point as in my text, as follows: "For a chaotic system, a predictability horizon may be defined beyond which for a given accuracy in the specification of the initial conditions the behavior of the system cannot be predicted. The predictability horizon only moves away linearly as the number of decimal places in the initial specification increases . . . initial conditions can never be specified absolutely precisely so that the predictability horizon represents a fundamental limit to our ability to predict"(pp. 48-49). Houghton, "New Ideas of Chaos in Physics," *Science and Christian Belief* 1 (1989): 41-51.

[9]Michael Berry, "Breaking the Paradigms of Classical Physics from Within," 1983 Cercy Symposium, *Logique et Théorie des Catastrophes*.

variables characterizing the initial conditions even in these "Newtonian" systems. Does this limitation on our knowledge of these variables pertaining to individual "particles" (whether atoms, molecules, or billiard balls) in an assembly reduce the period of time within which the trajectory of any individual "particle" can be traced? Is there an ultimate upper limit to the predictability horizon set by the irreducible Heisenberg "fuzziness" in the initial values of those key initial parameters to which the eventual states of these systems are so sensitive? Does "eventual unpredictability" prevail—at least with respect to the values of those same parameters which characterized the initial conditions and to which the HUP applied? The physicist J.T. Houghton thinks so[10] and so apparently does another, Joseph Ford, who (to anticipate the next section somewhat) having argued that "chaos is merely a synonym for randomness" goes on to affirm that "we may legitimately inquire if there is any chaos (randomness) in quantum dynamics *beyond that of the type contained in the probability density* $\psi^*\psi$."[11] Houghton's contention does not appear to go beyond the limitations of the italicized condition and so can be provisionally accepted,[12] since it does not seem to be subject to the outcome of the debate on quantum chaos (see end of section 2.4 below).

2.4 *"Chaotic" and "Dissipative" Systems*

One of the striking developments in science in recent years has been the increasing recognition that there are many other systems (physical, chemical, biological and

---

[10]Houghton in "New Ideas," expresses the situation thus: ". . . as soon as the required specification of the initial conditions [in chaotic systems, defined as in n. 8 above] involves details of the movements of individual electrons or atomic nuclei [one could add—any material entity], the Heisenberg uncertainty principle becomes relevant. We then come up against an inability, not only in practice but in principle, to specify with perfect precision the state of the system at any given time"(p. 49). The use of "in principle" in this last phrase is disputed, but whether or not there will prove to be hidden variables, physicists have to accept that the HUP has to be applied to our knowledge of initial conditions.

[11]Joseph Ford, "What is Chaos, that we should be mindful of it?" in *The New Physics*, ed. Paul Davies (Cambridge: Cambridge University Press, 1989), 365. What Ford is referring to in the italicized phrase can be expounded as follows based on an editorial comment by Russell:

Quantum mechanical systems display an inherent statistical character, represented by the wave function (ψ) and computed from its absolute value ($\psi^*\psi$). However, the wave function itself evolves deterministically, according to the Schrödinger equation (in non-relativistic quantum physics), and does not display the kind of chaotic behavior typified by the time evolution of even simple (classical) systems such as those represented by the logistic equation. Whether or not quantum systems do actually display additional statistical behavior beyond that represented by the wave function—that is, whether or not there is a quantum version of classical chaos (called "quantum chaos")—is still an open question, as discussed in the last paragraph of this section.

[12]A position I myself adopted in *Theology for a Scientific Age*.

indeed neurological) which can also in practice become unpredictable in their macroscopically observable behavior, even when this is governed by deterministic equations. Basically the unpredictability arises for us in practice because two states of the systems in question which differ, even only slightly, in their initial conditions eventually generate radically different subsequent states. Broadly this is often denoted as 'chaotic behavior' (e.g., see n.8); or 'chaos' is taken as equivalent to 'randomness' (as with Ford, section 2.3 above and n.11). More precise definitions of 'chaos,' are possible in particular systems and this may be illustrated mathematically by the sensitivity of the eventual reiterated solutions of the "logistic equation":[13]

$$x_{n+1} = k\, x_n\, (1 - x_n)$$

to values of the initial conditions (x_0) when the parameter k (which in real systems often represents external conditions) is such as to locate the system in the "bifurcation" regime (k=3.0 to 3.58), or in the "chaotic/random" regime (k=3.58 to 4). This particular equation has proved to be significant for certain natural systems (predator-prey patterns, yearly variation in insect and other populations, etc.) and this combination of an overall influence of environmental factors, "external" conditions, and sensitivity to initial conditions has transpired to be operative in a very wide range of natural systems. In some of these systems there can occur an amplification of a fluctuation of the values of particular variables (e.g., pressure, concentration of a key substance, etc.) pertaining at a micro-level within the system so that the macroscopic state of the system as a whole undergoes a marked transition to a new regime of new patterns of, for example, pressure, concentrations, etc. The state of such systems is then critically dependent on the initial conditions which prevail within the key, transitory fluctuation which is subsequently amplified. A well-known example is the "butterfly effect" of Edward Lorenz whereby a butterfly disturbing the air here today is said to affect what weather occurs on the other side of the world in a month's time through amplifications cascading through a chain of complex interactions. Another is the transition to turbulent flow in liquids at certain combinations of speed of flow and external conditions. Yet another is the appearance, consequent upon localized fluctuations in reactant concentrations, of spatial and temporal patterns of concentration of the reactants in certain, otherwise homogenous systems which involve autocatalytic steps—as is often the case, significantly, in key biochemical processes in living organisms.

One common feature of all these dynamic systems is that they are "non-linear" with respect to the relation between certain key variables of the system, for example the fluxes of material or energy and the "forces" controlling them. It is now realized that the time-sequence of such complex dynamical systems can take many forms: "limit cycles" such as those which govern the motion of the planets or the pendulum of a grandfather clock; regular oscillations in time and space with

---

[13]For a detailed discussion of the logistic equation, see Wildman and Russell, "Chaos."

respect to the concentrations of key constituents in those chemical systems already mentioned or with respect to biological populations of predators and prey; systems that "flip" periodically between two or more allowed states. As a key parameter increases, all kinds of further complexities can occur with successive bifurcations, and so forth, leading eventually into "chaotic/random" regimes.

In the real world most systems do not conserve energy: they are usually *dissipative systems* through which energy and matter flow, and so are also open in the thermodynamic sense. Such systems can often, when non-linear and far removed from the state of equilibrium, give rise to the kind of sequence just mentioned. Moreover, non-equilibrium physics and the discovery of new properties of matter in far-from-equilibrium conditions has also led to the recognition that, in such change-overs to temporal and spatial patterns of system behavior we have examples of "symmetry-breaking," and of macroscopic-scale correlations, that is, of self-organization."[14] Ilya Prigogine and his colleagues at Brussels have called this "order through fluctuations" and "order out of chaos,"[15] perhaps somewhat misleadingly in view of the current usage of the term 'chaos.' The point is that new patterns of the constituents of the system in space and time become established when a key parameter passes a critical value.

Explicit awareness of all this is only relatively recent in science and necessitates a re-assessment of the potentialities of the stuff of the world, in which pattern formation had previously been thought to be confined only to the macroscopically static, equilibrium state. In these far-from-equilibrium, non-linear, open systems, matter displays its potential to be self-organizing and thereby bring to existence new forms entirely by the operation of forces and the manifestation of properties we thought we already understood. As Crutchfield, *et al.* put it: "Through amplification of small fluctuations it [nature] can provide natural systems with access to novelty."[16]

How do the notions of causality and predictability relate to our new awareness of these phenomena? We shall discuss this initially without taking account of quantum uncertainties. Causality, as usually understood, is clearly evidenced in the systems just discussed. Nevertheless the identification of the causal event(s) now has to be extended to include unobservable fluctuations at the micro-level, whose effects in certain systems may extend through the whole system so as to produce correlations between the values of characteristic parameters which

---

[14]See Gregoire Nicolis, "Physics of Far-From-Equilibrium Systems and Self-Organization," in *The New Physics*, ed. Paul Davies (Cambridge: Cambridge University Press, 1989), 316-47; and Peacocke, *The Physical Chemistry of Biological Organization* (Oxford: Clarendon Press, 1989), chaps. 2 and 4.

[15]See, *inter alia*, Ilya Prigogine, *From Being to Becoming: Time and Complexity in the Physical Sciences* (San Francisco: Freeman, 1980); *idem* and Isabelle Stengers, *Order Out of Chaos: Man's New Dialogue with Nature* (London: Heinemann, 1984); and Peacocke, *Biological Organization*, chap. 2.

[16]J.P. Crutchfield, et al., "Chaos" (in this volume).

extend over a spatial scale many orders of magnitude greater then that of the original region of fluctuation itself.

The equations governing all these systems are deterministic, which means that *if* the initial conditions were known with infinite precision, prediction would be valid into the infinite future, as famously postulated by Laplace. But it is of the *nature of our knowledge of the real numbers* used to represent initial conditions that they have an infinite decimal representation and we can know only their representation up to a certain limit. Hence there will always be, for systems whose states are sensitive to the values of the parameters describing their initial conditions, an "eventual unpredictability" *by us* of their future states beyond a certain point.[17] The "butterfly effect," the amplification of micro-fluctuations, turns out to be but one example of this and we also note that in such cases the provoking fluctuations would anyway be inaccessible to us experimentally. Hence, although deterministic equations govern the time-course of these systems, they are unpredictable beyond certain limits because of the inevitably limited accuracy of *our* knowledge, making them in practice irreducibly unpredictable.

For those dissipative systems that are in regimes in which they could move to a finite number of possible future states, our lack of precise knowledge of the parameters prevailing in the crucial fluctuation(s) which could provoke symmetry-breaking at the critical value of a key parameter, allows us to affirm only the probability of future occupancy of any of the possible states. That is, *which* state will be occupied is unpredictable, as well as uncontrollable, and may be said by us to be a matter of "pure chance." Only a detailed knowledge of the particular, individual fluctuation initiating the symmetry-breaking would allow prediction and that is never going to be observable by us. Hence, in spite of the excitement generated by our new awareness of the character of the systems described in this section, the basis of the unpredictability *in practice* of their overall states is, after all, no different from that of the eventual unpredictability at the micro-level of description of the Newtonian systems discussed in the previous section (2.3).

What will be the effect of quantum (HUP) uncertainties on the predictability of the systems displaying "chaotic behavior," that is, having an immense sensitivity to initial conditions (to use Houghton's definition)? The effects of any uncertainty within "the type contained in the probability density $\psi^*\psi$" would also establish an irremovable horizon of "eventual unpredictability" for these "chaotic" systems—at least with respect to the level of description of the parameters about which there was this initial uncertainty. This quantum limit to predictability would also extend, for systems that amplify the effects of fluctuations, to macroscopic states, allowing only a probabilistic knowledge of which among possible states will actually occur.

But a further question remains, expressed by Ford in the following terms: "If we exclude from consideration the innate randomness of $\psi^*\psi$ and the like [see previous paragraph], what opportunities for chaos remain?" He continues, "There are two: randomness in the Schrödinger time evolution or flow of the wave

[17]E.g., the direction of rotation of a Bénard cell in a convecting fluid, or that of a rotatory wave pattern in the Belousov-Zhabotinsky reaction.

function $\psi^*\psi$ or randomness in the quantum eigenvalues, eigenfunctions and matrices." And, he tells us, a "major battle . . . is even now being waged over the existence of [such] quantum chaos." That being so, it would be unwise to prejudge the outcome of what Ford also says would be "an earthquake in the foundations of physics,"[18] but concludes that "the evidence weighs heavily against quantum chaos [as so defined]."[19]

3 *Whole-Part Constraint (or "Downward/Top-Down" Causation)*

3.1 *General*

The notion of causality, when applied to systems, has usually been assumed to describe "bottom-up" causation—that is, the effect on the properties and behavior of the system of the properties and behavior of its constituent units. However, an influence of the state of the system as a whole on the behavior of its component units—a constraint exercised by the whole on its parts—has to be recognized. D. Campbell[20] and R.W. Sperry,[21] called this "downward" (or "top-down") causation,[22] but it will usually be referred to here as "whole-part constraint." For, to take the example of the Bénard phenomenon, beyond the critical point, individual molecules in a hexagonal "cell," over a wide range in the fluid, move with a common component of velocity in a coordinated way, having previously

---

[18]Ford, "What is Chaos?," 366.

[19]Ibid., 370. Ford goes on to point out that—even if quantum chaos, as he has just defined it, does not occur—because "quantum mechanics is firmly rooted in the incomputable number continuum," it follows that "if chaos is ubiquitous in nature at the microscopic level, *whether or not in quantum mechanics*, then there is an incomputable and uncontrollable noise level on the smallest scale" (emphasis added). An irreducible, unpredictability would then arise from the same considerations as prevailed with Newtonian systems, namely on the basis of the infinite decimal representation of real numbers (see section 2.3, and earlier in this section) and, more generally, of algorithmic complexity theory. (Ibid., 365-70.)

[20]Donald T. Campbell, "'Downward Causation' in Hierarchically Organized Systems," in *Studies in the Philosophy of Biology: Reduction and Related Problems*, ed. Francisco J. Ayala and Theodosius Dobzhansky (London: Macmillan, 1974), 179-86.

[21]Roger Sperry, *Science and Moral Priority: Merging Mind, Brain, and Human Values* (Oxford: Blackwell, 1983), chap. 6.

[22]The "downward/top-down causation" terminology can, I have found, be misleading since it is actually meant to denote an effect of the state of the system as a whole on its constituent parts, such as the constraints on the parts of the boundary conditions of the system as a whole—more broadly, the constraints of actually being in the interacting, cooperative network of that particular, whole system. The word 'causation' is not really appropriate for describing such situations, so—in this paper—I have preferred the usage of 'whole-part constraint,' rather than that of 'downward-' or 'top-down causation,' which I used more generally in *Theology for a Scientific Age*.

manifested only entirely random motions with respect to each other. In such instances,[23] the changes at the micro-level, that of the constituent units, are what they are because of their incorporation into the system as a whole, which is exerting specific constraints on its units, making them behave otherwise than they would in isolation. Using "boundary conditions" language,[24] one could say that the set of relationships between the constituent units in the complex whole is a *new* set of boundary conditions for those units. There is also, of course, the effect on a system of its total environment (ultimately, the whole universe), since no system is ever truly isolable though the particular *system* effects can usually be distinguished from these.

It is important to emphasize again that recognition of the role of such whole-part constraint in no way derogates from the continued recognition of the effects of its components on the state of the system as a whole (i.e., of "bottom-up" effects). But the need for recognition of the former is greater since hardly anyone since the rise of reductionist scientific methodologies doubts the significance of the latter. Indeed this lack of a proper recognition of whole-part constraint has unfortunately often inhibited the development of concepts appropriate to the more complex levels of the hierarchy of natural systems.

On a critical-realist view of the epistemology of the sciences,[25] this implies that the entities to which the theories and experimental laws refer in our analyses correspond, however inadequately and provisionally, to epistemologically non-reducible features of reality, which have to be taken into account when the system-as-a-whole is interacting both with its parts and with other systems (including human observers). These new features may be deemed putatively to exist at the various levels being studied; that is, they can also have an ontological reference, however tentative.[26] It would then be legitimate to envisage the postulated reality which constitutes a complex system-as-a-whole (the "top" of the "top-down"

---

[23]Another example from another area of science, a computer programmed to rearrange its own circuitry through a robot that it itself controls, has been proposed by Paul Davies in *The Cosmic Blueprint* ([London: Heinemann, 1987], 172-4, fig. 32) as an instance of what he called "downward causation." In this hypothetical (but not at all impossible) system, changes in the information encoded in the computer's software (usually taken as the "higher" level) downwardly cause modifications in the computer's hardware (the "lower" level)—an example of software-hardware feedback.

[24]As does M. Polanyi in *inter alia* "Life's Irreducible Structure," *Science* 160 (1968): 1308-12. See the account in Peacocke, *God and the New Biology*, 23-5.

[25]Peacocke, *Intimations of Reality: Critical Realism in Science and Religion* (Notre Dame, IN: University of Notre Dame Press, 1984), chap. 1. See also *idem*, *Theology for a Scientific Age*, 11-19, for references to other authors.

[26]This over-brief comment refers, of course, to the much debated issue of reductionism and the possibility of emergent realities at higher levels of complexity. My own views are expressed *inter alia* in *Theology for a Scientific Age*, 39-41; and *God and the New Biology*, chaps. 1-2, where references to the extensive literature may be found. See also the next section.

terminology) as exerting a constraint upon its component parts, the realities postulated as existing at those lower levels—while continuing, of course, to recognize the often provisional nature of our attempted depictions of realities at all levels.

3.2 *Evolution*

The pattern of "causal" relationships in biological evolution is interesting in this connection. We are dealing with a process in which a selective system "edits," as it were, the products of direct physico-chemical causation (i.e., changes in DNA) over periods of time covering several reproductive generations. D. Campbell gives an example of this:[27] the surfaces and muscle attachments of the jaws of a worker termite are mechanically highly efficient and their operation depends on the properties of the particular proteins of which the jaws are made. These have been optimized by natural selection. Any particular organism is only one in a series of generations of populations of termites and it is the increasing efficacy of the proteins in constituting efficient jaws that is operative in selection and thereby determines the sequences of the DNA units. Yet when one looks at the development of a *single* organism, one observes only, with the molecular biologists, the biochemical processes whereby protein sequences, and so structures, are "read out" from the DNA sequences. Hence the network of relationships that constitute the temporal evolutionary development and the behavior pattern of the whole organism is determining what particular DNA sequence is present at the controlling point in its genetic material in the evolved organism. This is what Campbell called "downward causation."

It is not adequate to describe such complex interlocking networks of events and changes operating at different levels as *causally* connected in a sequential, constant conjunction of events. We seem rather to have here a determination of form through a *flow of information*, as distinct from a transmission of energy, where "information" is conceived of in a broad enough sense[28] to include the

---

[27]Campbell, "Downward Causation," 181-2.

[28]The relation between the different usages of 'information' has been usefully clarified by J.C. Puddefoot in "Information and Creation," in *The Science and Theology of Information*, ed. C. Wassermann, R. Kirby, and B. Rordoff ([Geneva: Labor et Fides, 1992], 15). He distinguishes three senses relevant to the present context (my numberings):

(i)"Information" in the physicists', communication engineers', and brain scientists' sense, that of C.E. Shannon—the sense in which "information" is related to the probability of one outcome, or case selected, out of many, equally probable, outcomes or cases. In this sense it is, in certain circumstances, the negative of entropy.

(ii)"Information" in the sense of the Latin *informo,-are*, meaning "to give shape or form to." Thus, "information" is "the action of informing with some active or essential quality [sense ii]," as the noun corresponding to the transitive verb 'to inform', in the sense (ii) of "To give 'form' or formative principle to; hence to stamp, impress, or imbue *with* some specific quality or attribute" (quotations from the *Shorter Oxford English Dictionary*

selective input from the environment towards molecular structures—for example, the DNA sequences in the termite jaw example. Such determinative relations may operate between two different kinds of "level" in nature. The determination of form by form requires a flow of information, in this case, between levels.[29]

3.3 *The Brain, Mental Events, and Consciousness*

It is in terms such as these, relevant to our later considerations of God's interaction with the world (section 4) that some neuro-scientists and philosophers have come to speak of the relation between mental events experienced as consciousness and the physico-chemical changes at neurons that are the triggers of observable actions in those living organisms whose brains are sufficiently developed that it is appropriate to attribute to them some kind of consciousness. As John Searle has recently put it:

> Consciousness . . . is a real property of the brain that can cause things to happen. My conscious attempt to perform an action such as raising my arm causes the movement of the arm. At the higher level of description, the intention to raise my arm causes the movement of the arm. At the lower level of description, a series of neuron firings starts a chain of events that results in the contraction of the muscles...the same sequence of events has two levels of description. *Both of them are causally real*, and the higher level causal features are both caused by and realized in the structure of the lower level elements.[30]

For Roger Sperry and Donald MacKay, "mental events" for human beings are the internal descriptions we offer of an actual *total* state of the brain. The total brain state acts as a constraint on what happens at the more micro-level of the individual neurons; thus what occurs at this micro-level is what it is because of the prevailing state of the whole. There is operative here, it is being suggested, a whole-part constraint of one "level" upon another, from that of the brain state as

---

on Historical Principles).

(iii)"Information" in the ordinary sense of "that of which one is apprised or told" (*Shorter Oxford*, sense I.3).

Puddefoot points out that "information (i)" is necessary to shape or give form, as "information (ii)," to a receptor. If that receptor is the brain of a human being, then "information (iii)" is conveyed. In this paper the term "information" (and its associates) is being broadly used to represent this whole *process* of (i) becoming (iii)—and only modulating to (iii) when there is a specific reference to human brain processes.

Although in actual natural systems, there is never a flow of information without some transfer of energy, however small, the *concept* of *information* is clearly distinguishable from that of *energy*.

[29]D.M. MacKay, "The Interdependence of Mind and Brain," *Neuroscience* 5 (1980): 1389-91.

[30]John Searle, *Minds, Brains, and Science* (Cambridge, MA: Harvard University Press, 1984), 26 (emphasis added).

a whole to that of the individual neurons. Descriptions of the total brain state in purely neurological terms would be exceedingly complex and, indeed, considering the complexity of the brain, may never be forthcoming in anything other than broad terms. The causal effectiveness of the whole brain state on the actual states of its component nerves and neurons is probably better conceived of in terms of the transfer of information rather than of energy, in the way a program representing a certain equation, say, controls the chips in a computer—but this whole area of investigation is still very much *sub judice*. (For example, is there a 1:1 correlation between brain states and mental states? Can a mental state be "realized" in a number of different brain states?)

It seems that, with the evolution of brains, this kind of whole-part constraint has become more and more significant in the evolutionary development, as the whole state and behavior of the individual organism itself plays an increasing role. This has also, as we saw, introduced an element of flexibility into the evolutionary process. Furthermore, since the brain-in-the-body is a dissipative system, it now becomes possible to envisage that the actual succession of states of the brain may prove in practice not to be describable in terms of currently available scientific concepts. This would then point to the need for some higher-level concepts (those called "mental"?) to denote and explicate sequences of events in the brain and the "whole-part constraints" operating from this level. Furthermore, as Nancey Murphy has written: "We attribute freedom to the person insofar as the states of the organism are attributable to the person as a whole, involving intentions, desires, etc. So if the brain states are not predictable [I would say "describable"] when considered solely at that [holistic] level, we have evidence that higher-level (free) processes are the determinative factor."[31]

4 *God's Interaction with the World in the Light of these Scientific Perspectives*

4.1 *Unpredictability, Open-Endedness, and Flexibility*[32]

The world appears to us less and less to possess the predictability that has been the presupposition of much theological reflection on God's interaction with the world since Newton.[33] We now observe it to possess a degree of openness and flexibility

---

[31]In a personal note, January 24, 1994. See also, Peacocke, "Natural Being and Becoming."

[32]This section represents an elucidation of *Theology for a Scientific Age*, 152-6, and a tightening up of the argument, by making more careful distinctions between "in principle" and "in practice" unpredictability with respect to deterministic chaos and quantum theory and to whether or not God or ourselves are under consideration.

[33]Here, and in what follows, by both 'world' and 'all-that-is' I intend to refer to everything apart from God, both terrestrially and cosmologically. I regard the world/all-that-is, in this context, as an interconnected and interdependent system, but with, of course, great variation in the strength of mutual inter-coupling. All wave functions, e.g., only go

within a lawlike framework, so that certain developments are genuinely unpredictable *by us* on the basis of any conceivable science. We have good reasons for saying, from the relevant science and mathematics, that this unpredictability will, in practice, continue.

The history of the relation between the natural sciences and the Christian religion affords many instances of a human inability to predict being exploited by theists postulating the presence and activity of God to fill the explanatory gap. However, as these gaps were filled by new knowledge, "God" as an explanation became otiose. Do we now have to take account of, as it were, *permanent* gaps in our ability to predict events in the natural world. Does this imply there is a "God of the (to us) *uncloseable* gaps"? There would then be no possibility of such a God being squeezed out by increases in scientific knowledge. This raises two theological questions: (1) "Does God know the outcome of these situations/ systems that are unpredictable by us?" and (2) "Does God act within such situations/systems to effect the divine will?"

4.1.1 *Non-Quantum Considerations*

We will first respond to these questions *excluding quantum theory considerations*. With respect to (1), an omniscient God may be presumed to know, not only all the relevant, deterministic laws which apply to any system (such as those discussed in sections 2.3 and 2.4), but also all the relevant initial conditions of the determining variables to the degree of precision required to predict its state at any future time, however far ahead, together with the effects of any external influences from anywhere else in the universe, however small. So there could be no "eventual unpredictability" with respect to such systems for an infinite, omniscient God, even though there is such a limiting horizon for finite human beings—because of the nature of our knowledge of real numbers and because of ineluctable observational limitations. To take a particularly significant example, divine omniscience must be conceived to be such that God would know and be able to track the minutiae of the triggering fluctuations in dissipative systems, unpredictable and unobservable by us, whose amplification leads at the macroscopic level to one particular, macroscopic outcome (e.g., a symmetry-breaking) rather than another—consequently also unpredictable by us.

Only if we thus answered (1) affirmatively, could we then postulate that God might choose to influence events in deterministic systems in the world by changing the initial conditions so as to bring about a macroscopic consequence conforming to the divine will and purposes—that is, also to answer (2) affirmatively. God would then be conceived of as acting, as it were, "within" the flexibility we find in these (to us) unpredictable situations in a way that could never be detected by us. Such a mode of divine action would never be inconsistent with our scientific

---

asymptotically to zero at infinity—and recall the effects of gravity from distant galaxies even on the collision of billiard balls (section 2.3), not to mention the ecological connectedness of terrestrial life within itself and with the state of the earth. (See also section 4.2 and n. 42.)

knowledge of the situation. In the case of those dissipative systems whose macro-states (often involving symmetry-breaking) arise from the amplification of fluctuations at the micro-level that are unpredictable and unobservable by us, God would have to be conceived of as actually manipulating micro-events (at the atomic, molecular and, according to some,[34] quantum levels) in these initiating fluctuations in the natural world in order to produce the results at the macroscopic level which God wills.

But such a conception of God's action in these, to us, unpredictable situations would then be no different in principle from that of God *intervening* in the order of nature with all the problems that that evokes for a rationally coherent belief in God as the creator of that order. The only difference in this proposal from that of earlier ones postulating divine intervention would be that, given our recent recognition of the actual unpredictability, on our part, of many natural systems, God's intervention would always be hidden from us.

Thus, although at first sight this introduction of unpredictability, open-endedness and flexibility into *our* picture of the natural world seems to help us to suggest in new terminology how God might act in the world in now uncloseable "gaps," the above considerations indicate that such divine action would be just as much "intervention" as it was when postulated before we were aware of these features of the world. This analysis has, it must be stressed, been grounded on the

---

[34]In a paper in the previous volume in this series ("Divine Action, Freedom, and the Laws of Nature," in *Quantum Cosmology and the Laws of Nature: Scientific Perspectives on Divine Action*, ed. Robert John Russell, Nancey Murphy, and C.J. Isham [Vatican City State: Vatican Observatory, 1993; Berkeley, CA: Center for Theology and the Natural Sciences, 1993], 185-207), W.P. Alston reverted to an earlier idea of William Pollard, in *Chance and Providence: God's Action in a World Governed by Scientific Law* (New York: Scribner, 1958), that God can act at the quantum level because quantum theory predicts only probabilities of particular outcomes of a given situation. He wrote, "God can, consistent with quantum theory, do something to bring about a physically improbable outcome in one or more instances without any violation of physical law" (p. 189). For reasons outside the scope of this particular paper I am skeptical of this proposal, as I am of the similar ones also advanced in this volume by George Ellis ("Ordinary and Extraordinary Divine Action: The Nexus of Interaction") and, possibly, also those of Nancey Murphy ("Divine Action in the Natural Order: Buridan's Ass and Schrödinger' Cat").

Later (p. 190) Alston argues that at levels other than the quantum one—i.e., at levels where there are genuinely deterministic laws relating outcomes to initial situations—God could be an additional factor, "a divine outside force," not previously allowed for by the science involved. But this is vulnerable to the criticism that such a "force" cannot but be in the natural order to be effective—and therefore amenable to scientific investigation. But it is these very investigations which unveil the deterministic laws which demonstrate that such arbitrary occurrences do not occur. So such "intervention" is an incoherent idea, as well as casting doubts on the foundations on which the very existence of a divine creator can be postulated, namely the regularity of that ordering of the natural world which the sciences reveal and from which is inferred that it manifests and expresses an endowed rationality.

assumption that God *does* know the outcome of natural situations that are unpredictable by us; that is, on an affirmative answer to (1). It assumes total divine omniscience about all actual, natural events.

4.1.2 *Quantum Theory Considerations*

Consideration of the foregoing in the light of quantum theory cannot avoid the continuing current disagreements concerning the basis and significance of the quantum uncertainties which are expressed in the HUP, which qualifies total predictability, however interpreted. The broad possibilities may be delineated as follows:

(i) *There are "hidden variables"*—that is, there are underlying deterministic laws, unknown to us, which govern the time-course of the precise values of the variables (momentum, position, energy, time, etc.) appearing in the HUP. The uncertainties, the "fuzziness," in our knowledge of the values of these variables is purely an *epistemological* limitation on our part, which would not also be one for an omniscient God. Such a God would know both these laws and the relevant initial conditions. Hence the conclusions about how God might interact with the world which were drawn in the previous section (4.1.1) would still apply.

(ii) *No "hidden variables"*[35]—that is, the epistemological limitations expressed in the HUP can never be obviated, not only in practice but also in theory, and represent a fundamental uncertainty that inherently exists in the values of the variables in question—an ontological claim that there is indeterminism with respect to these variables. The future trajectory of *any* system will always inherently have that unavoidable lack of precise predictability, given by the HUP relations, with respect to these variables—it is genuinely *in*deterministic in these respects (if not in all, e.g., in the statistical properties of the ensemble). Only a probabilistic knowledge of these variables is possible for us, but this limitation is insurmountable. It represents an "in principle" limitation. This is the majority view of physicists.

If this is so, we would have to conclude that this inherent unpredictability also represents a limitation of the knowledge even an omniscient God could have[36] of the values of these variables and so of the future trajectory, in those respects, of the system. That is to say, God has so made the quantum world that God has allowed God's own *possible* knowledge to be thus limited. In this regard, then, God's

---

[35]The positive evidence for the absence of, at least, *local* "hidden variables" is discussed in a way accessible to the general reader by John C. Polkinghorne, *The Quantum World* (Harmondsworth: Penguin Books, 1986), chap. 7. Whether or not such "hidden variables" can, more generally, be regarded as absent is still a much discussed question, though the majority view is against their existence.

[36]See also Peacocke, *Theology for a Scientific Age*, 122, for a discussion of God's "self-limitation."

omniscience is "self-limited."[37] God's knowledge with respect to HUP variables in future states would be the maximum it could be compared with ours, but would nevertheless still be only probabilistic. Moreover, if the future, as I and others have argued[38] has no ontological status—that is, does not exist in any sense— then it has no content of events *for* God to know, so it logically cannot be known even to an omniscient God, who knows all that it is possible to know. According to this view, God knows the future *definitively* only by prediction on the basis of God's omniscient knowledge of all determining laws and an infinitely precise knowledge of all initial relevant conditions (as in 4.1.1, above); or *probabilistically* in the case of quantum-dependent events (for God cannot predict in detail the outcome of in-principle unpredictable situations, on the no-hidden-variables assumption). This conclusion about the basis of God's foreknowledge would still apply, even if it is thought that God acts in the world by altering the initial conditions of a train of events (including possibly individual quantum events, as other authors in this volume have proposed [see n.33]) to obtain the outcome God wills and so must foresee.

An easily envisaged example, related to the relations expressed in the HUP, is afforded by radioactive decay, in which a quantum event has an observable macroscopic outcome in the decay of the atoms. In this case, the foregoing is arguing that God does not know *which* of a million radium atoms will be the next to disintegrate in, say, the next 10^{-3} seconds, but only (as we ourselves) what the average number will be that will break up in that period of time. There is no fact of the matter about which atom will decompose at a particular future moment *for* God to know. The proposal of "self-limiting" omniscience means that God has so made the natural order that it is, in principle, impossible, even for God, as it is for us, to predict the precise, future values of certain variables—which is what I take "in principle" to mean in this context. God's omniscient knowledge of the *probabilities* of these future values will, of course, always be maximal.

Hence, in the case of systems sensitive to initial conditions, the introduction of quantum uncertainty introduces an upper limit to predictability with respect to

---

[37]See ibid., 91-94 and 121-23. In the latter passage (p. 122), I included in the category of systems whose future states cannot be definitively known even to God (since they are in principle unknowable) not only the operation of human free will and (quantum) systems in the "Heisenberg" range, but also "certain non-linear systems at the microscopic level." I had in mind here (and was insufficiently precise in specifying) those non-linear systems in which effects of *quantum* events can be amplified to the macroscopic level so that, in principle, knowledge only of the *probabilities* of occurrence of particular macroscopic states is possible—knowledge maximally available to God. (Such systems were mentioned also in the penultimate paragraph of section 2.4.) If "quantum chaos," in the strict sense defined by Ford (end of section 2.4, and n. 11) proves to be possible, further reflection on God's knowledge of the future states of systems in which it occurs will be needed in the light of its precise epistemological character.

[38]See Peacocke, *Theology for a Scientific Age*, 128-133, which gives references to other authors who also hold this view. See also n. 45 below.

certain parameters which cannot be avoided.[39] This limit on total predictability applies to God as well as to ourselves, if there are no hidden variables. So the answer to the theological question (1) then has to be, in the light of such quantum considerations, that *God* also cannot know, beyond real limits, the outcomes of those situations, the trajectories of those systems which are also in principle (if no hidden variables exist) unpredictable for us beyond those same limits. God, of course, knows maximally what it is possible to know, namely the *probabilities* of the outcomes of these situations, the various possible trajectories of such systems. But this does not suffice for us to give a clear affirmative answer to question (2) to the effect that God could act in such situations or systems to implement the divine will.

On this, to some no doubt revisionary, view God bestows a certain autonomy not only on human beings, as Christian theology has long recognized, but also on the natural order as such to develop in ways that God chooses not to control in detail. God allows a degree of open-endedness and flexibility in nature, and this becomes the natural, structural basis for the flexibility of conscious organisms and, in due course and more speculatively, possibly for the freedom of the human-brain-in-the-human-body, that is, of persons (see above, end of section 3.3). So it does help us to perceive the natural world as a matrix within which openness and flexibility and, in humanity, perhaps even freedom could naturally emerge.

4.1.3 *Implications of "Chaotic" Determinism*

One set of considerations (section 4.1.1, concerned with the infinite decimal representation of real numbers and algorithmic complexity) only reinforces the rejection of an already long-rejected interventionism[40] as the basis for God's interaction with the world to influence events. The other set of considerations (section 4.1.2 [ii], concerned with the HUP and its consequences if there are no hidden variables) implies that God cannot know precisely the future outcome of quantum dependent situations,[41] so cannot act *directly* to influence them in order to implement the divine purpose and will, as we may be tempted to postulate. It should be noted that this does not derogate at all from God having purposes which are being implemented through the propensities (to complexity, self-organization, information-processing, and consciousness) that load, as it were, the dice whose throws shape the course of natural events.

The above discussion leads us to infer that this newly-won awareness of the unpredictability, open-endedness, and flexibility inherent in many natural processes and systems does not, of itself, help directly to illuminate the "causal

---

[39]Note that the argument being pursued here does not depend on establishing that "quantum chaos," in the strict sense of Ford's definition (section 2.4, last par., and n. 12) actually occurs.

[40]See the comments on "divine intervention" at the end of n. 33 above.

[41]These would also include the future trajectories of systems displaying "chaotic" behavior if "quantum chaos" proves to be theoretically feasible (see end of section 2.4).

joint" of how God acts in the world, that is, the nature of the interface between God and all-that-is—much as it alters our interpretation of the meaning of what is actually going on in the world. Defining the problem (à la Austin Farrer) as that of the "causal joint" between God and the world is inappropriate for it does not do justice to the many levels in which causality operates in a world of complex systems interlocking in many ways at many levels. It is to this major feature of the world as perceived by the sciences that we must now turn.

4.2 *Whole-Part Constraint as a Model for God's Interaction with the World*

In a number of natural situations, interactions within complex systems constituted of complex sub-systems at various levels of interlocking organization can best be understood as a two-way process. Real features of the total system-as-a-whole are constraints upon events happening within the sub-systems at lower levels—events, which, it must be stressed, in themselves are describable in terms of the sciences pertinent to that lower level. In the light of this it is suggested that we can properly regard the world-as-a-whole as a total system so that its general state can be a holistic constraint[42] upon what goes on at the myriad levels that comprise it. For all-that-is displays, with wide variations in the degree of coupling, a real interconnectedness and interdependence at the quantum, biological, and cosmological levels and this would, of course, be totally and luminously clear to God in all its ramifications and degrees of coupling.[43]

I want now to explore the possibility that these new perceptions of the way in which levels within this world-system interact with each other (from higher to lower and *vice versa*) might provide a new resource for thinking about how God interacts with that world-as-a-whole. In making such a suggestion I am not postulating that the world *is*, as it were "God's body," since, although the world is not organized in the way a human body is, it is nevertheless a "system." The world-as-a-whole, the total world system, may be regarded as "in God,"[44] though ontologically distinct from God. For God is uniquely present to it all, all its individual component entities, in and at all spaces and all times (in whatever

---

[42]As pointed out by Russell in this context, the notion of whole-part constraint does not depend critically on the idea of "boundary conditions" of a complex system, with its implicit phase-spatial connotations—and indeed the universe does not have a boundary in *that* sense. I am also indebted to him for some of the wording in the sentences that follow.

[43]The interconnectedness and interdependence of all-that-is would be infinitely more apparent to God. For God holds it all in existence; is present to all space and time frameworks of reference; and all-that-is is "in God," on the, properly qualified, "pan-en-theistic" model which I espouse (see Peacocke, *Theology for a Scientific Age*, 370-72, n. 75, and associated text, as well as figure 1).

[44]A "pan-en-theistic" view the sense of which I have explicated further, with references to other authors, in *Theology for a Scientific Age* (references in n. 42 above).

relativistic frame of reference[45]) and has an unsurpassed awareness of its interconnected and interdependent unity—even *more* that we can have of the unity of our own bodies. If God interacts with the "world" at a supervenient level of totality, then God, by affecting the state of the world-as-a-whole, could, on the model of whole-part constraint relationships in complex systems, be envisaged as able to exercise constraints upon events in the myriad sub-levels of existence that constitute that "world" without abrogating the laws and regularities that specifically pertain to them—and this without "intervening" within the unpredictabilities we have noted.[46] *Particular* events might occur in the world and be what they are because God intends them to be so, without at any point any contravention of the laws of physics, biology, psychology, sociology, or whatever is the pertinent science for the level of description in question.

In thus speaking of God, it has not been possible to avoid talk of God "intending," and so using the language of personal agency. For these ideas of whole-part constraint by God cannot be expounded without relating them to the concept of God as, in some sense, an agent, least misleadingly described as personal. In thus speaking, we are focusing upon *particular* events, or patterns of events, as expressive of the "purposes" (e.g., of communication) of God, who is thereby conceived of as in some sense personal. Such *particular* intentions of God must be distinguished from that perennial sustaining in existence of the entities, structures, and dynamic processes of the world, which is an inherent component of all concepts of God as creator. This sustaining is properly regarded as "continuous," an aspect of God as *semper creator* with respect to the *creatio continua*. What is being further suggested here is that we have to envisage God as at *any* time (and in this sense only, "all the time") being able to exert constraints upon the world-as-a-whole, so that *particular* events and patterns of events can occur, which otherwise would not have done so. This is usually regarded as God's "providential" action, unhelpful as the distinction between creation and providence often proves to be.

---

[45]For further elaboration of the understanding of God's relation to time assumed in this point, see *Theology for a Scientific Age*, 128-33. I there summarized my position in the following terms: "God is not 'timeless'; God is temporal in the sense that the divine life is successive in its relation to us—God is temporally related to us; God creates and is present to each instant of the (physical and, derivatively, psychological) time of the created world; God transcends past and present created time: God is eternal," (p. 132) in the sense that there is no time at which God does not exist nor will there ever be a future time at which God does not exist.

For a discussion of the issues concerning the "block universe" model used in relativistic physics and their implications for the God-time relation, see the illuminating discussion of Isham and Polkinghorne, "The Debate over the Block Universe," in *Quantum Cosmology*, 135-44. I take the view of the latter in this debate.

[46]Both of the (for us) practical kind (e.g., chaotic systems) and of the in-principle, inherent kinds (e.g., certain quantum events).

4.3 *Personal Agents as Psychosomatic Unities*

The way in which, in the preceding, we have found ourselves drawn towards the model of *personal* agency in attempting to explicate God's interaction with the world is intriguing in the contemporary context—and not only because of its biblical and traditional role. For in one particular instance of a system manifesting whole-part constraint, the human-brain-in-the human-body, we have an immediate sense of the non-reducibility of the whole—in our "consciousness," as folk psychology calls it.[47] For, over recent decades, the pressure from the relevant sciences has been inexorably towards viewing the processes that occur in the human brain and nervous system, on the one hand, and the content of consciousness, our personal, mental experience, on the other, as two facets or functions of one total unitive process and activity.[48] We have already seen that combining a non-dualist account of the human person and of the mind-body relation with the idea of whole-part constraint illuminates the way in which states of the brain-as-a-whole could have effects at the level of neurons, and so of bodily action; and could actually also be holistic states of the brain-as-a-whole. Such states could be legitimately referred to in non-reducible mentalist language as a real modality of the total unitive event which is the activity of thinking that is accomplished by the human-brain-in-the-human-body.

This invoking of the notion of whole-part constraints of brain states as a whole upon the states of the "lower" level of its constituent neurons in giving an account of human agency, affords, I would suggest, a new insight into the nature of human agency very pertinent to the problem of how to model God's interaction with the world. My suggestion is that a combination of the recognition of the way whole-part constraints operate in complexly interconnected and interdependent systems with the recognition of the unity of the human mind/brain/body event, together provide a fruitful model for illuminating how we might think of God's interaction with the world. According to this suggestion the state of the totality of the world-as-a-whole (all-that-is) would be known maximally only to the omniscience of God and would be the field of the exercise of the divine omniscience at God's omnicompetent level of comprehensiveness and comprehension.[49] When we act as personal agents, there is a unitive, unifying, centered constraint on the activity of our human bodies, which we experience as the content of our

---

[47]I am indebted, at this point especially in this paper, to the wording of some illuminating comments by Philip Clayton.

[48]See, *inter alia*, Colin Blakemore, *The Mechanics of the Mind* (Cambridge: Cambridge University Press, 1976); *The Brain: Readings from* Scientific American *Magazine* (San Francisco: Freeman, 1979); *Mind and Brain: Readings from* Scientific American *Magazine* (Oxford: Freeman, 1992); and the comprehensive information in *The Oxford Companion to the Mind*, ed. Richard L. Gregory (Oxford: Oxford University Press, 1987).

[49]With the qualification of the self-limitations of divine omniscience and divine omnipotence already made (See Peacocke, *Theology for a Scientific Age*, 121-3).

personal subjectivity (the sense if being an "I") in its mode of willing action. God is here being conceived of as a unifying, unitive source and centered influence on events on the world.[50]

We are here courting the notion that the succession of the states of the system of the world-as-a-whole is also experienced as a succession by God, who is present to it all; and that this might be modeled after the way we presume a succession of brain states constitutes a succession in our thoughts. God would then be regarded as exerting a continuous holistic constraint on the world-as-a-whole in a way akin to that whereby in our thinking we influence our bodies to implement our intentions. This suggestion is, for me at least, entirely metaphorical, providing only a model for God's interaction with the world and thereby enabling us to conceive coherently and intelligibly how God might be conceived of as interacting with the world consistently with what we know of its nature and with the character of God already inferred on other grounds. As such, therefore, it has its limitations, indeed—with all such attempts—an inevitably negative aspect. For, in a human being, the "I" does not transcend the body ontologically in the way that God transcends the world and must therefore be an influence on the world-state from "outside" in the sense of having a distinctively different ontological status.[51] But at least the suggested model helps us to conceive how God's transcendence and immanence might be held coherently together as a transcendence-in-immanence.

This now affords a further clue to how that continuing interaction of God with the world-as-a-whole which implements particular divine purposes might best be envisaged—namely as analogous to an input, a flow of information, rather than of energy.[52] For different, equally probable, macroscopic states of a system—and so, in the model, of the world-as-a-whole—can possess the same energy but differ in form and pattern, that is, in information content (cf. n.27). Moreover, since God is properly regarded by most theists as in some sense "personal," this "flow of information" may be more appropriately envisaged as a means of *communication* by God of divine purposes and intentions when it is directed towards that level in the hierarchy of complexity which is uniquely capable of perceiving and recognizing it, namely, humanity.[53]

---

[50]Do we have here one aspect of humanity as *imago dei*?

[51]The "ontological gap at the causal joint." See n. 51 below.

[52]To the best of my knowledge the first application of the concept of information to the interaction of God and the world was made by John Bowker in *The Sense of God: Sociological, Anthropological, and Psychological* (Oxford: Clarendon Press, 1973), chap. 5: "Structural Accounts of Religion." See also the expression in similar terms of this idea, of God "informing" the world in a "downward/top-down" manner, by Polkinghorne, *Science and Providence*, 32 ff.; *idem, Reason and Reality: The Relationship between Science and Theology* (London: SPCK Press, 1991), chap. 3; and Peacocke, *Theology for a Scientific Age*, 161, 164, 179, and 207.

[53]The consequences of this I have tried to develop in *Theology for a Scientific Age* (1993 ed.), Part 3.

5 *Conclusion*

The foregoing suggests a way in which we could think of divine constraints (properly called "influences," to cohere with the model of *personal* agency) making a difference in the world, yet not in any way contrary to those regularities and laws operative within the observed universe, which are explicated by the sciences applicable to their appropriate levels of complexity and organization. This holistic mode of action on and influence in the world is God's alone and distinctive of God. God's interaction with the whole and the constraints God exerts upon it could thereby shape and direct events at lesser levels so that the divine purposes are not ultimately frustrated. Such interaction could occur without ever abrogating at any point any of the natural relationships and inbuilt flexibilities and freedoms operative at all of the lower levels, and discerned by the sciences and ordinary human experience.

Only God in the mode of transcendence is present to the totality of all-that-is, as well as, in the mode of immanence, to the individual entities that comprise created existence. Accordingly, God's experience is of the world-as-a-whole as well as of individual entities and events within it. Only God could be aware of the distinctiveness of any state of that totality and which of its states might or might not succeed it in time (or whatever is the appropriate dimension for referring to "succession in God"). This divine knowledge would always be hidden from and eternally opaque to us, existing as we do at levels at which the conceptual language will never be available for apprehending God's own "inner" life. The best we can do, as we have already urged, is to stretch the language of personal experience as the least misleading option available to us. According to this approach, we are free to describe any particular events at our own level of existence in the natural terms available to us (e.g., in those of the sciences explaining both the "bottom-up" and whole-part effects within the natural order); and at the same time we are free to regard at least some of those events, whether private and internal to us or public and external to all, as putatively and partially manifesting God's intentions, God's providence, and so as being communications from God. For God could have brought it about that these particular events are what they are and not something else by that overall comprehensive constraining influence which only God can exert (but does not *necessarily* do so) in a whole-part manner upon any lower-level event occurring in the totality of existing entities in order to implement divine intentions, such as communicating with humanity.

God, I am suggesting, is thus to be conceived of as all the time the continuing supra-personal, unifying, unitive Agent acting, often selectively, upon all-that-is, as God's own self purposes. We must go on recognizing—and this is essential to the whole proposal— that, in the light of our earlier discussion, it is God who has chosen to allow a degree of unpredictability, open-endedness, and flexibility in the world God continues to hold in existence and through whose processes God continues to create; and that God, so conceived, does not intervene to break the causal chains that go from "bottom-up," from the micro- to the macro-levels.

What does this imply about the "causal joint" between God and the world? As already mentioned (n.27), in the world we observe through the sciences, we know of no transfers of information without some exchange of matter and/or energy, however minimal. So to speak of God as "informing" the world-as-a-whole without such inputs of matter/energy (that is, as not being "intervention") is but to accept the ultimate, ontological gap between the nature of God's own being and that of the created world, all-that-is apart from God. Hence the present exercise could be regarded essentially as an attempt, as it were, to ascertain where this ontological gap, across which God transmits "information" (i.e., communicates) is most coherently "located," consistently with God's interaction with everything else having *particular* effects and without abrogating those regular relationships to which God's own self continues to give an existence which the sciences increasingly discover.

I would want to emphasize, with Gordon Kaufman[54] and Maurice Wiles,[55] that God's action is on the world-as-a-whole, but to stress more strongly than they do that this maintaining and supporting interaction is a continuing as well as an initial one, and can be general *and particular* in its effects. The freedom of God to affect the world is indeed reinforced and protected in this model. For the notion of whole-part constraint now allows us to understand how initiating divine action on the state of the world-as-a-whole can itself have consequences for individual events and entities within that world. Moreover such divine causative, constraining influence would never be observed by us as a divine "intervention," that is, as an interference with the course of nature and as a setting aside of its natural, regular relationships.

The proposed model allows the effects of natural events, including the unpredictable ones and the outcome of freely-willed human decisions, to work their way up through the hierarchy of complexity and so to contribute to the state of the world-as-a-whole. It therefore also helps us to model more convincingly the interaction, dialogue even, between human decisions and actions, on the one hand, and divine intentions and purposes, on the other. It is in such a context that the notion of God communicating with humanity can be developed in which the significance of religious experience, revelation, the incarnation, prayer, worship, and the sacraments may be grounded.[56]

In conclusion, it would seem that, the unpredictabilities of non-linear dynamic systems do not as such help us in the problem of articulating more coherently and intelligibly how God interacts with the world. Nevertheless recent insights of the natural sciences into the processes of the world, especially those on whole-part constraint in complex systems and on the unity of the human-brain-in-the-human-body, have provided not only a new context for the debate about how God might be conceived to interact with, and influence, events in the world but have also afforded new conceptual resources for modeling it.

---

[54]Gordon D. Kaufman, *God the Problem* (Cambridge, MA: Harvard University Press, 1972).

[55]Maurice Wiles, *God's Action in the World:* The Bampton Lecture for 1986 (London: SCM Press, 1986).

[56]See Peacocke, *Theology for a Scientific Age* (1993 ed.), Part 3.

PARTICULAR PROVIDENCE AND THE GOD OF THE GAPS

Thomas F. Tracy

1 *Posing the Question*

The "God of the gaps" has not fared well in modern theology. This vivid pejorative phrase has come to characterize the result of a theology that eagerly (perhaps desperately) rushes forward to insert God at those points where human understanding reaches its limits. The shortcomings of this theological maneuver are well-expressed in some of Dietrich Bonhoeffer's later prison letters. Commenting on the experience of reading a new book on modern physics, he remarked that:

> It has again brought home to me quite clearly how wrong it is to use God as a stop-gap for the incompleteness of our knowledge. If in fact the frontiers of knowledge are being pushed further and further back (and that is bound to be the case), then God is being pushed back with them, and is therefore continually in retreat. We are to find God in what we know, not in what we don't know. . . . That is true of the relationship between God and scientific knowledge, but it is also true of the wider human problems of death, suffering, and guilt.[1]

Bonhoeffer contended that the appeal to God at the limits of human comprehension (whether scientific or existential) is part of an attack by "religion" on human maturity and strength, and it is bound to end in embarrassing failure in a world increasingly come of age.

Bonhoeffer's warning against motivating religious faith from human existential boundaries has been largely ignored by contemporary theologians, but his critique of the God who fills scientific gaps has been almost universally embraced. It clearly has not been a winning strategy for theologians to pounce upon points at which scientific explanations are currently incomplete and insist that here at last we see the hand of God at work in the world. The all but inevitable result of this maneuver is nicely illustrated by the fate of Newton's contention that God steps in occasionally to correct accumulating inaccuracies in the orbits of the planets. A century later, and with better calculations in hand, Laplace was able to make his famous remark that he had "no need of that hypothesis." It makes no sense to persist in planting the banner of theology at the boundaries of the known world, only to be perpetually forced into retreat as new territory is opened up to human understanding.

---

[1]Dietrich Bonhoeffer, *Letters and Papers from Prison*, ed. Eberhard Bethge, enlarged ed. (New York: Macmillan, 1979), 311.

Contemporary theologians, therefore, have been understandably reluctant to adopt an account of God's relation to the world that appeals to the incompleteness of scientific explanations. Any theological reliance upon gaps in our understanding of the natural order may look like a return to the discredited apologetic strategy that Bonhoeffer described. It is important, on both scientific and theological grounds, not to embrace this conclusion without further reflection, however. We need to think a bit more about the sorts of "gaps" that can occur in scientific understandings of the world. And we need to consider the conceptual requirements of a theology which speaks of the God who acts in history.

1.1 *Explanatory and Causal Gaps*

The first point to note is that not all gaps are the same, and it is important to make some distinctions. In the broadest sense, "*explanatory gaps*" occur whenever we are unable to give a complete account of the sufficient conditions for an event. Defined in this way, such gaps are a commonplace of inquiry; indeed, given our cognitive limits, we cannot fully spell out the sufficient conditions for anything that happens in the world. The explanatory gaps that are of principal interest for my purposes are those that occur either when (1) we must acknowledge that we do not yet have theories or explanatory strategies adequate to answer questions raised by our scientific inquiries, or when (2) our theories entail that human knowers will not in principle be able to give a sufficient explanation of some of the events that fall within the theory's scope. In the discussion that follows, I will refer to gaps that arise in these two circumstances as, respectively, explanatory limits in practice and in principle. The question of how to classify any particular instance of explanatory incompleteness may, of course, be a matter of scientific controversy. It might be argued, for example, that a putative instance of the second type (a limit in principle) is really just an instance of the first (a limit in practice).

Explanatory gaps of this second type *may*, but need not always, be understood to reflect "*causal gaps*." Causal gaps are breaks in the order of what is commonly called "event causation" (or "transeunt causation") which occur when events are not uniquely determined by their antecedents. Where causal gaps occur, later events cannot be deduced from a description of their antecedents and deterministic laws of nature.[2] As a result, causal gaps entail a particular sort of explanatory gap.[3] It is important to emphasize that the entailment does *not* run in

---

[2]On covering law, or deductive nomological, explanation see Carl Hempel, *Aspects of Scientific Explanation, and Other Essays in the Philosophy of Science* (New York: Free Press, 1965).

[3]Note that an event which cannot be explained in terms of a deterministic covering law may nonetheless have some other explanation that *is* sufficient. If the event is an incompatibilist free action, for example, it may have an adequate explanation in terms of the agent's reasons for acting. Further, if one subscribes to a concept of "agent causation" which is irreducible to event causation, then although there cannot be a complete explanation of the act in terms of antecedent events and the laws of nature, the act will have a cause (i.e., the

the other direction; not all explanatory gaps (even those that reflect limits in principle) will indicate the presence of causal gaps, and we must be very cautious about concluding from the former to the latter. Hasty inferences of this kind are one of the fatal shortcomings of a God of the gaps strategy in theology.

It now appears that explanatory gaps in principle occur within some of the most powerful contemporary physical theories. Quantum mechanics and, more recently, chaos theory provide the two leading candidates. The Heisenberg uncertainty principle acknowledges intrinsic limits on our ability to specify the state (e.g., both position and momentum) of quantum systems, and therefore to explain how later states arise from their antecedents. This explanatory incompleteness is subject to various interpretations, but it cannot be avoided within current theoretical structures, as J. von Neumann has shown. Many physicists are skeptical about the prospects for new theoretical proposals that would eliminate indeterminacy and show that it was a function of human ignorance (i.e., that it reflected an explanatory limit in practice but not in principle).[1] Further, one well-established interpretation of this explanatory incompleteness takes it to reflect a causal gap in the order of nature. On this view, there is no underlying deterministic process that would explain the behavior of the quantum mechanical system; the probability equations tell us as much as can be known about the relation of past, present, and future states of the system. This was Heisenberg's own later understanding of his uncertainty principle.[2]

Chaos theory presents a significantly different example of explanatory incompleteness. A non-linear dynamic system operating according to the laws of classical mechanics can display an extraordinary sensitivity to its initial, or boundary, conditions. Initial conditions that are only minutely different can produce widely divergent states of the system. Our ability to predict these future states, therefore, is determined in part by the accuracy of our specification of its initial conditions. In order to apply the relevant deterministic laws and predict states of the system that are significantly distant from the starting point, we would need to specify the initial conditions with an accuracy that exceeds our epistemic limits. Once the system has moved through a specific number of "generations"

---

agent). The idea of agent causation is both ancient and controversial. For recent discussions of it, see Roderick M. Chisholm, "The Agent as Cause," in *Action Theory: Proceedings of the Winnipeg Conference of Human Action*, ed. Myles Brand and Douglas Walton (Dordrecht: D. Reidel, 1976), 199-211; and Peter van Inwagen, *An Essay on Free Will* (Oxford: Oxford University Press, 1983), 152.

[1]David Bohm's hidden variable theory is the best known of such proposals. See Bohm, *Causality and Chance in Modern Physics* (Princeton, NJ: D. Van Nostrand, 1957); and *idem*, *Wholeness and the Implicate Order* (London: Routledge & Kegan Paul, 1980).

[2]For an overview of quantum physics and a description of leading interpretations of indeterminacy see Robert John Russell, "Quantum Physics in Philosophical and Theological Perspective," in *Physics, Philosophy, and Theology: A Common Quest for Understanding*, ed. Robert John Russell, William R. Stoeger, and George V. Coyne (Vatican City State: Vatican Observatory, 1988), 343-74.

(fixed by the accuracy of our description of the starting point), it ceases to be predictable for us. Chaos theory, therefore, generates an explanatory gap of the second type (i.e., a gap that cannot in principle be closed). Note that it does so, however, precisely by carrying a thoroughly deterministic analysis as far as possible, given our cognitive limits. Because the chaotic system is described by a deterministic equation, we have no grounds for claiming that there are underlying causal gaps in the system. Thus, while chaos theory and quantum theory both involve explanatory gaps in principle, they differ in that the gaps in chaos theory do *not* indicate the presence of causal gaps in chaotic systems, whereas the gaps in quantum theory indicate the *possible* presence of causal gaps in quantum systems.[3]

It remains to be seen whether either of these contemporary sources of incompleteness in our understanding of the world has any significance for theology. That question hinges upon a set of theological considerations to which we are about to turn. But it is worth noting two preliminary points. First, if these scientific theories establish themselves as integral elements in our contemporary understanding of nature, then theologians who wish to develop an account of God's activity in the world will have to take these gaps into account. Second, when a theologian does so, s/he will not simply be returning to the earlier "God of the gaps" theology, which sought to exploit explanatory gaps of the first type and claimed without scientific warrant that these represented causal gaps. Rather, the theologian will be working with explanatory gaps of the second type, which (*ex hypothesi*) represent fundamental limitations upon what we may know (within, of course, the terms of the theories that recognize these gaps).

Recall Bonhoeffer's remark that "we are to find God in what we know, rather than what we don't know." One effect of quantum mechanics and chaos theory, if they turn out to be correct in their essentials, is that they add to our knowledge of the world precisely by pointing out limits in principle on what we may know and by explaining why we cannot know it. We cannot draw theological conclusions directly from this scientific development. But if we are concerned to construct, on theological grounds, a theistic understanding of the world, it is legitimate to incorporate (along with the rest of what we currently think we know) this knowledge of our ignorance.

1.2 *Theology without Gaps?*

There is a second, and crucial, reason contemporary theologians ought not summarily to dismiss appeal to gaps in scientific accounts of the world: theology may not be able to get along altogether without gaps. The considerations here are

---

[3]On chaos theory see, e.g., James P. Crutchfield, et. al., "Chaos" (in this volume); and Joseph Ford, "What Is Chaos, that we should be mindful of it?," in *The New Physics*, ed. Paul Davies (Cambridge: Cambridge University Press, 1989). See also James Gleick, *Chaos: Making a New Science* (London: Heinemann, 1988); and Ian Stewart, *Does God Play Dice? The Mathematics of Chaos* (Oxford: Basil Blackwell, 1989).

explicitly theological, rather than scientific, and they center on the traditional claim that God acts in the world.

The scriptural sources of Judaism and Christianity (and Islam) are rich in stories of divine action. The God depicted in the biblical narratives does not, like an Aristotelian First Mover, rest within a self-contained perfection unmoved by the world. Rather this God has purposes for the world and acts at particular times and places to advance those purposes. As the biblical stories unfold, God plays a strikingly intimate role in history, engaging individuals and communities in a dramatic, unfolding relationship grounded in God's initiatives and promises. The authors of the Hebrew Bible present Yahweh as the one who freed their people from slavery in Egypt and gave them the law, the same one who had earlier entered into a covenant relation with Abraham and his descendants and who remains faithful in judgment and mercy throughout the hardships of Israel's history. This history of divine action plays a central role in identifying who God is and in shaping the language and imagination of the biblical religions.

Notwithstanding the prominence of this theme at the foundation of these religious traditions, the idea that God acts in history has had an uncertain career in modern theology. There are many reasons for this: for example, the emergence of historical criticism of the biblical stories, the persuasiveness of Enlightenment moral challenges to the self-interest and special pleading that is too often at work in talk of divine providence. Of principal interest for my purposes, however, are those misgivings about divine action that are especially tied to the rise of the natural sciences. The sciences have taught us to look at the world as a law-governed natural order, and they have achieved such impressive results precisely because they seek to explain events in terms of their place within this natural order, without reference to any supernatural agency.

How, then, can we understand God to act within such a world? Do scientific and theological descriptions of the world compete, so that if we explain an event scientifically we cannot understand it as a particular divine action? This conclusion may appear inevitable if we take universal causal determinism to be an axiom of "the modern, scientific worldview," as a succession of influential contemporary theologians have supposed. In such a world there is, in Rudolph Bultmann's phrase, "no room for God's working," and God can act within the causal series that constitutes the history of the universe only by interrupting it in miraculous interventions.[4] But then, as David Hume saw clearly, we will have powerful epistemic grounds for rejecting claims about divine action in history, since it will (always, said Hume) be more likely that a miracle report is mistaken or fraudulent than that the otherwise uniform order of the world has been temporarily suspended.

Many contemporary theologians have been convinced, therefore, that (1) traditional ways of speaking about divine action in the world require that there be

---

[4]Rudolf Bultmann, *Jesus Christ and Mythology* (New York: Scribner, 1958), 65. See also Gordon Kaufman, "On the Meaning of 'Act of God,'" in *God the Problem* (Cambridge, MA: Harvard University Press, 1972); and John Macquarrie, *Principles of Christian Theology*, 2d ed. (New York: Scribner, 1977).

gaps in the natural order, and that (2) such gaps are at least (a) epistemically problematic, if not (b) incompatible with the methods or assumptions of the natural sciences, or even (c) conceptually incoherent.[5] I have argued elsewhere that theologians should not concede either 2(b) or 2(c), and others have effectively made the case that we need not accept the strong general version of 2(a) that Hume put forward.[6] But these responses are not alone sufficient, though they are helpful and important in overcoming theological timidity in the face of faulty philosophy of science or preemptive epistemic vetoes. Questions remain about the relation between scientific descriptions of the world as a natural order and theological claims regarding divine action within that world. Do these theological affirmations require incompleteness in scientific explanations, and if so, can it be claimed with any plausibility that our world is incomplete in the right ways?

2 *Divine Action and the Order of Nature*

I want to consider three important theological strategies for replying to these questions, without claiming that these strategies exhaust the alternatives. Two of them argue, in quite different ways, that talk about divine action does not require that there be any gaps in the natural order. The first insists that, strictly speaking, God does not act *in* the world at all, and it shifts the focus instead to God's enactment *of* history as a whole. God acts through, but never among or amidst, finite causes, so that the order of nature remains untouched. The second strategy, by contrast, holds that God does act *in* history, but argues that this is possible quite without gaps in the natural order. The third approach both affirms divine action in history and grants that this requires gaps in the natural order. But it contends that the structure of the world is open in ways that are relevant to theology. God's activity in the world, therefore, need not take the form a miraculous interruption of the ordinary course of events.

I will argue that the first two strategies face telling theological and conceptual objections. If we think it important to continue to speak of divine action in history, then we have good reason to consider whether the world displays an open and flexible structure that would allow ongoing divine action without disrupting the order of nature. In making such a world, God would be the creator not only of natural law but also of the indeterministic gaps through which the world remains open to possibilities not exhaustively specified by its past.

---

[5]This last, and strongest, claim is made by Kaufman, "Meaning of 'Act of God.'"

[6]See Thomas Tracy, "Enacting History: Ogden and Kaufman on God's Mighty Acts," *The Journal of Religion*, 64, no. 1 (January 1984). On Hume's argument, see e.g., William Alston, "Divine Action: Shadow or Substance?," in *The God Who Acts: Philosophical and Theological Explorations*, ed. Tracy (University Park, PA: Pennsylvania State University Press, 1994).

2.1 *The Enactment of History*

The defining move of the first strategy is to shift the emphasis in talk of divine action from providence to creation, locating God's activity solely at the foundations of the world. One simple way of carrying out this procedure is exemplified by the Deists. Their God brings the world system into being, establishing its initial conditions and the natural laws which structure its development. Having gotten the world going, however, God plays no ongoing role within its history. Though Deists often claimed that this, along with a set of moral teachings, constitutes the "essence" of Christianity, they generally were not concerned with whether their views represented an adequate expression of the historic Christian faith. There are, however, theologically much richer ways of carrying out this absorption of providence into creation, and we should not jump to the conclusion that whenever we see this move we are witnessing a version of Deism. A particularly thoroughgoing and subtle development of this approach can be found in Friedrich Schleiermacher, whose theological proposals provide the starting point and paradigm for liberal Protestant theology. It is worthwhile to consider Schleiermacher's position on divine action in some detail, since it nicely illustrates the strengths and weaknesses of this overall strategy.

For Schleiermacher, the whole of God's action is contained in the creation and preservation of the world, and these two forms of divine action are united in a developmental conception of God's continuous creation. On Schleiermacher's principles, of course, statements about divine action must either be direct expressions of religious feeling or be the result of logically ordered reflection upon it.[7] At the heart of his understanding of religion generally and Christianity in particular lies the claim that, at its deepest level, human self-consciousness coincides with consciousness of God as the whence from which we and our world continuously receive our being.[8] As he develops this central idea, Schleiermacher makes two further claims that are particularly pertinent for my purposes.

First, he insists that no distinction can be made "between the mediate and immediate activities of God . . . without bringing the Supreme Being within the sphere of limitation."[9] We must not pick out certain events as direct acts of God in the world and contrast them with ordinary events in which God acts through finite causes (i.e., indirectly).[10] Rather, God acts in the same way in every event, namely,

---

[7]Friedrich Schleiermacher, *The Christian Faith,* ed. H. R. Macintosh and J. S. Stewart (New York: Harper & Row, 1963), 81.

[8]Note that the claim about the world's dependence upon God must be derived, on Schleiermacher's principles, from religious self-consciousness. See his argument, ibid., 151-52.

[9]Ibid., 179.

[10]Another way of making this point is to say that, for Schleiermacher, every act of God is a *basic*, or *simple*, act. A basic act is one that an agent undertakes without having to perform any prior intentional action as the means to this end. In a game of pool, for example, I may intentionally cause one of the balls to drop into a corner pocket by striking it with the

as its absolute ground and source, as the power which posits it in its entirety. This divine creative activity is always direct; each member of a causal series, and not just the first, will depend equally and immediately upon God's action (hence, Schleiermacher can say that theological talk of creation has nothing at all at stake in claims about the beginnings of the world). The same must be said of the operation of "free causes" (i.e., human agents). In the very act of making a free choice I depend absolutely upon God, and God acts unconditionally to constitute me as a free finite agent who makes that choice.[11] Every finite entity at every moment of its being and activity is the immediate result of God's creative agency. Any other way of understanding God's active relation to the world, Schleiermacher alleges, ends up treating God as one limited agent among others.

Second, not only does each event in the world's history bear a relation of immediate and absolute dependence upon God, it also is integrated into a complete (gap free) system of natural relations. It is interesting to note that Schleiermacher provides *theological grounds* for this claim; he does not present it as a conclusion forced upon theology by natural science, though he clearly has the sciences in view.

> It is . . . an expedient often adopted by human indolence to attribute what is not understood to the supernatural immediately; but this does not at all belong to the tendency to piety. . . . Where a pious feeling is actually existent, there the interdependence of nature is always posited.[12]

Not only does the pious mind have no need to find gaps among natural causes in which God can act, it has its own reasons to deny that there are such gaps. Schleiermacher insists that religious consciousness achieves its fullest development when we see the world as a single, integrated "nature-system" to which we belong in a universal solidarity of absolute dependence upon God. "Divine preservation, as the absolute dependence of all events and changes on God, and natural causation, as the complete determination of all events by the universal nexus, are one and the same thing simply from different points of view."[13]

The intention of these two moves is clear: they are meant to ensure that no conflict can occur between scientific (or historical) descriptions of the world and

---

cue ball. Any complex intentional action of this sort must originate in an act which I perform without having to do anything else (as an intentional action) in order to bring it about. This basic intentional action (perhaps a movement of my body) will be what I do *directly*, or *immediately*, in contrast to its various *indirect*, or *mediated*, results (e.g., sinking the 8-ball, winning the game, winning a bet, etc.). We can make use of this distinction in discussing divine action: God acts directly, or immediately, when God brings something about as a basic divine action, and God acts indirectly when God does one thing by doing another (e.g., when God acts through created causes).

[11]Schleiermacher, *Christian Faith*, 179-80.

[12]Ibid., 172 and 174.

[13]Ibid., 174. Compare Stoeger's statement that "the direct causal nexus *is* the active richly differentiated, profoundly immanent presence of God in created beings and in their interrelationships. . . ." ("Describing God's Action," section 9).

theological talk about God's active relation to created things. Schleiermacher's key strategy in achieving this result is to deploy his own version of the classical distinction between primary and secondary causes. On the one hand, there is the "horizontal" order of created (secondary) causes, that is, the individuals that in their activities and relations across time jointly make up our world. On the other hand, there is the vertical order of divine (primary) causation, God's continuous creation/preservation of all finite things. Every event can be understood *both* in terms of its place within the nature-system of the world and in terms of its relation to God's agency.

Schleiermacher insists that these two orders of causation can and must be kept distinct; it is always a mistake to offer explanations of events that mix or cross them. On the horizontal level, our explanations must exclude reference to God and appeal only to other events in the created order. Schleiermacher affirms here a principle of explanatory autonomy among events in nature and history. On the vertical axis, by contrast, we must (a) attribute *all* events to God's direct creative agency, and (b) treat them as proceeding from the divine agency in the *same way*. We cannot, that is, appeal to divine action to explain any particular event in contrast to any other. Rather, we must strictly maintain the universality and uniformity of God's action. The metaphysical relation of God to created things differs not at all from case to case, though the particular content of God's action varies with all the diversity of the created things that God brings to be.[14]

---

[14]In a helpful overview of major theological positions regarding divine action, Owen Thomas suggests that "the liberal view of the divine activity in the world was merely a simplified form of the traditional doctrine of primary and secondary causes with miracles deleted" (Owen Thomas, "Introduction," in *God's Activity in the World* [Chico, CA: Scholars Press, 1983], 4). It is worth noting a crucial difference between Schleiermacher and Aquinas, however, in addition to their views on miracles. Aquinas attempts to sustain the distinction between what God does directly and through intermediaries. God acts directly in causing and conserving the existence of finite things and in "concurring" with their operation. But God acts indirectly through the finite cause in producing its effect. Secondary causes are the instruments of God's action, rather in the way a tool is the instrument of the sculptor's art. The effect can be attributed to both the tool and the sculptor, but on different levels. Note Stoeger's use of this distinction in his essay in this volume, "Describing God's Action in the World in Light of Scientific Knowledge of Reality." Like Schleiermacher, Stoeger contends that we must not think of God as "intervening" in the natural world, but that we can "conceive God's continuing creative action as being realized through the natural unfolding of nature's potentialities. . . ."(section 7) In contrast to Schleiermacher, however, Stoeger holds both that events in the world are indirect divine acts and that we never have experience of God's direct creative action (sections 8-9). Without a distinction between direct and indirect divine action, as I argue below, it is difficult to see how one can avoid the conclusion that God is the only metaphysically real cause. Cf. Nancey Murphy on causation, "Divine Action in the Natural Order: Buridan's Ass and Schrödinger's Cat" (in this volume), section 2.1.3.

On primary and secondary causation in Aquinas see, e.g., Etienne Gilson, "The

On this account, every event is God's act, and no event or set of events can be set apart as flowing from the divine agency in a distinctive way. As Schleiermacher put it in the *Speeches*, "Miracle is simply the religious name for event."[15] Is there any sense, therefore, in which certain events can be singled out as God's acts? There are at least two carefully circumscribed ways in which this can be done. First, we can pick out events as divine actions insofar as they play a special role in *our apprehension* of God's universal agency. Here the distinctiveness of these events consists in their subjective significance to us in awakening God-consciousness. No claim need be made that there is anything objectively noteworthy about them, although there may be. Their importance lies in the function they perform within an historical community; they are for us the preeminent acts of God because they happen to evoke in us a particularly vivid recognition of God's action. That action, however, is in fact present in *all* events.

Second, an event or set of events may play a distinctive role in *God's purposes* for the world.[16] The crucial instance of this for Schleiermacher is found in the life of Jesus. Given Schleiermacher's founding theological principles, he must formulate his Christology in terms of the content and import of Jesus' consciousness of God; Jesus is the Christ because he possessed a complete and unimpeded God-consciousness.[17] Note that although this perfect God-consciousness is unique to Jesus alone, it emerges naturally out of human history and represents the realization of an inherent human possibility. We cannot say that God acts in the life of Jesus Christ in a unique way (e.g., as incarnate deity or by raising him from the dead); any such claim is excluded both by Schleiermacher's assertion of the explanatory autonomy of the created order and by his insistence on the metaphysical uniformity of divine action. But we can say that the life of Jesus plays

---

Corporeal World and the Efficacy of Second Causes," in *God's Activity in the World*, ed. Thomas. For a use of this distinction to separate the realms of science and religion, see Ernan McMullin, "Natural Science and Christian Theology," in *Religion, Science, and the Search for Wisdom: Proceedings of a Conference on Religion and Science*, ed. David Byers (Washington D.C.: National Conference of Catholic Bishops, 1987).

[15]Schleiermacher, *On Religion: Speeches to its Cultured Despisers*, trans. John Oman (New York: Harper & Row, 1958), 88.

[16]This is a point that David Griffin overlooks in his critique of the liberal theologians' version of primary and secondary causation. He supposes that the *only* sense in which an event can be a special act of God, if miracles are excluded, is in the first of the two senses I discuss here. But one can affirm, as Schleiermacher does, both that (1) God bears the *same* relation as agent to every event, and (2) God so designs and determines the course of history that certain events within it make an objectively distinctive contribution to the realization of God's purposes. This avoids a purely subjective, and therefore relative, understanding of revelation. Liberal theologians after Schleiermacher, however, have not always recognized this possibility themselves. For Griffin's critique see his "Relativism, Divine Causation, and Biblical Theology," in *God's Activity in the World*, ed. Thomas, 117-36.

[17]For all his efforts to avoid exposing theology to any risk of empirical contradiction, Schleiermacher at this point is forced to make strong claims about the historical Jesus.

a specific salvific role in God's purposes for history (the whole of which, of course, is enacted by God).[18]

It becomes evident here that a theological price must be paid for this strategy of interpreting all talk of God's action in the world in terms of a continuous divine enactment of history. This procedure has the advantage of eliminating any possibility of conflict between Christian claims about divine action and our expanding knowledge of the world. But this immunity from empirical risk is purchased at the cost of significant limitations on what we are able to say about a number of central theological topics.

First, we have just noted that Schleiermacher's account of divine action restricts the range of options available to us in Christology. His strategy rules out a "high" Christology which would affirm that God enters history in Jesus Christ, joining divinity to humanity in a unique form of divine action. Christian talk about the divinity of Christ must signify, rather, the perfection of his God-consciousness, which makes God available to us through him.[19] These alternatives paths in Christology are matters of longstanding and momentous theological debate, and we may hesitate to adopt a general account of divine action that settles this question in advance.

Second, within the terms of Schleiermacher's theology, there can be no interaction between God and finite persons. We cannot say, for example, that God sometimes acts in response to our actions, for that would violate the cardinal tenet of Schleiermacher's theological system, namely, that relation to God is distinguished from relation of finite things precisely by the utter absence in the former of any reciprocal influence or initiative on our part.[20] Further, if the world's history is to constitute an unbroken continuum of natural causes, then God cannot modify the course of events once that history is under way. Whatever God may intend for any individual (say, that she should ask for in prayer and receive in practice a heightened sense of God's presence) must be "built into" the causal destiny of history from the outset.

Third, it follows that there can be no opposition between the divine and the human wills. God's action is the unconditional ground of all our choices, and there

---

[18]More specifically, God elects that (a) history shall be ordered as a progressive development from a "lower" consciousness submerged in the immediate vitalities of animal life to a fully emergent "higher" consciousness of God, and that (b) this higher consciousness shall radiate outward in history from its first full appearance in Jesus. Langdon Gilkey in *Reaping the Whirlwind: A Christian Interpretation of History* (New York: Seabury, 1976), 210-213, reads Schleiermacher as putting forward a developmental scheme of salvation. Along similar lines, Michael Root provides an illuminating account of the doctrine of election in Schleiermacher's thought. See his article, "Schleiermacher as Innovator and Inheritor: God, Dependence, and Election," *Scottish Journal of Theology* 43: 87-110.

[19]Schleiermacher, *Christian Faith*, 385-89.

[20]See Michael Root, "Innovator and Inheritor," 96ff., for a discussion of the impact of this on narrative structures in Christian theology.

is nothing in the choice that can be attributed to us but not to God. Schleiermacher seems to hold that human agents freely choose just what God, in the continuous creation of each individual, "enacts" us as choosing. But this leaves Schleiermacher in a difficult position in accounting for sin and moral evil, which must now be incorporated into the divine purposes not simply as something God *permits* but as something God *causes*. Sin becomes an element in the developmental progression that God undertakes as the history of creation, an element destined from the outset to be overcome through Christ.

It is worth mentioning two more general difficulties with the conceptual underpinnings of Schleiermacher's proposals. Although Schleiermacher affirms human freedom, it is uncertain just what this amounts to, either in relation to finite causes or in relation to God. If our actions are integrated into a system of nature that constitutes a deterministic whole, then we can at most be free in the sense that we are at liberty to do what we choose, though our choices are determined by the nexus of antecedent natural causes. And if our actions are themselves ordained by God, can we any longer regard them as freely undertaken?[21]

This difficulty, however, is quickly swallowed up within a more pervasive problem. It is not at all clear that, on Schleiermacher's account, there can be active, or effective, finite causes at all. For he insists that God is always the *direct* cause of the entire existence, activity, and attributes of each individual. We must not, for example, distinguish between God's action of preserving the existence of finite things and God's "cooperation" with them in the exercise of their causal powers.[22] It is not altogether clear how to interpret Schleiermacher on these matters, since on his principles statements that look like metaphysical claims must always be read as expressions of religious feeling. But he appears to be saying that God does not merely sustain the existence and causal properties of, say, a flame and a kettle of water, bringing about the boiling of the water by means of the heat of the flame. Rather, God directly brings about the event of the flame being hot and the event, a short time later, of the water boiling. But if each event in a causal series is entirely determined by the direct action of God, then there appears to be no causal work done by the finite events themselves; they can be causes and effects, but only in the sense that they stand in certain regular relations to each other (e.g., as proposed by Hume and by contemporary counterfactual analyses of causality).

---

[21]Schleiermacher's various remarks on human freedom are difficult to interpret. He speaks of human agents as "free causes" who move themselves (see *Christian Faith*, para. 49). The power of self-movement is essential to our humanity and God-consciousness, and the causality of other living things should be understood as a diminished form of the "self-activity" that is present in human beings. As we have seen, however, he also insists on the "complete determination of all events by the universal nexus" (p. 174) and contends that the activities of nature and of free causes are "completely ordained by God" (p. 189). These claims can be consistent only if freedom is understood in a restricted sense (e.g., as action in accordance with our beliefs and desires, which are themselves determined by the causal nexus).

[22]Ibid., 176.

It appears, therefore, that Schleiermacher's position generates an "occasionalism" in which the only effective agent is the divine agent.[23] He could avoid this problem by making a distinction between God's direct (or basic) action in causing the continued existence of finite things, and God's indirect action in bringing about certain events through secondary causes. But Schleiermacher is quite clear about refusing to make this distinction.

There are several good reasons, then, not to pursue Schleiermacher's particular version of the strategy that substitutes God's creative enactment of history for talk of divine action in history. This strategy can be put into play in other ways, of course. But Schleiermacher's theological position was worked out with remarkable imagination and skill, and if his proposals face deep objections, then this does not bode well for the approach his work exemplifies. A number of contemporary theologians have taken positions of this type, and while they avoid occasionalism and usually deny universal determinism, they are somewhat less successful than Schleiermacher in generating a coherent account of divine action. It will be helpful here to consider briefly one recent proposal.

In a widely influential essay, Gordon Kaufman has presented an account of what it means to say that God acts.[24] He begins, as contemporary discussions of this topic often do, by explaining what this notion cannot mean, namely, that God performs particular actions which affect the course of events in the world. The natural sciences have taught us, he says, to see the world as a unified and autonomous whole. If I read him right, the relevant autonomy here is explanatory; that is, we have learned to explain what happens in terms of causal relations to other entities and events in the world, and not by appeal to any supernatural personal agency. The idea of a direct act of God is unacceptable for us because such an event would involve a gap in the order of nature; it could not be sufficiently explained in terms of antecedent finite events, and so would constitute "an absolute beginning point" for a novel causal series. Warming to the subject, Kaufman declares that such an event is not just epistemically problematic, it is "literally inconceivable," for the notion of an event without "adequate finite causes" is "quite as self-contradictory as the notion of a square-circle."[25]

We must, therefore, rigorously avoid talk of divine action *in* history. Nonetheless, it is open to us to think of history as a whole as God's act. In working out this proposal, Kaufman draws upon a distinction in action theory between an agent's guiding intention in action, her "master act," and the particular "subacts" by which that intention is carried out. When we speak of divine action, we refer first and foremost to God's master act, which is "*the whole course of history*."[26] Understood this way, God's action does not break in upon the world and disrupt

---

[23]Occasionalism is the view that finite entities or events are only "occasional," rather than real, causes; that is, their activity is merely the occasion for a divine action that produces the effect.

[24]Kaufman, "Meaning of 'Act of God,'" chap. 6. Also see chaps. 3 and 4.

[25]Ibid., 130-31, n. 11.

[26]Ibid., 137.

it, but rather gives history its structure and direction. This theological vision of history as a single divine action can readily accommodate modern theories of cosmic and biological evolution, which recognize the progressive emergence over time of higher levels of organization out of simpler structures. Theological and scientific descriptions of the world coincide here in a way that is reminiscent of Schleiermacher's insistence that religious consciousness itself requires that we regard the world as a complete and harmonious nature-system. Kaufman follows Schleiermacher as well in arguing that, although history as a whole is the primary unit of divine action, we can nonetheless pick out certain events as having special significance both in moving history toward the ends that God intends and in making the divine purposes visible to us.

This proposal faces some of the same difficulties that Schleiermacher encountered, and it generates several new ones of its own. Unlike Schleiermacher, Kaufman does not develop an account of God's relation to the particular events that constitute the unfolding course of history, and this leaves us with a series of puzzling questions. There can, of course, be no master act without sub-acts to carry it out. What, on Kaufman's account, should we regard as a divine sub-act, and what exactly is God's relation to these events—in what way are they *God's* acts? One possible answer to this question would affirm that every event is a divine sub-act.[27] One could claim that each event in the world's history can be attributed to God's agency by virtue of proceeding from the original act of creation which established the fundamental structures and direction of cosmic history.[28] Just as a human agent can act indirectly by means of a series of instrumental causes, so God may be understood to act in every event that occurs within unbroken causal chains which flow from the initial creation. The event of the sun warming the earth today is (indirectly) God's act, even though this event is linked to God's (direct) agency through a staggeringly complex and extended series of intermediate events (each of which also counts as God's act). We can add that God not only initiates the world's history but also sustains it at every moment. We then could say (*contra* Schleiermacher, who rejects the direct/indirect distinction) that God acts in every event both directly (to sustain all finite things in existence) and indirectly (to bring about events here and now through a series of secondary causes stretching back to God's original creative act).

The difficulty with this construal of Kaufman's proposal is that he explicitly denies that he is endorsing universal causal determinism; he considers determinism to be incompatible with genuine human freedom.[29] This generates two acute

---

[27]Kaufman mixes together the ideas of a divine subact with the idea of an event that we single out as a special act of God, and so he denies that basic natural processes should be regarded as subacts of God (see, e.g., ibid., 144). He affirms, however, that these processes flow from God's creative initiative and constitute the basic structures that support the realization of God's purposes at higher levels of organization.

[28]"It is God's master act that gives the world the structure it has and gives natural and historical processes their direction," ibid., 138.

[29]Ibid., 133.

difficulties for his position, however. First, it contradicts his key argument for eliminating all talk of divine action in the world. He had contended that every event *must* have adequate causal antecedents in the natural order. But indeterministic events, including free human acts, will not have such antecedents; they will, in fact, be precisely the "beginning points for chains of events . . . *within* ongoing natural and historical processes" which Kaufman had declared to be inconceivable.[30] If such events are admissible in considering free human action, then Kaufman no longer has any general conceptual grounds for ruling them out in talk of divine action (though both divine action and free human action may be epistemically problematic).

Second, if the world includes events that are not determined by their historical antecedents, what relation to they bear to God's agency? Each of the options here presents problems for Kaufman's position. On the one hand, if God determines these naturally undetermined events, then God does after all act *in* history. On the other hand, if God does not determine them (as Kaufman affirms is the case for free human actions), then it is not clear what place these events have within God's master-act. History is not, in this case, entirely God's act, but rather is a complex texture of divine initiative, natural indeterminacies, and creaturely freedom. Given the virulent presence of moral evil in the world, it certainly appears that the acts of free creatures can run counter to the divine intention for history. And these free acts will set causal series in motion which echo (and perhaps amplify) throughout history, as theological reflections on 'original' sin suggest. This provides a powerful motive for affirming that God acts in history (viz., that God does so *in response* to human actions); one common telling of the Christian story understands the significance of Christ in just these terms.

Kaufman need not concede this point, however. It is open to him to argue, for example, that (1) God foreknows the choices of free creatures and builds in appropriate correctives through those parts of the world that are determined by God's initial creative act, or more generally, that (2) God sets the parameters of history in such a way that, whatever the details of free human action, God's purposes will be realized.[31] Kaufman does not in fact work out these sorts of details in his position, and it is worth noting that he finds it difficult entirely to rid his discussion of reference to divine action in history. In commenting on God's relation to individual persons, Kaufman remarks that "certain of [God's] subacts are responsive to our acts; in this we may rejoice, finding meaning for our lives and

---

[30]Ibid.,134.

[31]The first of these two options requires, I believe, that God have "middle knowledge," i.e., that God know what every possible free created agent would freely choose to do under any set of circumstances in which it might be placed. Only if God has such knowledge would God be able to design a world that takes our specific free actions into account. There are good reasons to doubt, however, that such knowledge is possible (viz., because there may be no true counterfactuals about free acts). William Alston explores these issues in "Divine Action, Human Freedom, and the Laws of Nature," in *Quantum Cosmology*, 191-99.

comfort for our souls."[32] But if God acts only to initiate and sustain the world, then God cannot respond to our actions, though it may be possible for God to *anticipate* them in the design of history. And if God does not act in response to human actions, then only created free agents will make any novel contribution to the direction of history.

Once again, it is clear that a theological price must be paid for avoiding all talk of God acting in history. The precise character of that cost will be determined by the particular way in which this overall strategy is carried out, and my brief remarks here do not, of course, exhaust the possibilities.[33] It is safe to say, however, that this approach will limit one's options in Christology, constrict what can be said about God's interaction with human beings, and present certain puzzles in soteriology. It is possible to do creative theological work within these constraints, as Schleiermacher illustrates. But it may not be necessary to accept them. Certainly it can be shown, for example, that Kaufman claims too much (even, as we have seen, for his own purposes) in insisting that an act of God in the world is inconceivable. And it is not difficult to make a case that universal determinism is neither a necessary presupposition nor a well-established conclusion of the natural sciences. I will return to this point in considering the third strategy below. It is worth entertaining the question, however, of whether we could speak of particular divine action *in* the world quite apart from there being any gaps in its natural structure. It is to this question that I now turn.

2.2 *Particular Providence without Gaps*

We have already noted one way in which there can be divine action in a thoroughly deterministic world, namely, in such a world each event (after the initial state of the world system) can be considered an *indirect* divine action, brought about by means of an extended series of created causes. We have also seen that there is a limited sense in which particular events in a deterministic world can be especially identified as God's acts, namely, if they have a notably important role in advancing and/or revealing God's purposes. The question I wish to raise here, however, is whether there is any stronger sense in which particular events in such a world can be singled out as God's acts. I will consider two proposals that appear to make this claim, and argue that each of them is unsuccessful.

In an illuminating essay, Brian Hebblethwaite sets out to defend the idea of particular providence against the charge, brought by Kaufman and Wiles, that it

---

[32]Kaufman, "Meaning of 'Act of God,'" 146-47.

[33]Maurice Wiles, in *God's Action in the World*, ed. Thomas, works out a proposal of this type, drawing upon Kaufman's work. Schubert Ogden also pursues this basic strategy in his discussion of divine action, though he makes use of the rather different resources provided by Bultmann and Charles Hartshorne. See Ogden, "What Sense Does it Make to Say, 'God Acts in History'?" in *The Reality of God, and Other Essays* (New York: Harper & Row, 1966).

involves an untenable commitment to divine intervention in the world.[34] Hebblethwaite wants both to affirm that God realizes particular purposes in nature and history, and to deny that this divine action requires that there be gaps in the natural or historical explanations of events. In making a case for this position, he draws upon the rather enigmatic remarks of Austin Farrer about the paradox of "double agency," namely, that a single event can be attributed both to God and to a created cause. Hebblethwaite suggests that we understand "God's particular action in relation both to individual lives and to historical developments as taking place in and through regular natural and historical processes."[35] This is not simply to say that God initiates and sustains the world's history; that, of course, is just what Kaufman and Wiles propose. In opposition to this absorption of providence into creation, Hebblethwaite maintains that God is also at work throughout the course of the world's history, shaping the development of events in specific ways that realize the divine purposes. We should not, however, think of this providential activity as a matter of occasional supernatural incursions into natural processes. On the contrary, Hebblethwaite insists that God acts continuously "in and through" the law-governed structures of the world. God's activity is universal and immanent, shaping events from within the network of interacting finite causes. Rather than overriding the structures of nature, God's guidance of events takes effect through the exercise of the creatures' own powers of operation, so that God "[makes] the creature make itself at each level of complexity without faking or forcing the story."[36] In this way, "the whole web of creaturely events is to be construed as pliable or flexible to the providential hand of God."[37] No gaps are needed here, he contends, and although there may in fact be gaps in the causal structures of the world, he endorses the view that "indeterminacy in nature is *not* to be thought of as the specific latching-on point for divine action."[38]

Hebblethwaite presents a theologically appealing picture of God's pervasive creative role in the world's unfolding history, and it seems to me that we would do well to aim for something like this in our account of providence. It is not at all clear, however, that his account can entirely dispense with gaps in the order of natural causes. On the one hand, Hebblethwaite wants to say that God's action *makes a difference* in the way things go in the world; because God acts with this or that particular purpose, events develop differently than they would have if God had not so acted. But once this is said, we no longer can claim that a *complete* account could be given of these events strictly in terms of creaturely agencies. If God, in addition to creating and sustaining secondary causes, acts to affect the course of events in the world, then God's agency makes a contribution to the

[34]Brian Hebblethwaite, "Providence and Divine Action," *Religious Studies*, 14, no. 2: 223-36.

[35]Ibid., 234.

[36]Ibid., 228.

[37]Ibid., 226.

[38]Ibid., 233.

causal nexus, and any account of the relevant events that is couched strictly in terms of finite agencies must contain some gaps.

This will be the case even if we envision God to act upon the finite causal nexus as a whole, rather than affecting it at particular points. There is a helpful distinction to be made here between (a) overriding the lawful operation of a causal system (say, in the miraculous healing of a withered human limb), and (b) setting that causal system as a whole to work in some particular way (e.g., in using one's restored legs to jump for joy). In the latter case, the processes that constitute the physiology of the human body operate without causal irregularity, but they are directed to the agent's purpose; these processes are, as Hebblethwaite suggests, "flexible or pliable" to the will of the agent. So perhaps Hebblethwaite is proposing that we think of God's action on the latter model rather than the former, namely, that God acts by continuously influencing the causal nexus of the world as a whole, setting it to work in the service of God's purposes but doing so in accordance with the natural structures that God has established for it.

This is an inviting model for divine action, but it does not altogether eliminate explanatory gaps. How is this divine "influence," this "setting to work," to be understood? The claim that God acts in and through, rather than between or among, created agencies, does not by itself advance the discussion. We still must consider the question of what these prepositional phrases *mean* in this context. Note that in the original analogy, the person who leaps for joy *does* have a quite definite effect within the causal history of the body. If we think of this person as possessing free will in the strong, or libertarian, sense, then the agent's free choice will have necessary but not sufficient conditions in the events leading up to it; that is, it will constitute a gap in the causal history of the psychophysical system. If, on the other hand, that causal history remains unbroken, then the system is flexible only in the sense that (a) it could, if it received different triggering inputs, produce a different output, and (b) there are a wide range of such input-output combinations.

If God's action in the world is analogous to the human agent's bodily action, then particular divine initiatives will involve some causal contribution which brings it about that the world-system develops in this way rather than that, realizing one of the possible operations permitted by the structure of the system. This divine influence will be part of the causal history of the world, and a description of that history will be incomplete if God's agency is left out of the account. Hebblethwaite is quick to disavow any view that has God act at points of causal indeterminacy in the world, but it is difficult to see another way to sustain his claims about God's ongoing providential activity.[39] If each successive state of the world system is

[39]Cf. Arthur Peacocke's, "God's Interaction with the World" (in this volume); and *idem*, *Theology for a Scientific Age: Being and Becoming—Natural and Divine* (Oxford: Blackwell, 1990), chaps. 3 and 9. Peacocke invokes the concept of top-down, or whole-part, causation in developing a similar model of divine action. It is important to remember that this concept is suggested by the structure of *explanation* frequently employed in analyzing the behavior of individual constituents in organized physical systems. In order to explain

uniquely determined by its antecedents, then the system will be flexible only in the sense that it may be organized in the way we have just discussed, namely, that it could, depending upon the triggering inputs it receives, function in a variety of ways without breaking down. God's creative action will initiate and sustain the world's history and determine its direction, but God will not affect events once the are underway.

I want to consider a second, quite intriguing, suggestion about how God might be understood to act in the world without producing or exploiting gaps in its causal structures. In an essay on "Science and God's Action in Nature," John Compton suggests a particularly interesting use of the familiar analogy between God's relation to the world and a human agent's relation to her body.[40] Compton points out that some movements of the body can be described simultaneously in two different ways, namely, both as a component physiological process within a complex organism and as an action undertaken by an agent to realize her purposes. Considered strictly as a physiological process, the bodily movement (e.g., a movement of one's arm) can be analyzed in detail as a causal sequence which, given sufficiently good science, may be "explained as completely as is desired."[41] At the same time, this bodily movement is also an intentional action (e.g., of trying to throw a strike past the batter) which can be described and explained in terms of the agent's reasons and purposes.

The key to Compton's proposal is the claim that these two descriptions/ explanations do not in any way compete with each other. Rather, they constitute coordinated but independent forms of discourse about human beings.

---

their behavior, we turn not simply to their immediate causal antecedents, but also to the organization and operation of the system as a whole, which constrains the behavior of the parts. Peacocke moves from this explanatory procedure to the suggestion that here we see an *ontological* structure: namely, one in which "the real nature of the system as a whole" affects its parts *not* through an energetic causal relation but rather through a flow of information. This provides the model for God's action upon the world; the world is "in God," who as a higher-order whole constrains the world's development without entering into its causal history, even at points of indeterminacy.

Peacocke's development of this proposal is richer both scientifically and theologically than I can convey here, but it nonetheless hinges on a series of problematic claims: e.g., the supposition that top-down explanations cannot be analyzed in terms of structures of bottom-up explanation; the move from whole-part explanation to treating the whole (or the nature of the system) as a cause; the idea that the state of a system can be affected by information without an energetic interaction. Each of these claims must be explicated and defended if we are to go not to assert that God can have particular affects on the world yet leave its causal history untouched.

[40]John Compton, "Science and God's Action in Nature," in *Earth Might Be Fair: Reflections on Ethics, Religion, and Ecology*, ed. Ian Barbour (Englewood Cliffs, NJ: Prentice-Hall, 1972).

[41]Ibid., 37.

> The analysis of actions reveals a different structure and system of relations from the analysis of bodily events. . . . Whereas bodily events sustain causal connections in space and time, the same events, viewed as an action, sustain relations in a means-ends continuum, with other more inclusive actions. Similarly, whereas events simply *are* what they are, actions are always subject to appraisal and criticism as justified or unjustified, . . . The entire logic of bodily events and the logic of actions—each equally applicable to me and my behavior—are different.[42]

Each has a complete cast of characters, without the need for interaction with the other story, but quite compatible with it.[43]

Compton seems to have something like the following in mind. Descriptions of human beings as biological organisms and as agents of intentional actions cannot conflict because the vocabularies they use are at every point logically distinct. The physiological description speaks of events that can be analyzed into simpler components and explained in terms of efficient causation within a law-governed structure. Teleological and evaluative language has no place here. The agentic description, on the other hand, speaks instead of acts that have a unity defined by the agent's intention, that are related to other acts in means/ends relations, and that are subject to normative assessments. These two vocabularies cannot conflict because they do not make claims that can be aligned with one another as contradictories. They employ logically distinct subject and predicate concepts, with the result that we cannot generate subject-predicate assertions in the one vocabulary that deny what is affirmed in the other. Conceptual confusion, rather than a conflict of truth claims, is generated when we cross these vocabularies and try to offer a physiological explanation of events described as intentional action or an intentional explanation of events described as a physiological process. The solution to the problems this generates is not to abandon one vocabulary or the other, but rather to pay attention to their logical differences.

This opens up some promising possibilities in theology. Just as efficient-causal and intentional-teleological vocabularies offer distinct and non-competing ways of talking about human beings, so too will these vocabularies both be available in describing the world around us. We can regard events as belonging to a law governed natural order and also as enacting God's intentions. This frees theology from any need to claim that events which the faithful have come to regard as special divine actions must necessarily, on the physical level, lie beyond the scope of adequate scientific explanation. "God does not need a 'gap' in nature in order to act, any more than you or I need a similar interstice in our body chemistry."[44] A single set of events (say, a strong east wind at the Sea of Reeds) can both be explained in terms of deterministic (though unpredictable) meteorological conditions and be described as a mighty act of God in freeing the Jews from slavery in Egypt. Note that this account allows us not only to regard history as a

---

[42]Ibid., 38.

[43]Ibid., 39.

[44]Ibid., 39 and 44.

whole as a divine action (God's "master act," as Kaufman put it), but also to say that specific events enact God's particular purposes, so that God acts in history as well. In his own development of these ideas, Compton follows Kaufman in focusing on God's enactment of history as a whole, but I see nothing in his position that requires this emphasis.[45]

This approach is ingenious, and its theological applications are enticing. The fundamental difficulty with it, however, resides in the thoroughness with it disconnects the two vocabularies that it distinguishes. While there are important logical differences between the description of a bodily movement as a physiological process and as an intentional action, these modes of description and explanation do interact in important ways and can come into conflict. This is evident in the fact that certain sorts of causal explanations of a bodily movement are *incompatible* with also describing it as an intentional action. If, for example, the movement of a person's limb has as its proximate cause a disorder of the nervous system, then that behavior cannot also be described as an intentional action. The motion of the limb is something that happens to the agent, not something that s/he does. If bodily motion is to be intentional action, then the agent must exert intentional control over it, initiating and regulating it for the attainment of her purposes. This is not to say, however, that if we describe the movement as an intentional action, we cannot also offer a complete physiological-causal explanation of it. We need not deny that intentional actions might be constituted by certain deterministic causal series (though whether such actions can be regarded as free is, of course, a matter of ancient and intractable debate.) But there will, in any case, be some causal series that cannot be intentional actions, given the structure and history of the organism in which they occur.

As it happens, Compton himself acknowledges this interaction of causal and intentional vocabularies, and so undercuts the foundations of his position. In commenting on the shortcomings of mind-body dualism, he remarks that:

> consciously held reasons and intentions *are* causes of our action, and they do effectively modify bodily conditions . . . these reasons and intentions have a basis in certain complex states of our brains and nervous systems that may, in principle, be analyzed.[46]

Compton appears to be joining philosophers like Donald Davidson in holding that the agent's reasons explain action because they cause it, and they are able to cause it because mental states are realized in brain states that are the proximal cause of the resulting bodily motion.[47] This is a plausible and widely held view, and Compton is right that this account of intentional action does not require (as we noted above) that there be any gaps in the physiological explanation of action. But,

---

[45]Compton also follows some process theologians in suggesting that we think of the world as God's body. One could make use of his account of causal and teleological vocabularies, however, without adopting this wider analogy.

[46]Compton, "Science and God's Action," 41.

[47]See Donald Davidson, "Actions, Reasons, and Causes," *Journal of Philosophy* 60 (1963).

in the first place, positions of this sort directly link physiological and intentional descriptions of action; indeed, the latter depend upon the former. Understood this way, Compton's analysis of human action no longer provides a conceptual model that releases us from puzzling over the relation between descriptions of events as divine actions and scientific descriptions of the causal order of the world. And, in the second place, action theories of this type are thoroughly deterministic, though they may be coupled with a compatibilist position on human freedom. If this is taken over into theology, Compton's position will reduce to one that we have already considered, namely, a divine determinism in which, after the initial creative act, every event will be an indirect act of God, but no event will be a direct divine act in history.

There may, of course, be other ways of defending the claim that particular divine action in history is possible even in a world whose causal structure is closed and complete. As I noted in concluding the first section, the difficulties faced by particular deployments of a theological strategy do not show that the strategy as a whole is doomed to failure. But in this instance the outlook is bleak. If by an "act of God in history" we mean a divine initiative (beyond creation and conservation) that affects the course of events in the world, then it is at least very difficult to see how such an action could leave a closed causal structure untouched.

2.3 *Divine Action in an Open World*

The upshot of my argument thus far is this. If we wish to affirm not only that God enacts history (as the first strategy claimed) but also that God acts in history, then there are good reasons to think that the world God has made will have an open ("gappy") structure. Note that the motivation for this claim is entirely theological and conceptual. Nothing has yet been said, in section 2 of this discussion, about the findings of the contemporary natural sciences. Our task so far has been to see what claims we would be led to make about the world if we consistently pursue the implications of the claim that God acts in history. We can then go on to ask whether these theologically motivated claims are consonant with descriptions of the world currently being given by the natural sciences.

I want to turn, then, to the third strategy for understanding divine action. This approach affirms particular divine action in history and acknowledges that this requires gaps in the natural order, but it contends that the structures of the world are "open" in ways that accommodate divine action without disruption. We need to consider two questions here. First, in what respects must the structures of nature be open on this view? Second, do the contemporary natural sciences provide support for the claim that the natural order is open in this way? I noted earlier (section 1.1) two instances of explanatory incompleteness in principle (rather than merely in practice) in contemporary scientific theory, namely, chaos theory and quantum mechanics. We need to ask whether either or both of these explanatory gaps proves helpful in working out this third strategy for conceiving of particular divine action.

The world will have an open structure in the required sense if (a) the lock-step of complete causal determination (by antecedent conditions together with the laws of nature) is broken at certain points, and (b) these departures from perfect determination are an integral part of the order of nature, rather than being disruptions of it, and (c) these under determined events can make a difference in the course of events which follows them. The second condition provides that God's particular actions need not take the form of miraculous deviations from the ordinary patterns of nature. Miracle is not excluded as a possibility; the creator of heaven and earth can act beyond the causal powers granted to created things. But under these conditions, God can both bring about particular effects in the world and preserve the immanent structures of nature.

There are at least two different kinds of events which could satisfy these conditions: namely, indeterministic chance and free intentional action, where the latter is understood as incompatibilist, or libertarian, freedom.[48] Events of both these types are not exhaustively determined, though they will be constrained, by the conditions under which they occur. There is more to free action, however, than mere non-determination. Freedom requires *self*-determination; the agent must be able to decide which of the alternatives open to her will be realized in her action. So non-determination is a necessary but not sufficient condition for freedom, and advocates of this sort of freedom bear an additional burden of argument (i.e., beyond that born by someone who asserts simply the presence of structured chance in the world).

In reflecting upon the openness of the world to God's action, I will focus primarily on the theological relevance of chance in the structures of nature. It is interesting to note, however, that one could make human freedom the key to understanding the world as a structure open to novel development. This is Alfred North Whitehead's strategy in working out a metaphysical scheme that builds an element of at least rudimentary self-definition into the history of each entity (i.e., each actual entity receives its data from proximate entities and its initial aim from God, but the entity unites these in its own distinctive way). Whitehead's resistance to mechanism and reductionism, and his insistence on giving creative freedom a founding role in his understanding of the world, contribute to the theological usefulness of his system.

---

[48]Incompatibilists hold that freedom and determinism *cannot* both be true. Some incompatibilists are "hard determinists," who hold that determinism is true, and therefore deny that we are free. Other incompatibilists affirm that we are free and deny determinism; those who take this view are typically called "libertarians." In contrast to both hard determinists and libertarians, compatibilists (also sometimes called "soft determinists") hold that freedom and determinism *can* both be true; some contend, in fact, that we can act freely only if we are determined. Well known proponents of incompatibilist freedom include Roderick Chisholm, *Person and Object: A Metaphysical Study* (La Salle, IL: Open Court, 1976); Richard Taylor, *Action and Purpose* (Englewood Cliffs, NJ: Prentice-Hall, 1966); and Peter van Inwagen, *An Essay on Free Will* (Oxford: Oxford University Press, 1983). Prominent compatibilists include Davidson, *Essays on Actions and Events* (Oxford: Clarendon Press, 1980); and Daniel Dennett, *Elbow Room: The Varieties of Free Will Worth Wanting* (Cambridge, MA: MIT Press, 1985).

There is not space here to pursue this Whiteheadian alternative in detail, but it is worth noting one important feature of the generalization of freedom into a fundamental metaphysical category. The concept of libertarian freedom, as we just noted, involves weightier philosophical commitments than the concept of indeterministic chance: namely, the former adds to the latter the notion of agent self-determination. This self-determination cannot be explained in terms of the causal role of events (say, desires and beliefs) leading up to the choice; the agent's free choice will have necessary but not sufficient conditions in these events. But neither is the agent's free choice merely a chance event. Some libertarians have suggested that here we find a distinctive type of causal relation: "agent causation" rather than "event, or transeunt, causation." But no one knows quite what to say about agent causation (who or what is the "agent" who exercises it? how is it that individuals of this type have this power?), and critics suggest that "agent causation" is merely the name of the central puzzle about freedom, not a solution to it. Whitehead takes this sort of freedom as the key to understanding all becoming, so that free self-determination (greatly diminished though it might be) is part of the explanation of every event. In effect, agent causation is made a more basic category than event causation, and the latter is understood as an abstraction from the former. This is a powerful idea, but it also leaves the mystery of agent causation unaddressed, and it risks the charge of generalizing a perplexity rather than resolving it. For if we find it difficult to explicate the notion of indeterministic self-determination in the instance that is most familiar to us (viz., in our own free action), then these difficulties are compounded when we employ this concept as a metaphysical analogy and extrapolate a modified version of it to the primitive constituents of all things.

This is not, of course, an exhaustive analysis of process philosophies. My point in making these remarks is simply to suggest that there may be some advantage to seeing what can be done theologically with the concept of chance; this will involve more modest metaphysical claims and carry a somewhat lighter burden of argument than the appeal to freedom as a general metaphysical category.

In exploring the place of chance in the structures of nature, it is important to distinguish (a) causal chance and (b) various forms of chaotic unpredictability from (c) indeterministic chance. By causal chance I mean the intersection of causal chains that were previously unrelated and whose particular convergence is improbable (as when a meteor crashes through someone's roof). These are the sorts of events that insurance companies (which otherwise are not well-known for their sense of humor) call "acts of God." Classical physics would describe these events as thoroughly determined by the causal nexus, and though they usually catch us by surprise, we may be able to predict them if the relevant causal chains come to our attention in advance. It is their relative improbability, and therefore their unpredictability in practice, that makes them a suitable paradigm of one kind of chance.

If chance intersections of causal chains are usually unpredictable in practice, chaotic systems (after a surprisingly small number of cycles) are unpredictable in principle. These systems, as we noted above, are so extraordinarily sensitive to their initial conditions that arbitrarily close starting points for these processes can produce dramatically divergent outcomes. The results will be unpredictable in principle, since it will not be epistemically possible either (a) to specify the initial conditions with full accuracy or (b) to predict their result by considering the operation of the system under similar, yet slightly different, initial conditions. This epistemic unpredictability is vividly illustrated by noting the limits on our ability to anticipate accurately the behavior even of a rather simple system of masses in motion (e.g., the familiar idealized billiard balls) operating in complete accord with the laws of classical mechanics. A minuscule error in specifying the initial state of such a system (say, in the initial angle of impact) quickly escalates and, after a relatively small number of collisions, leaves us completely unable to predict the future state of the system. In order to predict each position of the billiard balls even for a short time we would need to specify the initial conditions with an accuracy that utterly exceeds our epistemic reach.[49]

This certainly should awaken us from Laplace's dream of a world that is entirely calculable, and it suggests the astonishing fecundity of the processes that structure our world. But it does not, by itself, generate the ontological openness that would be fruitful for theology. The reign of causal determinism persists unbroken here; the non-linear equations describing chaotic systems fix exactly what each new state shall be, given the immediately preceding state. There is nothing here to suggest that the world's history is not exhaustively specified by past events conjoined with deterministic covering laws. Chaotic processes generate richness and variety in the world, as well as continued surprises for the human observers of nature, but they do not allow for novelty, for a future that is not entirely written into the past. Rather, the presence of such systems guarantees only that we cannot read the future out of the past.

Chaos theory, therefore, reveals that we are behind a "veil of ignorance" with regard to the determining conditions of many events in our world. It is certainly possible for the theologian to claim that God is at work on the other side of this veil. But such divine actions are just as much a matter of miraculous intervention in natural causal chains as any overtly astonishing mighty act of God in history.[50] The difference between parting the Red Sea (at least as Cecil B. DeMille depicted it) and making a infinitesimal adjustment in the initial conditions of a chaotic

[49]"If the player ignored an effect even as minuscule as the gravitational attraction of an electron at the edge of the galaxy, the prediction would become wrong after one minute!" (Crutchfield, et al., "Chaos").

[50]Cf. Peacocke, "God's Interaction," section 6.1., where he makes this same point about divine intervention which take place under the cover of unpredictability. He overlooks the distinction, however, between the deterministic unpredictability of chaotic systems and the indeterministic unpredictability of quantum system. Divine action through quantum indeterminacy need not, as I will argue shortly, involve divine "intervention," in the sense that Peacocke discusses.

system is that human observers are unlikely to overlook the former and utterly unable to detect the latter.[51]

John Polkinghorne takes a considerably more positive view of the significance of chaos theory for efforts to develop a theological view of the world as open, without disruption, to divine action within it. "The flexibility-within-regularity which chaotic dynamics suggests, does appear appropriate to a world held in being by the God of love and faithfulness, whose twin gifts to his creation will be openness and reliability."[52] This is well-said. But why we should think that chaotic unpredictability suggests ontological openness? Nancey Murphy points out the fallacy of employing a "critical realist" principle (viz., that our knowledge models reality) to license the inference from (a) "We cannot predict the outcome of this system given our limited knowledge of its initial conditions" to (b) "The outcome of this system is not determined by its initial conditions."[53] Polkinghorne does appear to make just this argument.[54] It is worth noting, however, that Polkinghorne's claims about chaos and openness in nature are embedded in his larger program of exploring the idea of "downward emergence" in which "the laws of physics are but an asymptotic approximation to a more subtle (and more supple) whole."[55] On this view, deterministic laws would turn out to be abstractions that have limited applicability, and the deeper story about the natural order would involve the openness of which Polkinghorne speaks. If this view were vindicated, then chaos theory could be regarded as an instance of it. But it is hard to see why we should run the inference in the other direction, that is, from chaos theory to a more holistic view. For chaos theory, as we have noted, is quite prepared to explain unpredictability entirely as a function of its deterministic equations.

We need to look beyond chaotic unpredictability toward indeterministic chance, then, if the world is to be "open" in the requisite sense. And here quantum mechanics holds great interest.[56] If, by virtue of claiming that God acts in history,

---

[51]Perhaps God has good reasons to act in both these ways. The wondrous act in nature commands our attention and provides a means of self-revelation for the divine agent. The undetectable contribution to a chaotic system allows God to shape the direction of events in the world without undercutting our efforts to understand the world as a lawful natural order. Murphy discusses the latter point in "Divine Action."

[52]John Polkinghorne, "A Note on Chaotic Dynamics," *Science and Christian Belief* 1, no. 2 (1989): 126.

[53]Murphy, "Divine Action."

[54]Polkinghorne, *Science and Providence: God's Interaction with the World* (London: SPCK Press, 1989), 29; and *idem*, "The Laws of Nature and the Laws of Physics" in *Quantum Cosmology*, 441.

[55]Polkinghorne, "The Laws of Nature and the Laws of Physics," 2.

[56]It is worth noting that we should not assume that if indeterministic chance appears anywhere in the world, it will do so only deep within the underlying structure of things. If, as appears likely, physical systems at higher levels of organization display properties that cannot be reduced to (i.e., explained exhaustively in terms of) the properties of their simpler constituents, then it is at least possible that indeterminisms may appear at these higher levels

we have some stake in the idea that the world displays an open structure, then quantum mechanics may appear to be the good luck of theology. I want to emphasize, however, that physical theory at this level does not provide us with a ready-made metaphysic that theologians can borrow. On the contrary, quantum mechanics challenges our ability to generate any ontological scheme that satisfies the mathematics of this science.[57]

Among physicists and philosophers of science there are many competing interpretations of the statistical character of the regularities at this level of nature. It has been argued, for example, that this feature of quantum physics might reflect (a) the presence of hidden determining variables or (b) epistemic limits in principle on scientific observation/interaction with the world or (c) genuine indeterministic chance in nature.[58] The first of these alternatives, however, has been undercut by Bell's theorem, which shows that the spin correlations predicted by quantum mechanics are incompatible with deterministic hidden-variable theories that deny instantaneous action at a distance. With accumulating experimental results confirming the predictions of quantum mechanics, there is a growing consensus that classical deterministic explanations for these systems are no longer a viable alternative.[59] In the familiar Newtonian world at the macroscopic level we find discrete entities with determinate properties interacting in absolute space and time. Now at the quantum level we find a world of non-locality and non-separability, where particles that have once interacted continue to constitute a single system of linked probabilities even when vastly separated. In contemplating the world

---

as well. These questions about the location and causal role of chance involve a significant empirical component which must wait upon developments in the relevant scientific disciplines. But it is safe to say that current scientific thinking is recognizing a more subtle and extensive interplay of order and chance, regular structure and flexibility, than had been imagined even two decades ago, must less in the era of unchallenged Newtonian determinism. See, e.g., George F. R. Ellis, "Ordinary and Extraordinary Divine Action: The Nexus of Interaction" (in this volume), section 2.

[57]This point is well-made by Russell in, "Quantum Physics," 353-54.

[58]See Barbour, *Issues in Science and Religion* (New York: Harper & Row, 1977), chap. 10, section 3; and *idem*, *Religion in an Age of Science*, The Gifford Lectures, 1989-1991, vol. 1 (San Francisco: HarperSanFrancisco, 1990), chap. 4, section 1. See also Russell, "Quantum Physics," 348-50; and Max Jammer, *The Philosophy of Quantum Mechanics: The Interpretation of Quantum Mechanics in Historical Perspective* (New York: Wiley, 1974).

[59]See John S. Bell, *Speakable and Unspeakable in Quantum Mechanics: Collected Papers on Quantum Philosophy* (Cambridge: Cambridge University Press, 1987); James T. Cushing and Ernan McMullin, eds., *Philosophical Consequences of Quantum Theory: Reflections on Bell's Theorem* (Notre Dame, IN: University of Notre Dame Press, 1989); R.I.G. Hughes, *The Structure and Interpretation of Quantum Mechanics* (Cambridge, MA: Harvard University Press, 1989); and Michael L.G. Redhead, *Incompleteness, Nonlocality, and Realism: A Prolegomenon to the Philosophy of Quantum Mechanics* (Oxford: Clarendon Press, 1987).

according to quantum mechanics, we quickly encounter the limits of our (non-mathematical) imaginations, for familiar (deterministic) macroscopic relationships are simply unable to provide coherent models of nature at this level.

The open-endedness and conceptual strangeness of the scientific discussion should keep us cautious in our theological uses of this science: theological metaphysics should not rush in where physicists fear to tread, although in the discussion of these matters the boundary between physics and metaphysics has also been crossed repeatedly from the scientific side.[60] But if, as I have suggested, theology has an interest in an open-structured world, and if at least one widespread interpretation of current physical theory describes such a world, then it is legitimate to explore ways in which these theological and scientific descriptions of the world might cohere. Without attempting to discuss current scientific theory in detail, I want to comment on two conditions, in addition to indeterminism, that would need to be satisfied if quantum mechanics is to provide the sort of openness that is sought by our third theological approach to divine action.[61]

First, the indeterministic chance that appears in quantum mechanics would need to be embedded in an ordered structure. The presence of chance events in the world will be of limited help to theology if these events simply represent random disruptions of otherwise orderly natural processes. Events of this sort could not be given a coherent place in a scientific description of the world; the world would not so much have an open structure as an incomplete one. If we are to understand God to have an ongoing and pervasive role in contributing to the direction of events, then the world must be structured in a way that is both open and ordered, smoothly integrating chance and law.[62] This condition does appear to be satisfied by the probabilistic distribution of indeterministic chance in quantum mechanics.

---

[60]Consider, e.g., Einstein's insistence that quantum mechanics must be incomplete (since God does not play dice with the universe) or, on the side of indeterminism, Werner Heisenberg's later writings on the uncertainty principle (e.g., *Physics and Beyond: Encounters and Conversations* [New York: Harper & Row, 1971]). Physicists have now become quite bold in their forays into metaphysics: e.g., see John Barrow and Frank Tipler, *The Anthropic Cosmological Principle* (Oxford: Oxford University Press, 1986). In Tipler's latest book, theology becomes "a branch of physics" (*The Physics of Immortality: Modern Cosmology, God, and the Resurrection of the Dead* [New York: Doubleday, 1994], ix.).

[61]These are the second and third conditions that I noted above in defining the sort of openness in nature that is required by a theology which affirms that God acts in the world without disrupting its immanent order.

[62]It is notable that creation myths often portray chaotic disorder as the power that must be subdued by the gods if a world is to be formed in which human beings can live. At the same time, however, human attempts to seek divine guidance have often involved the use of some randomizing process which creates an "opening in the world," beyond the reach of immediate human control, in which the gods can act. (Consider, e.g., the remark in Proverbs 16: 33, "The lots may be cast into the lap, but the issue depends wholly on the LORD.") On the account I am exploring, both of these ancient religious impulses contain some wisdom; God creates a world of ordered indeterminacies, a world of bounded but open possibilities.

Quantum events are not uniquely specified by antecedent conditions, but their probabilities are lawfully ordered and predictable.

Second, indeterministic chance at the quantum level would need to make a difference in the way events unfold in the world. Chance will be irrelevant to history if its effects, when taken together in probabilistic patterns, disappear altogether into wider deterministic regularities. It is commonly said that this is the case with quantum indeterminacies, since the statistical patterns of these events give rise to the deterministic structures of macroscopic processes.

Does this entirely undercut the usefulness of quantum chance for a theological proposal of the third type? Remember that the theological strategy we have been considering seeks to say both that (a) God performs particular actions which turn events in a direction that they would not have gone otherwise, and that (b) these actions need not suspend or disrupt the order of nature. The impulse here is to portray God as having an ongoing providential role in shaping the world's history, over and above creating it, sustaining it, and establishing its lawful structures and initial conditions. But it looks as though a defender of this view who turns to quantum mechanics for scientific help will confront a dilemma. If God's action at the quantum level remains within the bounds of probabilistic law, then God's activity adds nothing to the course of events beyond contributing to higher level deterministic regularities.[63] If, on the other hand, God's action in determining quantum indeterminacies departs from the lawful probabilities, then we are back to "interventionist" disruptions of the natural order. So it look as though we are left, after all, with just two possibilities: God acts by establishing and sustaining a lawful natural order and God may act by intervening to bring about particular effects that "violate" those laws.

Theological appeal to quantum openness can, I think, slip out of this dilemma. It will be possible to open up a third alternative *if* quantum events which fall within the expected probabilities sometimes have significant macroscopic effects over and above contributing to the stable properties and lawful relations of macroscopic entities. Here the discussion turns on scientific questions that I cannot presume to answer. It is apparent, however, that quantum events can set in motion causal chains at the macroscopic level. We are able, for example, to design devices whose outputs are contingent upon chance events at the quantum level (as the uncertain fate of Schrödinger's cat vividly illustrates). Here we arrange an apparatus that has a quantum event as its "trigger," and which amplifies that event through a simple causal mechanism. There may be mechanisms of this sort, though

---

[63]We could say that the effect of God's activity at the quantum level is simply to produce the statistical patterns which account for the stable properties of higher order entities. Murphy takes this view in section 5 of "Divine Action." However, Murphy does not restrict God's action to the bounds of these statistical laws; God can act outside the quantum mechanical regularities to bring about extraordinary events (section 5.2.2).

vastly more complex, at work in neurophysiology and genetics.[64] More generally, chaos theory indicates that nature contains delicately balanced processes that can amplify minute differences in initial inputs into dramatically divergent outcomes. It has been suggested that these systems can be determined by quantum phenomena, whose effects they would magnify with a double unpredictability: (1) the unpredictability of indeterministic chance at the quantum level, and (2) chaotic unpredictability in the system that conveys the quantum effect. The physics involved here, however, is relatively new and notably controversial. For example, it appears that the wave equations at the heart of quantum mechanics do not generate chaotic behavior, and it remains to be seen whether a single coherent account can be given of quantum physics and chaos that would allow for a chaotic amplification of quantum events.[65] Whatever the resolution of this issue, however, there is reason to think that chance events at the lowest reaches of the world's organization can have significant effects on the development of events at the macroscopic level.

If these two conditions are satisfied (and I have done no more here than to suggest that they might be), then it is open to us to propose that one way in which God may act in history is by determining at least some events at the quantum level. This provides a means by which God might (as Hebblethwaite suggested) act to guide the direction of events without overriding the structures of nature that God has established; God's work in creation will then include not only initiating and sustaining the world, but also making an ongoing contribution to the development of events within it. This providential determination of otherwise undetermined events will not transgress natural law; as long as this divine action operates within the statistical regularities described by the relevant sciences, it will remain hidden within the order of nature. Nor will God's activity at the quantum level require that God act as a quasi-physical force, manipulating sub-atomic "particles" as though they were determinate entities of the sort envisioned in Bohr's initial (and quickly abandoned) planetary model of the atom. Rather, God will realize one of the several potentials in the quantum system (the "wave packet"), which is defined as a probability distribution.[66] There is no competition with or displacement of finite causes here, since there is no sufficient finite cause that could explain why the

---

[64]In "Theistic Evolution: Does God Really Act in Nature?" (in *Genes, Religion, and Society: Theology and Ethical Questions Raised by the Human Genome Project*, ed. Ted Peters, forthcoming, 1996) Russell points out that quantum events are a source of genetic mutation, and he argues that these mutations play an important role in providing for novelty in nature. The biological assembly of the organism (the phenotype) and the processes of natural selection between organisms can then function as a highly significant amplifier of indeterministic quantum events. In this way, God may be understood to affect the course of evolution by determining events at the quantum level.

[65]Ford, "What Is Chaos?"

[66]I am indebted to Barbour for this point in his written comments on an earlier version of this paper. Also see his comments on William Pollard's proposals (Barbour, *Issues in Science and Religion*, 428-30.).

probability function collapses as it does. The world at the quantum level is structured in such a way that God can continuously affect events without disturbing the immanent order of nature.

This represents one significant way in which God can act, but it is important to emphasize that I am *not* saying that God acts *only* through the gaps in the causal order of nature. God creates and conserves all finite things, establishing the initial conditions and laws of nature that structure the world's unfolding history. In reflecting upon God's relation to particular events in the world, therefore, we can say that: (1) God acts directly in every event to sustain the existence of each entity that has a part in it, (2) God can act directly to determine various events which occur by chance on the finite level, (3) God acts indirectly through causal chains that extend from God's initiating direct actions, (4) God acts indirectly in and through the free acts of persons whose choices have been shaped by the rest of God's activity in the world (including God's interaction with those persons).[67] The picture that emerges here is of pervasive divine action which establishes, sustains, and continuously shapes a world of secondary causes and created free agents with an integrity of their own. All of this divine activity can take place without in any way disrupting the natural order of the world. That said, it should be added that God can also (5) act directly to bring about events that exceed the natural powers of creatures, events which not only are undetermined on the finite level, but which also fall outside the prevailing patterns and regular structures of the natural order.

It might be objected that this account of divine action ends up portraying God as one agent jostling for influence in the midst of many others on the cosmic scene. If we go beyond the first mode of action just noted (i.e., God's activity as creator/sustainer) and allow the other four, do we reduce the infinite divine creator to the status of a cosmic demiurge? In particular, is this a special problem for the second mode of divine action, that is, God's activity in the interstices of the natural order so as to affect the ongoing course of events? Recall here Schleiermacher's claim that if God is said to bear any relation to events in the world other than that of being their absolute ground, then God's action has been placed among finite agencies and must share in their limitations.

It is important to point out that this conclusion simply does *not* follow from the premises given. God can be the unique, transcendent creator of all finite things and also act directly among the secondary causes that God has brought into being. God's status as first cause—the primary agent whose action founds and sustains all other agencies—is not jeopardized here. In choosing to create a world of finite agencies with causal powers of their own, God freely constrains the uses of divine power out of regard for the integrity of creatures and the intelligibility of their world. I see no reason in general to hold that it is somehow unfitting for God to act at points of causal indeterminacy in nature. Given the Jewish and Christian affirmations of God's ongoing involvement with human destiny, we might expect that the world God has made will turn out to have an intelligible and consistent

[67]On the distinction between direct, or basic, and indirect, or mediated, divine actions, see n. 13 above.

structure, on the one hand, and be open to a continuous divine engagement which is consistent with that structure, on the other. And it will not do to object that this view simply insinuates God into the gaps in physical theory. If we understand God to be the creator of all that is, and if the world God has made includes indeterministic chance in its structures, then *God is the God of the gaps as well as the God of causal connection*. It is part of the theologian's task to try to understand God's relation to the former as well as the latter.

There are theologically interesting puzzles about God's relation to chance events, however. If we say that God might act to determine events that are undetermined by their finite antecedents, should we say that God does so selectively, sometimes leaving the course of events to chance? If so, then (*contra* Einstein) God really does play dice with the universe. Or should we say that God determines every otherwise undetermined finite event, for example, every quantum indeterminacy? Murphy takes this position and affirms that the statistical regularities of quantum mechanics are "nothing but summaries of patterns in God's action upon quantum entities and processes" (sec. 5.2.2). This claim leaves us with difficult questions about how to avoid universal divine determinism throughout the higher levels of organization in nature. William Pollard, in his earlier development of such a position, embraced this result as providing for a strong sense of divine sovereignty and providential control.[68] Murphy, on the other hand, resists the deterministic conclusion.[69]

It will be helpful, by contrast with Pollard and Murphy, to explore at least briefly the first alternative: namely, that God sometimes leaves quantum events undetermined. There is an apparent paradox in the idea that God might create a world that includes events which are absolutely undetermined (i.e., determined neither by God nor by created events or agents, though they occur under certain limiting conditions reflected in the relevant statistical laws). In this case, God would cause there to be events which have no (sufficient) cause. This phrase can be given a non-paradoxical meaning for human agents, since we are able to exploit already existing elements of indeterministic chance in the world. We might, for example, arrange for the random emission of a particle in radioactive decay to register macroscopically on a Geiger counter. In doing so, we generate an event that belongs to a relatively short causal series originating in an event that has no

---

[68]Pollard, *Chance and Providence: God's Action in a World Governed by Scientific Law* (New York: Scribner, 1958).

[69]This will require a strong notion of irreducible emergent properties and a developed account of top-down causation. The latter, Murphy contends, "explains how human free agency is possible within a highly deterministic universe" (Murphy, "Divine Action," section 5.2.5). Her discussion of top-down causation, however, makes the point that the effects of wholes on parts are mediated by the bottom-up interactions of the parts. This leaves it unclear how freedom can appear as a top-down effect within a system of deterministic bottom-up causal relationships.

finite sufficient cause.[70] But this maneuver is not available to God since, *ex hypothesi*, there is nothing at all (i.e., no non-divine actuality) that exists independent of God's creative activity. If there are to be events that are not determined by any event or agent, including the divine agent, we must nonetheless say that they exist by God's will.

In trying to make sense of this, it will be helpful to distinguish between (1) the act of causing or sustaining the existence (*ex nihilo*) of some individual *I* at some time *t* and (2) the act of causing *I* to possess the property *P* at *t*.[71] In distinguishing these two acts, I am not saying that it is possible for *I* to exist at *t* but have no properties at all. God must cause the existence of something in particular. Indeed, if *P* is an essential property of *I*, then God cannot cause *I* to exist at *t* if *I* is not *P* at *t*. But it does not follow from this that *God* directly causes *I* to have the property *P* at *t*. Some other event or agent could be the proximate cause of *I* being *P* at *t*, even though God is the direct cause of the existence of *I* at *t*.

This is the case, for example, if God chooses to bring about some event indirectly, that is, through secondary causes. Recall for a moment our earlier discussion of Schleiermacher's drift toward occasionalism. What was needed, I suggested, was a distinction between the direct divine action of causing the existence of a finite cause and effect (say a source of fire and a kettle of boiling water) and the indirect act of causing the water to be heated by means of the flame. God directly brings about the existence of all the entities involved in this causal transaction. But God does not directly cause the water to boil.

So too, God might cause the existence of entities (or of the linked systems of indeterminate proto-entities that quantum mechanics suggests to us) but leave the successive states of the entity (or system) up to probabilistically structured chance, so that not even God determines the next state of affairs.[72] God would both be the

---

[70]Note that if human beings are agent-causes, then when we act freely we cause there to be events which do not have sufficient antecedent event-causes. But these events nonetheless are determined, viz., by the agent.

[71]We must also distinguish two senses of "causing to exist." In one familiar sense, finite agencies can cause existence; i.e., they can *cause* changes in various constituents of the world which sometimes lead to the formation or destruction of contingent individuals, as in birth and death. God, by contrast, causes to be *ex nihilo*; this act is *not* a species of change, since there is nothing to change until God creates. It is in this distinctively theological sense that I am speaking of God causing existence. This can only be a direct divine act and, most theists have held, is required at every moment to sustain the existence of each finite thing. God can also cause existence in the first sense, and may do so indirectly, i.e., by working through creaturely intermediaries, as (once again) in birth and death.

[72]Van Inwagen, in "The Place of Chance in a World Sustained by God," in *Divine and Human Action*, ed. Thomas Morris (Ithaca, NY: Cornell University Press, 1988), explores a related but even more puzzling question. He considers whether God, if faced with an infinity of different but equally acceptable possible initial states for the universe, might simply decree "Let one of this set be" without determining which would be actualized. This is not simply to choose one of these alternatives arbitrarily, i.e., without having any reason

absolute ontological ground of every event and bring into being a world that includes within its structure an important place for indeterministic chance. God could then choose whether or not to determine these finite indeterminacies in light of their impact on the course of events in the world.[73] In this way, God's creative work would include a continuous involvement in history, and the open potentialities of nature would emerge and be elaborated within the ongoing providential care of God.

3 *Conclusion*

I have done no more here than to sketch the outline of such a proposal, and there is a great deal more to be said on both the theological and scientific sides of the matter.[74] My aim has been to carry the discussion far enough to illustrate the advantages and vulnerabilities of this third way of developing the idea of divine action. This approach avoids the crippling conceptual difficulties that became apparent when we took a closer look at two examples of the second strategy, particular providence "without gaps"; if we affirm that God performs particular

---

to prefer it to the others. Nor is it to resort to a random selection procedure, a divine coin toss. Rather it is to decree that one of them shall be actualized without specifying which. Van Inwagen thinks this is possible, but I am not at all sure that it is. After all, none of the alternatives will be actualized unless *God* actualizes it; there is no other agent on the scene to do the job.

[73]Peacocke suggests that the outcomes of quantum uncertainties would be unknown even to God before they actually occur, and therefore that God would not be in a position to use them to influence the course of events (Peacocke, "God's Interaction," section 6.1.). One classical response to the related puzzle about divine foreknowledge and human freedom applies to this problem as well: if God exists outside time, then the whole of history is available to God's knowledge simultaneously. See, e.g., Norman Kretzmann and Eleonore Stump, "Eternity," *Journal of Philosophy* 79 (1981): 429-58. If God exists in time, there are a variety of strategies of response. If nature includes amplifiers of quantum events in the way I suggested, I see no reason that God would not know this and understand the precise relation of quantum inputs to macroscopic outputs; even if God does not know what the input will be in cases left to chance, God will certainly know the input whenever God determines it. Finally, it is also possible for God to determine enough quantum indeterminacies to guarantee that they will affect the course of events in the way God intends. Murphy, as we have seen, holds that God determines *all* such events.

[74]The best known presentation of a position that makes use of quantum indeterminacies is that of Pollard (*Chance and Providence*), who contends that divine providence is exercised though the determination of *all* quantum events. See Murphy's remarks on the relation of her proposal in *Chance and Providence* to Pollard's position (Murphy, "Divine Action," section 6.4.). See also Donald MacKay, *Science, Chance and Providence* (Oxford: Oxford University Press, 1978); David J. Bartholomew, *God of Chance* (London: SCM Press, 1984); and Keith Ward, *Divine Action* (London: Collins, 1990), chap. 5.

actions which affect the course of events in the world, then it certainly appears that we must also grant that there will be gaps in the explanation of these events in the sciences. Moreover, unlike the first strategy, God's "enactment *of* history," this third approach allows us to speak not only of God's enactment of history through the creation and preservation of the world, but also of divine action ***in*** history which shapes the unfolding course of events without disrupting the structures of nature. This is a theologically important result; talk of particular divine action in the world has played an undeniably important role in the history of Jewish, Christian, and Islamic thought, and it continues to figure prominently in religious life (e.g., the life of worship and prayer) within these traditions. If the third theological strategy is sustainable, then we need not conclude that our "modern, scientific worldview" requires us to abandon or radically reinterpret this language.[75]

These theological gains bring with them some vulnerabilities, however. For now we have tied theological reflection to interpretations of contemporary scientific work, and this means that we must give up the perfect neutrality on scientific matters that is the hallmark of the first strategy. I have argued that a theological proposal of the third type will find quantum mechanics more helpful than chaos theory. While both fields present us with explanatory gaps in principle, chaotic unpredictability emerges from precisely the sort of deterministic analysis which breaks down in quantum mechanics. Chaos, therefore, appears to be a source of epistemic but not ontological openness. By contrast, one widely accepted interpretation of the statistical character of quantum mechanics takes it to reflect indeterminacy in nature. Of course, quantum mechanics can be interpreted in more than one way, and the development of this science is by no means complete. In addition, the theological usefulness of quantum mechanics hinges in part on empirical questions about whether quantum indeterminacies are entirely "dampened out" in deterministic regularities at a higher level, or whether they are sometimes "amplified" by particular causal mechanisms in nature. We noted that chaotic processes may play a very significant role in this regard, but that the science involved here is new and quite uncertain.

It is clear, then, that the third approach exposes our account of God's active relation to the world to some degree of empirical risk. Modern theologians have been notably averse to running such risks, particularly in light of the embarrassing history of theological resistance to new scientific developments. As we saw in considering the first approach to divine action, however, there is also a price to be paid if we try to guarantee at the outset that no conflict can occur between

---

[75]There are, of course, important theological questions raised by the language of divine action in history, and these questions may lead to proposals for revision or to a preference for a theology of the first type. See, e.g., Maurice Wiles, "Divine Action: Some Moral Considerations," and James Gustafson, "Alternative Conceptions of God," both in *The God Who Acts*, ed. Tracy. I do *not* claim to have settled these theological questions, but only to have opened up some "conceptual space" in which they can be debated on ***theological grounds***, without facing an immediate scientific veto.

theological affirmations and scientific claims. This strategy of disengagement, after all, is still a form of theological response to natural science, and it too has often involved suppositions about the world described by the sciences which turn out to be false (e.g., the claim that science teaches universal determinism). Modern thinking about God's relation to the world inevitably takes place against a background of scientific theorizing about how that world is put together. It is best to be explicit, therefore, about what claims our theology entails, about what the sciences currently are saying, and about whether these can be brought into a mutually illuminating relation.[76] In doing so we must cope with the virtual certainty that scientific understandings of the world will change. There may be more gain than loss, however, in regarding our theological constructions as working proposals that must be continually rethought as we attempt *both* to remain faithful to the religious tradition that affirms God's action in the world *and* to appropriate fuller understandings of the world in which God acts.

---

[76]For an insightful discussion of the ways in which theological "control beliefs" interact with scientific beliefs as we form theoretical commitments, see Nicholas Wolterstorff, *Reason Within the Bounds of Religion* (Grand Rapids, MI: Eerdmans, 1976).

DIVINE ACTION IN THE NATURAL ORDER:
Buridan's Ass and Schrödinger's Cat

Nancey Murphy

1 *Introduction*

In the Medieval period, especially after the integration of the lost works of Aristotle into Western thought, God's action in the world could be explained in a way perfectly consistent with the scientific knowledge of the time. Heaven was a part of the "physical" cosmos. God's agents, the angels, controlled the movements of the "seven planets," which, in turn, gave nature its rhythms. But modern science has changed all that, primarily by its dependence on the notion of *laws of nature*. For Isaac Newton and other architects of the modern scientific worldview, the "laws of nature" were a direct expression of God's will—God's control of all physical processes. However, today they are generally granted a status independent of God, not only by those who deny the very existence of God, but also by many Christians, who seem to suppose that God, like a U.S. senator, must obey the laws once they are "on the books." Consequently, for modern thinkers, deism has been the most natural view of divine action: God creates in the beginning—and lays down the laws governing all changes after that—then takes a rest for the duration.

Not all modern theologians have opted for this deistic account, but in many cases the only difference has been in their additional claim that God sustains the universe in its existence. Those who have wanted (or who have believed Christianity needed) a more robust view of God's continued participation in the created order have been forced to think in terms of intervention: God occasionally acts to bring about a state of affairs different from that which would have occurred naturally.[1] It is an ironic bit of history: the laws that once served as an account of God's universal governance of nature have become a competing force, constraining the action of their very creator.

The series of conferences for which this essay was written involve a re-evaluation of the modern understanding of divine action in light of more recent science. Chaos theory has been proposed as an important avenue for a new view of divine action.[2] However, this essay grows out of a recognition that the turn to

---

[1]Authors represented in this volume are some of a small number of more recent thinkers who have sought *non-interventionist* accounts of special divine acts.

[2]John Polkinghorne is the most important proponent of this view. See, e.g., his *Science and Providence: God's Interaction with the World* (Boston: Shambhala, 1989); and *idem*, "Laws of Nature and Laws of Physics," in *Quantum Cosmology and the Laws of Nature: Scientific Perspectives on Divine Action*, ed. Robert John Russell, Nancey Murphy, and C. J. Isham (Vatican City State: Vatican Observatory, 1993; Berkeley, CA: Center for Theology and the Natural Sciences, 1993).

chaos and complexity has not solved the problem in the way it was intended. Furthermore, while the recognition of *top-down causation* is an important advance in our understanding of natural processes, as well as an important ingredient that must go into any new theory, it is not *in itself* an adequate account of divine action.[3] So the main goal of this paper is to provide an alternative account of causation and divine action that is both theologically adequate (consistent with Christian doctrine and adequate to Christian experience), and consistent with contemporary science.

1.1 *Preview of the Argument*

Following a brief critique of the most promising account of divine action based on chaos theory, I shall attempt to set out in advance the criteria a theory of divine action needs to meet. It is my contention that the problem of divine action is, at base, a metaphysical problem—one that cannot be solved by anything less radical than a revision of our understanding of natural causation. One way to understand the nature of metaphysics is as a set of interrelated theories about reality that are of the broadest possible scope, and thus descriptive or explanatory of the phenomena described by all other branches of knowledge. My goal, then, is to provide a theory of causation that takes account of phenomena germane to both science and theology. Thus, in section 2, I propose criteria of adequacy drawn from both theology and science.

Section 3 surveys relevant changes in metaphysical views of matter and causation, in particular contrasting the Aristotelian hylomorphic conception with the early modern corpuscular theory. This background is intended to put in question current metaphysical assumptions about the nature of matter and of natural causes. This section also considers consequences of recent developments in science for rethinking these metaphysical issues.

Section 4 advances a proposal. I shall argue that any adequate account of divine action must include a "bottom-up" approach: if God is to be active in all events, then God must be involved in the most basic of natural events. Current science suggests that this most basic level is that of quantum phenomena. Consequences of this proposal need to be spelled out regarding the character of natural laws and regarding God's action at the macroscopic level in general and the human level in particular.

In section 5 I attempt to answer some of the objections that have been raised against theories of divine action based on quantum indeterminacy, and also to show that this proposal meets the criteria of adequacy set out in section 2.

---

[3]Arthur Peacocke is to be credited with the most compelling accounts to date of the role of top-down causation in accounting for God's continuing action. See his *Theology for a Scientific Age: Being and Becoming—Natural and Divine*, 2d ed., enlarged, (Minneapolis, MN: Fortress Press, 1993). I owe a great debt to Peacocke's thought throughout this paper.

1.2 *Chaos Theory: The Road Not Taken*

One proposed solution of the problem of divine action in the natural world is John Polkinghorne's suggestion that God works within the indeterminacy of chaotic systems. Complex systems, being highly sensitive to initial conditions, are inherently unpredictable, since significant variations in initial conditions fall beneath the threshold of measurement. Polkinghorne argues from this fact to the claim that the futures of such systems are truly "open," and hence that God can operate within them without contravening the laws of nature.

I claim (1) that the argument from unpredictability to indeterminacy is fallacious; (2) that the attempt to find indeterminacy between the quantum and human levels is unnecessary if we have already made allowance for God's action at the most basic levels of organization; but (3) that the *unpredictability* recognized by chaos theorists is nonetheless extremely important for an account of divine action. If we begin with the hypothesis that God works at the quantum level, it is not necessary—in fact it is counterproductive—to argue for causal indeterminism at higher levels of organization (excluding the human level) since God's will is assumed to be exercised by means of the macro-effects of subatomic manipulations.

Polkinghorne, in speaking of chaotic systems, says:

> We are necessarily ignorant of how such systems will behave. If you are a realist and believe, as I believe, that what we know (epistemology) and what is the case (ontology) are closely linked to each other, it is natural to go on to interpret this state of affairs as reflecting an intrinsic openness in the behavior of these systems.[4]

Now, let us grant the realist thesis that what we know is (unproblematically) linked to what is the case. Let P stand for any proposition, then 'X knows that P' entails P. So far so good.

But Polkinghorne's argument is not from the *content* of some known proposition P to the character of the world; it is rather an argument from the character *of our knowledge of P* to the character of the world. Take any P that is a statement about the future (chaotic) state of a chaotic system: what the unpredictability amounts to is that for any person, X, and for any P, it is *not* the case that X knows that P. This implies nothing at all about the world's likeness to P.

To make such an argument is comparable to confusing a modal qualifier, which qualifies a proposition as a whole, with a property of an object described by that proposition. 'Possibly there are unicorns' does not entail that there are possible unicorns—that is, entities that are both unicorns and possible. Neither does 'The outcome of chaotic processes are inherently unpredictable' imply that there are outcomes that are indeterminate.

Is this move in Polkinghorne's thought simply an instance of using a bad argument for a position that may well be defensible on other grounds? I think not. The grounds upon which chaos theorists argue for the unpredictability of future

---

[4]Polkinghorne, *Science and Providence*, 29.

states depend upon the assumption that the future states *are determined by* initial conditions in so sensitive a manner that we cannot measure them. So the systems are presumed to be determined at a very precise level— small changes *produce* large effects.

So what chaos shows is not that there is genuine indeterminacy in the universe, but rather that we have to make *a more careful distinction* between predictability (an epistemological concept) and causal determinism (an ontological concept). In a similar way, the phenomenon of quantum indeterminacy forced earlier physicists to distinguish between ontological and epistemological indeterminacy. That the consensus now is in favor of an ontological interpretation does not obliterate the distinction; *a fortiori* it does not provide warrant for obliterating the distinction between ontological indeterminism and epistemological unpredictability in this case.

Furthermore, it is not clear to me that Polkinghorne's position would solve the problem even if the argument for indeterminacy were valid. Let us take a specific case. Suppose Father Murphy is playing billiards in a high-stakes game in the hope of winning enough to get his school out of debt. Let us also suppose that God intends him to win, and in order to do so must bring about his getting a particular ball in the pocket.

Murphy takes aim, hits the cue ball and the cue ball hits the #2 ball, which undergoes several more collisions. Polkinghorne rightly points out that we are unable to predict whether the ball will fall into the appropriate pocket. But what, exactly, could it *mean* to say that the outcome is open? Does it mean undetermined, *tout court*? Does it mean not uniquely determined by the laws of motion? I take this latter to be Polkinghorne's meaning, since I find it hard to imagine what it would mean to say that it is totally undetermined, and also because he sees such things as slight environmental influences as important to the outcome in such cases. So what we might better say is that there are a *range* of outcomes that are consistent with the laws of motion.

Now, how does God effect one of these possible outcomes? Polkinghorne suggests that in some cases God's input might be a non-energetic contribution of information. But to whom or what is the information contributed? How is it conveyed without any energy at all. And in what sense does this proposal avoid "turning God into a demiurge, acting as an agent among other agents?"[5] Polkinghorne quotes John V. Taylor with approval, when he writes:

> [I]f we think of a Creator at all, we are to find him always on the inside of creation. And if God is really on the inside, we must find him in the process, not in the gaps.[6]

I suggest that Polkinghorne has not provided a clear account of *how* God works on the inside, in the process.

---

[5]Ibid., 33.

[6]John V. Taylor, *The Go-Between God* (London: SCM Press, 1972), 28, quoted in Polkinghorne, *Science and Providence*, 31.

This raises the question of how God *could* work "on the inside." I take it that if God is to do so, the it is necessary that God work on the inside of *all* created entities—which must mean in turn that God works within the smallest constituents of macroscopic entities, since these smallest constituents are entities in their own right.[7] If we begin with this hypothesis, it is not necessary—in fact it is counterproductive—to argue for causal indeterminism at higher levels of organization (excluding the human level) since God's will is assumed to be exercised by means of the macro-effects of subatomic manipulations.

2 *Criteria of Adequacy for a Theory of Divine Action*

The theory of causation and divine action to be presented here might be construed as metaphysical—that is, metascientific and metatheological. As such, its primary confirmation should come from its consistency with both science and theology, and especially from the fact that it solves problems that have arisen at the interface between these two sorts of disciplines. To solve such problems is no small accomplishment, and insofar as this proposal could be shown to solve problems that its competitors cannot solve, it would have a high degree of acceptability.[8]

---

[7]It is interesting to speculate about the meaning of the distinction between God working "on the inside" versus "from the outside." We can give a clear sense to "from the inside" when we are speaking of macroscopic entities and God working within them by manipulating constituent quantum entities, since the quantum entities are "inside" of the macroscopic entity. But can we make sense of a distinction between the inside and outside of the quantum entities themselves? If God has no physical location, literally speaking, yet we say that God is omnipresent and immanent in all of creation, perhaps we are assuming that a disembodied agent's presence is to be defined in terms of the agent's causal efficacy—wherever God acts, there God is. Thus, to say that God works within quantum entities would be equivalent to saying that God affects quantum entities.

[8]Ideally, one would like to be able to show that such a proposal is progressive in the sense defined by Imre Lakatos. He proposed that a scientific research program is progressive if it can be developed in such a way that its theoretical content anticipates the discovery of novel facts. A similar criterion could be devised for metaphysical theories: that they anticipate and solve problems in other disciplines. That is, a metaphysical theory should be counted progressive if it turns out to contain resources for solving conceptual or empirical problems in or between other disciplines that it was not originally designed to solve. Lakatos's scientific methodology is found in "Falsification and the Methodology of Scientific Research Programmes," in *Criticism and the Growth of Knowledge*, ed. I. Lakatos and Alan Musgrave (Cambridge: Cambridge University Press, 1970), 91-196. See my adaptation of his work in "Evidence of Design in the Fine-Tuning of the Universe," in *Quantum Cosmology*.

2.1 *Theological Requirements*

To do justice to the Christian tradition, a theory of divine action ought to be consistent with widely accepted formulations of key Christian doctrines, and—this is at least as important—it must constitute suitable *presuppositions* for Christian *practice*.

2.1.1 *Doctrine*

I take it that one desideratum for theological construction is always to see what sense can be given in each age to traditional formulations. Only if the formulations of the past turn out to be hopelessly unintelligible should they be rejected or radically changed. God's continuing action in the created world has been spoken of in a number of different ways—as sustenance, providence, continuing creation. One traditional set of terms will turn out to be particularly useful: God's continuing work understood as *sustenance, governance,* and *cooperation.*[9] The sense that can be given to these terms by means of the proposal in this paper will become clear as we go along.

An additional doctrinal requirement, I suggest, is that an account of divine action throughout the hierarchy of levels of complexity must show forth God's *consistency*. Thus, if the paradigm of divine action for Christians is found in the story of Jesus, we should expect that same divine moral character to be manifested, analogously, in God's action within sub-human orders. I shall claim that the relevant feature of God's action in Christ, displayed analogously throughout the whole, is its non-coercive character.

2.1.2 *Presuppositions for Christian Practice*

The following seem to be required of any account of divine action that would be supportive of Christian belief and practice:

2.1.2.1 *Special Divine Acts*[10]

The first requirement is that we be able to distinguish in a meaningful way between events that are in some way special acts of God, and others that are not. This requirement is not met easily, since both doctrine and logic suggest that if God acts at all, God is acting in *everything* that happens.

There are at least three reasons for needing to distinguish special divine acts. First, our knowledge of a person comes primarily from the person's actions,

---

[9]These terms go back at least to Augustine, who formulated the discussion of grace and free will using the concepts of providence, sustaining activity, governance and cooperation. The terms have been used frequently in subsequent discussions of divine action.

[10]My use of 'special' here corresponds to that of "objectively special divine acts" as defined in Russell's "Introduction" to this volume.

including speech acts. Knowledge of God, therefore, must come primarily from seeing what God has done. However, it is well-recognized that the sum total of the events known to us so far (both natural and historical) provide at best an ambiguous testimony to the character of God.[11] So we need at least to be able to distinguish between God's acts and the actions of sinful creatures; ideally we ought to be able to make sense of recognizing certain historical events as actions of God that are especially revelatory of God's character, intentions, and providence.

A second reason for needing to distinguish between divine actions and natural events is to support the practice of petitionary prayer. If there is no sense in which God may be expected to bring about a state of affairs that would not otherwise have occurred, then the practice of petitionary prayer is groundless.[12]

An even more pressing reason for needing to distinguish a special class of divine actions is that to fail to do so makes God entirely responsible for every event, and thus exacerbates the problem of evil. As Polkinghorne argues, theodicy requires a "free-process defense," as well as a free-will defense.[13]

Notice, though, that a concept of the autonomy and regularity of natural processes is not merely a *parallel* to the theodicist's doctrine of free will; it is a *prerequisite* for it as well. In order to make intelligent, free decisions and take responsibility for our action we must live in a world where outcomes of our actions are often predictable, and this in turn requires that the universe exhibit law-like regularity.

2.1.2.2 *Extraordinary Divine Acts*

Many modern and contemporary Christians would be satisfied with an account of causation and divine action that met all of the above requirements. However, earlier Christians would have insisted as well that there be room in such an account for something on the order of miracles. I prefer *not* to use the term 'miracle' because it is now so closely associated with the idea of a violation of the laws of nature. I believe it could be shown that the primary reason for current rejection of miracles, in fact, has been this very definition.

So one reason for going against the Enlightened consensus and including as a second requirement for a theory of divine action that it leave room for what I shall call *extraordinary acts of God* is that the modern rejection of such acts was based on a mistaken view of the nature of miracles. A second is that elimination of all such events from Christian history leaves too little: the resurrection is an extraordinary act of God if ever there was one. Yet, as Paul asserts, if Christ is not

---

[11]See, e.g., David Hume's critiques of the argument from design in *Dialogues Concerning Natural Religion*, and John Wisdom's parable, "Gods," *Proceedings of the Aristotelian Society*, 1944-5.

[12]See my "Does Prayer Make a Difference?" in *Cosmos as Creation: Theology and Science in Consonance*, ed. Ted Peters (Nashville, TN: Abingdon Press, 1989), 235-45.

[13]See Polkinghorne, *Science and Providence*, 66-67.

raised, then Christian faith comes to nothing (cf. 1 Cor. 15:14,17,19). But if the resurrection is credible, then lesser signs cannot be ruled out *a priori*.

2.1.3 *Summary*

We can sum up the discussion of theological requirements by saying that an adequate account of divine action will have to avoid the opposite poles of deism and occasionalism. Occasionalism, as applied to theories of divine action, denies the causal interaction of created things: created entities only provide an "occasion" for the action of God, who is the sole cause of all effects. This position has been rejected on the grounds that it ultimately denies the reality of finite beings.

Schematic representations make clear the difference between these two extreme positions. Occasionalism can be represented as follows, where *G* stands for an act of God and *E* stands for an observable event:

$$\begin{array}{ccccc} G_1 & \rightarrow & G_2 & \rightarrow & G_n \\ \downarrow & & \downarrow & & \downarrow \\ E_1 & & E_2 & & E_n \end{array}$$

$$\xrightarrow[\text{time}]{\qquad\qquad}$$

Here, God is the sole actor, and any causal efficacy on the part of observable events is mere illusion.

The following sketch represents the deist option, where *L* represents a law of nature:

$$G \rightarrow \underbrace{E_1 \rightarrow E_2}_{L_1} \rightarrow \underbrace{\rightarrow E_n}_{L_n} \quad (G \searrow L_1)$$

$$\xrightarrow[\text{time}]{\qquad\qquad}$$

Here, God's action is restricted to an initial act of creation, which includes ordaining the laws that govern all successive changes.

Some modern accounts of divine action have sought to hold divine action and natural causation together: God acts in and through the entire created order. Thus, we get a combined picture:

$$\begin{array}{ccccc} G_1 & & G_2 & & G_n \\ \downarrow & & \downarrow & & \downarrow \\ E_1 & \rightarrow & E_2 & \rightarrow & E_n \end{array}$$

$$\xrightarrow[\text{time}]{\qquad\qquad}$$

This approach suffers from two defects. First, it leaves no room for any sort of special divine acts and, second, it seems impossible to do justice to both accounts of causation (the problem of double agency); one inevitably slides back into occasionalism or else assigns God the role of a mere "rubber stamp" approval of natural processes.

In short, we need a new picture of the relation of God's action to the world of natural causes that allows us to represent God's sustenance, governance, and cooperation in such a way that we can make sense of revelation, petitionary prayer, human responsibility, and of extraordinary acts such as the resurrection, without at the same time blowing the problem of evil up to unmanageable proportions.

2.2 *Scientific Requirements*

An adequate account of divine action must also be consistent with the sciences. Here, again, we can distinguish several types of consistency.

2.2.1 *The Results of Scientific Research*

An adequate account of causation in general and divine action in particular needs to "save the phenomena." That is, we are setting out to explain how God and natural causes conspire to bring about the world *as we know it*. The salient features seem to be, first, the general law-like behavior of macroscopic objects and events, qualified, however, by two major exceptions: the apparent randomness of individual events at the quantum level and human free actions.[14] The fact that the "rule of law" needs to be so qualified, however, suggests the value of recognizing as a second, equally important, feature of the world known by science its organization into a hierarchy of levels of complexity.[15] More on this below. It also suggests that in an account of divine action, attention needs to be given to three very different regions or "regimes" within the hierarchy: the quantum level, the realm of human freedom, and an intermediate regime wherein the behavior of entities is describable by means of deterministic laws.

2.2.2 *Presuppositions of the Practice of Science*

The law-like character of the natural world is not only a finding of science; it is a presupposition for engaging in scientific research in the first place. It has often been argued that the Christian (and Jewish) doctrines of God, stressing both God's freedom and God's rationality and reliability were crucial assumptions for the

---

[14]Perhaps the higher animals are also capable of free actions, but if philosophers are not agreed what it means to say that human actions are free, *a fortiori* we do not know what to say about the animals.

[15]See Peacocke, *Theology for a Scientific Age*, *inter alia*.

development of empirical science.[16] No revised account of divine action that undercuts the practice of science will be acceptable.

2.2.3 *Metascientific Factors*

I have been careful in the two preceding subsections to speak of "the law-like character" of the natural world, not of the existence of laws of nature. While many scientists assume that there must in some sense *be* such laws—that they must have some sort of *existence*[17]—I do not believe such a view is either a necessary prerequisite for doing science or necessarily supported by the findings of science.[18] Thus, I shall argue that the "de-ontologizing" of the "laws of nature" is a helpful move in understanding divine agency.

3 *Metaphysical Considerations*

I claimed above that nothing short of a revision of current metaphysical notions regarding the nature of matter and causation is likely to solve the problem of divine action. In this section we survey some important changes in the history of metaphysics as background, and then attempt to see where we are now and where we must go in our thinking about causes.

3.1 *From Aristotle to Newton*

One of the most striking changes from medieval (Aristotelian) hylomorphism to modern corpuscularism (á là Descartes and Newton) regards the "powers" of material things to move themselves or to change in other ways. Of course Aristotle and Newton would both agree that horses, for example, are material bodies, and horses, obviously, can move. So the question is a deeper one about the nature of matter itself.

For Aristotelians, all individual substances were constituted by two principles: matter and form. Individual substances could be arranged hierarchically with the more complex at the top. For the higher beings, the matter of which they were composed was already "en-formed" by the forms of lower realities. The lowest entities in the hierarchy of existents were the four elements: earth, water, air, and fire. But these elements were themselves constituted by *forms* (of earth, water, air, or fire) and "prime matter." Prime matter, however, was assumed to exist only

---

[16]See Ian Barbour, *Issues in Science and Religion* (New York: Harper & Row, 1966), 46.

[17]See, e.g., Paul Davies, *The Mind of God: Science and the Search for Ultimate Meaning* (New York: Simon & Schuster, 1992).

[18]For a discussion of this issue, see William Stoeger, "Contemporary Physics and the Ontological Status of the Laws of Nature," in *Quantum Cosmology*. See also Bas C. van Fraassen, *Laws and Symmetries* (Oxford: Clarendon Press, 1989).

as ingredient in the four elements (and hence as a basic ingredient in all higher substances), so it was only a theoretical construct within the system.

However, in Aristotle's system, prime matter, were it to exist independently of all forms, would be entirely passive since it is form that gives individual characteristics to existent beings, including whatever powers and actions are natural to that species of existent. Conversely, since all existent material beings are en- formed matter, all material beings have certain inherent powers and certain "motions" that are natural to them. Even stones, simple objects composed primarily of the element earth, have the intrinsic power to seek their natural position, which is at the center of the cosmos. That is why rocks fall when dropped, and sink when placed in water. So in this worldview, while prime matter is passive, it does not exist as such. All material *beings* ("primary substances"), in contrast, have inherent powers to act in their own characteristic ways. The self-moving capacities of animals and humans need no special explanation.

In contrast, René Descartes, Thomas Hobbes, Newton, and other early modern thinkers developed a worldview in which material bodies were inherently passive or inert. All macroscopic phenomena, including the movements of animals and human bodies, were manifestations of matter in motion. According to Hobbes, all that exist are "bodies." Bodies move. In doing so they move other bodies; that is all that happens.[19]

We can describe this change by making use of terms coined by Baruch Spinoza. He distinguished between immanent causes, which produce changes within themselves, and transeunt causes, which produce changes in something else. The change from the Aristotelian to the Newtonian world views included a change from a world filled with immanent causes to one in which all causes, when properly understood, are transeunt causes. According to Newton, all motion in the universe was introduced "from the outside" by God. The laws of nature were, in the first instance, laws of motion that determined the patterns of motion after that initial impetus.

It has been argued that Newton had theological motives for developing the inertial view of matter.[20] One motive was what might be called Calvinist theological maximalism—to give as much credit to God as possible for whatever happens. So Newton ascribed all motive power to God. Second, this view of the physical universe made an obvious argument for the existence of God: someone had to have set it in motion in the beginning.

So a second change in the understanding of causes, from Aristotle to Newton, regards the question of what it is that causes cause. For Aristotle causal analysis was given of *substances* and their modification (including locomotion). For Newton causal analyses are given of *changes*, and changes are ultimately changes in motion.

---

[19]This summary of Hobbes's materialism is Wallace Matson's, *A New History of Philosophy*, vol. 2 (San Diego, CA: Harcourt Brace Jovanovich, 1987), 288.

[20]See Eugene Klaaren, *Religious Origins of Modern Science: Belief in Creation in Seventeenth-century Thought* (Grand Rapids, MI: Eerdmans, 1977).

3.2 *Current Assumptions*

I submit that since the demise of the Newtonian worldview, philosophical accounts of causation have not kept pace with science. The question of the innate powers of matter is little addressed these days by either scientists or philosophers,[21] and it seems a crucial preliminary question for locating God's action in the physical universe. Yet, if scientists after Newton are willing to do without Newton's version of the Prime Mover, they must be assuming, *contra* Descartes and Newton, that matter is inherently active.[22] So what is the ultimate source of the world's processes? Do we look to the beginning, in the Big Bang; or do we look instead to the basis of all processes in the smallest constituent events? Are quantum events brought about by transeunt causes or by immanent causes?

Nor is it clear what answer is to be given today to the question of what it is that causes cause. It is more common now to speak of events or states of affairs, rather than objects, as the effects of causes. Suppose we describe an event as a change from one state of affairs S_1 to another S_2. Then, is it S_2 or the change from S_1 to S_2 that requires causal explanation? And is S_1 the cause, or merely a necessary condition? Scientific language is not consistent here. When there is a regular connection between states of type 1 and states of type 2 we are inclined to speak of S_1 as the cause of S_2. However, if there is no such regularity we have two options. The first is always to look for an additional factor to label as the cause. If none can be found we speak of S_2 as random—and in such cases some would say that the event is uncaused.

Furthermore, it is necessary to distinguish between entropy- increasing changes and entropy-decreasing changes. Entropy- increasing changes require no additional explanation; here S_1 is an adequate explanation of S_2 (or of the change to S_2). Entropy-decreasing changes require an exchange with the environment, which is sometimes designated the cause, and the status of S_1 is reduced to that of a necessary condition.

Another complication: it is also possible to treat the "laws of nature" as the most significant "ingredient" in a causal explanations, in which case S_1 is designated as the set of initial conditions. This tendency has been furthered by Carl Hempel's influential nomological account of explanation, wherein a causal

---

[21]However, Richard Taylor claims that there remain two important philosophical questions regarding causation that have not been satisfactorily resolved. One is whether the concept of power or causal efficacy is after all essential, and whether there is after all any kind of necessary connection between a cause and its effect. *The Encyclopedia of Philosophy*, 1967 ed., s.v. "Causation," by R. Taylor.

[22]See Wolfhart Pannenberg, *Toward a Theology of Nature: Essays on Science and Faith*, ed. Ted Peters (Louisville, KY: Westminster/John Knox Press, 1993), esp. chap.1, "Theological Questions to Scientists."

explanation takes the form of a law and a set of initial conditions from which the *explanandum* can be deduced.[23]

So with current physics and cosmology having displaced the simple clockwork model of the universe, we are left without a clear scientific answer to the question of the causal nature of matter. Neither do we seem to have an agreed-upon philosophical analysis of causal concepts.

3.3 *The Ontological Status of the Laws of Nature*

Another area of disagreement concerns the ontological status of the laws of nature. When Newton and his contemporaries spoke of "the laws of nature" they no doubt understood the term as a metaphorical extension of the notion of divine law from the realm of theological ethics.[24] The ontological status of the laws of nature was unproblematic: they were ideas in the mind of God (a move for which the way had already been paved by Christian Platonists, who "located" Plato's realm of the forms in the mind of God).

What status have the laws of nature in contemporary thought? Paul Davies notes that:

> As long as the laws of nature were rooted in God, their existence was no more remarkable than that of matter, which God also created. But if the divine underpinning of the laws is removed, their existence becomes a profound mystery. Where do they come from? Who "sent the message"? Who devised the code? Are the laws simply *there*—free floating, so to speak—or should we abandon the very notion of laws of nature as an unnecessary hangover from a religious past?[25]

Davies, along with a number of other scientists, opts for what I shall call a Platon*istic* account of the laws of nature, meaning that like Plato's eternal forms, the laws have an existence independent of the entities they govern.

However, no one, to my knowledge, has provided a suitable account of how (or where) the laws might "exist" and how they affect physical reality—the same problems that have led most philosophers to abandon Platonic metaphysics. Furthermore, William Stoeger has argued persuasively that no such account of the laws of nature is necessary. All one needs to recognize is that there are objective regularities and relationships in nature, which scientists describe in human language and with the aid of mathematics.[26]

Stoeger's view appears the most credible account of the status of the laws of nature, but even if his arguments were not persuasive, this would still be the most viable option, since there seems to be no intelligible answer to the question of *how*

---

[23]See Carl Hempel, *Aspects of Scientific Explanation: and Other Essays in the Philosophy of Science* (New York: Free Press, 1965).

[24]See Mary Hesse, "Lawlessness in Natural and Social Science," draft paper for conference on quantum cosmology and the laws of nature, Vatican Observatory, September, 1991, typescript, p. 1.

[25]Davies, *Mind of God*, 81.

[26]Stoeger, "Contemporary Physics and the Laws of Nature."

the laws of nature could "exist" independently of either the mind of God or of the reality that instantiates them. Still, Stoeger's account leaves unanswered the question of what accounts for the objective regularities and relations in nature *if not* pre-existent laws. To this question we return in section 4.

3.4 *Pointers Toward a New Metaphysic*

An absolutely crucial development in contemporary understandings of the nature of reality regards its non-reducible hierarchical ordering in terms of increasingly complex systems. In some ways this recognition represents a return to the Aristotelian view that the form (organization, functional capacities) of an entity is equally constitutive of reality as is the stuff of which a thing is made.

The recognition of top-down causation is integral to this view. The hierarchical conception of reality suggests that an investigation of causation and the role of divine action begin at either the top or the bottom of the hierarchy, or both. The present state of our knowledge gives primacy to bottom-up causation; it is not clear whether this is an accident resulting from the long dominance of reductionist thinking, or whether bottom-up influences do in fact play a more decisive role in events than do top-down influences. In any case, no account of what makes things happen can neglect what we now take to the lowest level of the hierarchy—the quantum level.

This level is odd from the point of view of a causal analysis: quantum events do not obey deterministic laws. Individual events violate the principle of sufficient reason, which expresses our expectation that things happen when and as they happen due to some specific cause; that we should be able to give a reason why this happened now, rather than later or not at all. So here is a radical incompleteness in our knowledge at this most basic level. There is a metaphysical gap that we hunger to fill—by means of hidden variables, or layers of the implicate order, or some other means.

4 *A Proposal*

Let me summarize the requirements and hints so far assembled for an account of divine action. We are looking for a way to make sense of the traditional claim that God not only sustains all things, but also cooperates with and governs all created entities. This account needs to be consistent with other church teaching; it needs to leave room for *special* divine acts for both doctrinal and practical reasons; and it must not exacerbate the problem of evil.

It also needs to be consistent with science in the sense of saving the phenomena, and must not undercut the practice of science. However, I claim that it need not be consistent with metaphysical assumptions about matter and causation, which seem at present to be in great disarray.

Finally, a revised metaphysical account of causation that includes divine action as an integral part needs to take into account the recent recognition of the non- reducible hierarchy of complexity; this suggests two likely starting points, based on either top- down or bottom-up causation.

4.1 *Top-Down Causation: Another Road Not Taken*

Peacocke has very helpfully explored the topic of top-down causation and its possibilities for a theory of divine action. The concept of top-down causation itself is crucially important for a number of reasons. It explains how human free agency is possible within a highly deterministic universe. Hence, it is an important element in any account of divine action, since God's influence on human consciousness would otherwise have no possible influence on the rest of the cosmos. It is also necessary simply to understand the relations among the various levels of complexity in the natural world.

However, I have serious reservations about the adequacy of an account of divine action in terms of top-down causation alone. I shall discuss top-down effects in the human realm below, so here my focus will be on purported top-down effects on the non-human world. The clearest account given so far of how God operates is by *analogy* to human (top-down) agency in the inanimate world. However, this analogy does not solve the problem because human agency is brought to bear on the natural world via bodily action. Since God has no body, we get no help with the question of *how* God brings it about that events obey his will. This pushes us to consider whether God's causal relation to the world is like the causal relation between a human mind and the body. But Peacocke rightly rejects any dualistic account; and if we understand mental events as a function of the operation of the organism at a high level of organization, we again have trouble applying the account to God—God would then be the world-mind or the world-soul.

Ordinarily we invoke the concept of top-down causation when we find processes that cannot be described or understood in abstraction from the whole system, comprised of the affected entity in its environment. However, in such cases, it appears that the effect of the environment is always mediated by specific changes in the entity itself. For example, team spirit only affects an individual insofar as sights and sounds emanating from the other people affect the individual's sensory organs. Environmental factors affect individual organisms by means of, say, food surpluses or shortages, which in turn affect an animal only insofar as it does or does not eat.

So top-down causation by God should also be expected to be mediated by specific changes in the affected entities, and this returns us to the original question of how and at what level of organization God provides causal input into the system. I suggest we turn to a bottom-up account as the most plausible supplement to Peacocke's top-down approach.

4.2 *Overview of Modes of Divine Action*

In brief, the following is my position. In addition to creation and sustenance, God has two modes of action within the created order: one at the quantum level (or whatever turns out to be the most basic level of reality) and the other through human intelligence and action. The apparently random events at the quantum level *all* involve (but are not exhausted by) specific, intentional acts of God. God's action at this level is limited by two factors. First, God respects the integrity of the entities with which he cooperates—there are some things that God can do with an electron, for instance, and many other things that he *cannot* (e.g., make it have the rest-mass of a proton, or a positive charge).[27] Second, within the wider range of effects allowed by God's action in and through sub-atomic entities, God restricts his action in order to produce a world that *for all we can tell* is orderly and law-like in its operation. The exact possibilities for God's action within higher reaches of the natural order by means of cooperation with and governance of sub-atomic entities are highly debatable and will be considered below. But I hope to show that by taking quantum events as the primary locus for divine action it will be possible to meet many of the theological needs placed upon such a theory without running into insuperable theological or scientific objections.

In the following sections I address each of three regimes mentioned above: the quantum level, the regime of law-like behavior, and the human realm, asking in each case what are the possibilities for divine governance and cooperation. Since the levels are interrelated, the position outlined for each regime will have consequences throughout.

Divine action in the regime of law needs to be understood in terms of both bottom-up and top-down components. For example, I shall suggest that there is an analogy between God's respecting the "natural rights" of humans and a similar respect for the inherent rights of lower entities to be what they are. So calculating the possibilities for divine interaction with macroscopic objects involves the interaction of top-down and bottom-up influences with the God-given characteristics and potentialities of those beings.

Most of what I shall have to say about God's mode of action at the human level will be non-controversial. Insofar as it presents anything new, it will be by applying the results of my proposal regarding bottom-up causation at this level.

4.3 *God's Action at the Quantum Level*

The first question to raise with regard to the quantum level is this: Does God produce solely and directly all the events (phenomena) at this level, or are the entities endowed with "powers" of their own? In other words, I am raising here the question introduced in section 3—the activity or passivity of matter—but relating it specifically to the most basic entities in the physical universe and their relation to God. There are two possible answers: either God is the sole actor at this level or the entities (also) have their own (God-given) powers to act.

I believe we can rather quickly dismiss the first option on theological grounds. To say that each sub-atomic event is solely an act of God would be a version of occasionalism, with all the attendant theological difficulties mentioned above: it exacerbates the problem of evil; it also comes close to pantheism, and conflicts with what I take to be an important aspect of the doctrine of creation—that what God creates has a measure of independent existence relative to God, notwithstanding the fact that God keeps all things in existence. To put the

---

[27]The sense in which God "cannot" do all things with an electron is explained in section 4.3.

point another way, if God were completely in control of each event, there would be no-*thing* for God to keep in existence. To create something, even so lowly a thing as an electron, is to grant it some measure of independence and a nature of its own, including inherent powers to do some things rather than others.

These considerations lead to the conclusion that it is necessary for theological reasons to grant that every created entity, however small and ephemeral, has an existence independent of God. To be is to be determinate, and to be determinate is to have certain innate properties, including actual or potential behaviors.

Now, the peculiarity of entities at the quantum level is that while specific particles have their distinguishing characteristics and specific possibilities for acting, it is not possible to predict *exactly when* they will do whatever they do. This allows us to raise another question: Is the *when*: (1) completely random and undetermined; is it (2) internally determined by the entity itself;[28] is it (3) externally determined by the entity's relations to something else in the physical system;[29] or, finally (4) is it determined by God?

To make sure that these four options are distinguished clearly, allow me to present an analogy. A medieval philosopher by the name of Buridan is supposed to have hypothesized that if a starving donkey were placed midway between two equal piles of hay it would starve to death for want of sufficient reason to choose one pile rather than the other. I am supposing that entities at the quantum level are miniature "Buridian" asses. The asses have the "power" to do one thing rather than another (walk to one of the piles of hay). The question is what induces them to take one course of action rather than the other (or to take a course of action at a particular time rather than another or not at all). By hypothesis, there is nothing external to determine the donkey's choice (no difference in the piles of hay). Also, by hypothesis, there is nothing internal (no sufficient reason) to determine the choice.

Insofar as epistemological interpretations of quantum theory and the quest for hidden variables are rejected, we are left with the conclusion that there is no "sufficient reason" either internal or external to the entities at this level to determine their behavior. While these issues are still open, many physicists have rejected both epistemological interpretations and at least local hidden variables.

By process of elimination, this leaves options 1 and 4: complete randomness or divine determination. The fact that the inventor of Buridan's ass believed the donkey would starve illustrates the philosophical assumption that all events must have a sufficient reason. This same intuition is what has made the apparent randomness of quantum events so difficult for the scientific community to accept. I shall argue that the better option is divine determination. While most of my argument will be for the advantages of this thesis for theology, it is important to

---

[28]In Spinoza's terms, is the entity itself an "immanent cause"?

[29]That is, moved by a "transeunt cause."

bear in mind that it has the further advantage of consistency with the principle of sufficient reason.[30] To put it crudely, God is the hidden variable.

4.3.1 *God's Governance, Cooperation, and Sustenance*

It is neither theologically nor scientifically problematic to maintain that God's creative activity involves the sustaining in existence of that which has been created. However, it poses an interesting question to see if we can find work for the terms 'cooperation' and 'governance'. These terms turn out to be quite valuable here. My proposal is that God's governance at the quantum level consists in activating or actualizing one or another of the quantum entity's innate powers at particular instants, and that these events are not possible without God's action. This is the manner and extent of God's governance at this level of reality.

I have already claimed that we need to maintain that all created entities, despite being sustained by God, are entities in their own right *vis-à-vis* God. Only in this way can we say that there is a created entity with which God can cooperate. God's action is thus limited by or constrained by the characteristic limitations of the entities with which he cooperates. This limitation is, in one sense, voluntary—God *could* cause an electron to attract another electron, but so far as we know has chosen not to do so. In another sense, though, this limitation is a logical necessity—an electron that attracts electrons is no longer (really) an electron.

This principle of God's respecting the integrity of the entities he has created is an important one. Proposing it is in line with Polkinghorne's speaking of "free processes" in nature. I further suggest, on the strength of a similar analogy with the human realm, that we speak of all created entities as having "natural rights," which God respects in his governance. This is the sense in which his governance is cooperation, not domination.

4.3.2 *God's Bottom-Up Causation*

The rationale for proposing this bottom-up account of divine governance is based upon what remains true about reductionism and determinism, even after recent criticisms of these positions are taken into account. The theological goal is to find a *modus operandi* for God at the macro-level—the level that most concerns us in our Christian lives. The ontological reductionist thesis seems undeniable—macroscopic objects are *composed of* the entities of atomic and subatomic physics.[31] This being the case, much (but not all) of the behavior of macro-level objects is determined by the behavior of their smallest constituents. Therefore, God's capacity to act at the macro-level must include the ability to act upon the most basic constituents. This is a conceptual claim, not theological or scientific.

---

[30]However, this is probably a minor point, since it not clear what the principle itself is based upon.

[31]This point stands even for those who want to add a mind or soul to the human body in order to get a living person: the *body* is still nothing but a complex organization of its most basic physical parts.

However, the theological question that arises immediately is whether God acts upon these parts-making-up-wholes only in rare instances, or whether God is constantly acting on or in everything. Over the long history of the tradition, I believe, the majority view has been that God acts in all things at all times, not just on rare occasions.

We can approach this question from the following angle: we object to interventionist accounts of divine action because it seems unreasonable that God should violate the laws he has established. We object to "God of the gaps" accounts of divine action for epistemological reasons—science will progress and close the gaps. But I think there is a more basic intuition behind the rejection of both of these views: God must not be made a competitor with processes that on other occasions are sufficient in and of themselves to bring about a given effect. In addition, if God's presence is identified with God's efficacy[32] then a God who acts only occasionally is a God who is usually absent.

So our theological intuitions urge upon us the view that, in *some* way, God must be a participant in *every* (macro-level) event. God is not one possible cause among the variety of natural causes; God's action is a necessary but not sufficient condition for every (post-creation) event. In addition, I claim that God's participation in each event is *by means of* his governance of the quantum events that *constitute* each macro-level event. There is no competition between God and natural determinants because, *ex hypothesi*, the efficient natural causes at this level are insufficient to determine all outcomes.[33]

4.3.3 *Conclusions*

In this section I have proposed a bottom-up account of divine action. God governs each event at the quantum level in a way that respects the "natural rights" of the entities involved. God's action is (and from the point of view of science, *must be*) such that, in general, these events "accumulate" in regular ways. However, within the limits provided by the natures of the quantum entities involved and by *our need* for an orderly and predictable world, God is free to bring about occasional extraordinary events at the macro-level—exceptions that suit God's own purposes. This account provides a *modus operandi* for God's constant and all-pervasive governance of the physical cosmos, but does not rule out special acts upon rare occasions.

Each event at the quantum level, then, needs to be represented as follows:

$$\begin{array}{c} G \searrow \\ S_1 \longrightarrow S_2 \end{array}$$

Here S_1 represents the prior state of the entity or system, and G represents an intentional act of God to actualize one of the possibilities inherent in S_1. Notice that

---

[30]See n. 6 above.

[33]My suspicion is that arguments based on quantum non-locality could also be used to reinforce the claim that if God works in any quantum event, God must work in all of them.

this is a radical revision of the meaning of 'cause' as it is used in science and everyday life, since on the view presented here no set of natural events or states of affairs is ever a sufficient condition for an event. One necessary condition will always be an element of divine direction; nothing ever happens without God's direct participation.

Notice, also, that this view splits the difference between Newton's view of the utter passivity of matter and Aristotle's view of substances possessing their own inherent powers to act. On this view, created entities have inherent powers, yet they are radically incomplete: they require God's cooperation in order to be actualized.

4.4 *God's Action in the Regime of Law*

By 'the regime of law' I mean to refer to the events occurring at all levels of complexity above the quantum level but below the level of free action. In this section I shall first mention the constraints placed upon our conclusions by the requirements of both science and theology. Second, I shall attempt to state the consequences that the proposal of the previous section has for a conception of the relationship of divine action to the laws of nature.

4.4.1 *Reconciling the Needs of Science and Religion*

We come to the crux of the problem of divine action when we address the regime of law. Science both presupposes for its very existence the strictly law-like behavior of all entities and processes, and constantly progresses in its quest to account for observable phenomena in terms of elegant sets of interrelated laws. As stated above, no account of divine action that undermines the practice of science or denies its major findings can be considered adequate.

However, the law-like regularity of nature has regularly been equated with causal determinism, with the result that God's action can be understood in one of three ways: God is *not* causally involved in the ongoing processes of the universe; God *is* involved, but only by intervention; or God's action amounts to supporting the ongoing regular processes.

Since none of these options seems an adequate account of divine action,[34] much of the previous work in this area has concentrated on finding respects in which the processes in this regime are not causally determined by prior conditions and natural laws. However, if we adopt a bottom-up view of divine causation, the problem of God's action at the macro-level reverses itself. The problem is not that of beginning with the law-governed character of macro-level phenomena, and then trying to find room for divine action. Rather, one begins with a strong measure of divine determinism at the most basic level of natural processes and then has to account for the observed regularity and the applicability of "laws of nature."

---

[34]For reasons described in section 2.

At this level we have to consider a similar set of questions as we did in considering God's relation to entities at the quantum level. But these questions are complicated by relations to answers given at the quantum level.

Let us take a very simple example of a macro-level entity: a billiard ball. The ball is composed of cellulose structures; cellulose in turn is composed of carbon, hydrogen, and oxygen. The stability of the elements and their ability to form this compound are effects of overall patterns in the behavior of the constituent sub-atomic entities. The characteristics of the wood itself (e.g., the grain and density) are the result of past biological processes, similarly grounded in (but not uniquely determined by) the behavior of the quantum-level entities. The characteristics of the wood give rise to some of the characteristics of the ball itself, such as its elasticity. Others, such as its shape and size are effects of the manufacturing process, and perhaps other accidents since then. So past environmental influences have interacted with influences determined in a bottom-up manner by characteristic behaviors of its constituents.

To be a billiard ball is to have a set of inherent properties that allow for a characteristic range of behaviors and interactions with other entities in the environment.

How does God act with respect to billiard balls? Is the rolling of a billiard ball always, sometimes, or never the result of divine action? To be consistent with the above analysis we must say that divine action is always a necessary condition, but never a sufficient condition for such an event. The continued existence of the ball is dependent upon God acting in regular ways at the quantum level (e.g., governing the movements of the electrons in its atoms). But such patterns of action give rise to an entity capable of interacting in some ways (but not others) with the environment. One of the "capacities" with which the ball is endowed by virtue of its constitution is to lie still until struck; another is to roll when struck by the cue stick. So the rolling of the ball (ordinarily) will be a joint effect of an impact and of God's sustaining the ball's characteristics "from below."

One might now ask how this account differs from a standard modern account of God sustaining entities in existence whose behavior is determined by the laws of nature, in particular the laws of motion. The differences are subtle.[35] First, God is not merely keeping the ball in existence; God is maintaining its typical characteristics through intentional manipulation of its smallest constituents. This fulfills the theological requirement that God be understood as acting within all macro-level events. Second, the behavior of the ball and its characteristic interactions with the environment are not determined externally by laws "out there," but are inherent characteristics, emergent from the behavior of its constituent parts.[36] And, third, within the limits provided by the "natural rights" of those constituents, God could effect extraordinary behaviors or interactions by governing the constituents in atypical ways. A philosopher once wrote that it is not impossible for all the atoms in a billiard ball to "go on a spree" so that the ball

---

[35]And of course they are intended to be subtle. The goal here is to produce an account of divine action that does not conflict with observations.

[36]There may be exceptions here, such as the law of gravity.

would suddenly move without any outside force. The account of divine action given here entails that such things are possible, but if they happen they are not the result of *chance* synchronization of random vibrations, but rather of *intentional orchestration* of the vastly many micro-events.[37]

4.4.2 *God and the Laws of Nature*

I have already suggested that we view the statistical laws of quantum mechanics as summaries of patterns in God's action upon quantum entities and processes. In light of this claim, what are we to say about the laws of nature above the quantum level but below the human level?

To say that these laws are nothing but summaries of individual acts of God is to ignore the fact that God's actions at the quantum level constitute macro-level entities that have their own distinct manners of operation. Mathematical description of the typical behaviors of these entities yields our "laws of nature."

Notice that at this point I am not saying anything new or unorthodox scientifically. I am simply assuming what has turned out to be true about reductionism. Macro-level objects are complex organizations of their most basic constituents (this is analytic). To a great extent,[38] the behavior of the whole is determined by the behavior of its parts. So the laws that describe the behavior of the macro-level entities are consequences of the regularities at the lowest level,[39] and are *indirect* though intended consequences of God's direct acts at the quantum level. What is unorthodox (scientifically) is the grounding I have given to the statistical regularities in the behavior at the quantum level.

Now, if the behavior of macro-level entities is dependent upon God's sustaining their specific characteristics by means of countless free and intentional acts, why do natural processes look so much like the effect of blind and wholly determinate forces? Since we have undermined the standard modern answer—determination by the laws of nature—a different account must be provided. The account to be given here is theological: one of God's chief purposes is (must have been) to produce a true cosmos—an orderly system. If we ask *why* God purposed an orderly universe we might speculate that it is for the intrinsic beauty and interest of such a cosmos; we could ground this speculation in our own aesthetic appreciation and in the supposition that our appreciation is an aspect of the *imago Dei*.

An equally significant explanation is the necessity for such order and regularity so that intelligent and responsible beings such as ourselves might exist to "know, love, and serve him." The law-like character of the universe is a necessary prerequisite for the physical existence of systems as complex as our

---

[37]The contentious point here has to do with the question whether or not quantum effects necessarily "wash out" at the macro-level. I am assuming that they need not. See section 5.4.

[38]That is, within the limits circumscribed by top-down causation.

[39]This is true even if the laws at higher levels cannot be derived mathematically from the laws of quantum mechanics.

bodies; it is also necessary for intelligence. There could not, of course, be brains capable of investigating the cosmos without the cosmos being orderly; but if, *per contra*, suitably complex beings did exist in a chaotic environment they would be unable to develop intelligence. *A fortiori* they would not be able to make responsible free choices. Free agency requires a background of law-like processes so that the effects of one's action can be predicted.

But how law-like does the cosmos have to be? There is a vast continuum between total chaos (which is actually unimaginable—think of the difficulty in producing a truly random series of numbers) and the absolute regularity (determinism) that has often been assumed since the rise of modern science. The assumptions upon which this paper is based require that at some level a principle of the uniformity of nature must prevail. Otherwise God's governance would not include intelligent use of cause-effect relations (any more than ours could), and we would be back to occasionalism. But this does not entail that *our* scientific laws could suffer no exceptions—in fact, I have just been arguing that by tampering with initial conditions at the quantum level, God can bring about extraordinary events; events out of keeping with the general regularities we observe.

So the question is: To what extent can God bring about such extraordinary events without defeating his own purposes? It is obvious that the whole cosmos does not fall into chaos if there are occasional exceptions. The more interesting question is how much disorder is possible without destroying *our* ability (or motivation) for intelligent appreciation of the cosmos or our ability to take responsibility for our actions.

John Hick has written that God withholds obvious signs of his existence in order to create epistemic distance, and hence to leave us free to believe or not in his existence.[40] This argument seems to have something right about it—it is certainly the case that God could act in such a way as to make it much more difficult to deny his existence. Yet the argument seems faulty, too. It seems to overlook the traditional account of disbelief as sin, and the fact that even in the face of the most astounding evidence given by and on behalf of Jesus, the crowds largely failed to believe. I suggest that God's action does remain largely hidden and is always ambiguous—when manifest it is always subject to other interpretations. But this is not because we would otherwise be forced to believe in God (as Hick claims) and then to obey him. Rather it is because we would lose our sense of the reliable behavior of the environment. When the environment is taken to behave in a set (and therefore predictable) manner, we can make responsible choices about how to act within it. If instead we saw the environment as a complex manifestation of divine action, we would lose our sense of being able to predict the consequences of our actions, and would also lose our sense of responsibility for them. So, for instance, if I carelessly allow my child to fall off a balcony, I would not see myself as responsible for his injuries since God was there with all sorts of opportunities for preventing them.

---

[40]See John Hick, *Evil and the God of Love* (London: Macmillan, 1966).

These psychological requirements for responsible action seem to require in turn that extraordinary acts of God be exceedingly rare (that we not have any adequate justification for expecting God to undo the consequences of our wrong choices) and that they normally be open to interpretation as (somehow) in accord with "the laws of nature." So God's relation with us requires a fine line between complete obviousness and complete hiddenness—the latter since we could not come to know God without special divine acts.[41] The difficulty in describing God's action is that we want to have it both ways: both that there be evidence for divine action—something that science cannot explain—*and* that there be no conflict with science. So a suitable theory of how God acts leaves everything as it was scientifically. But then there is no evidence upon which to argue that such a view ought to be accepted over a purely naturalistic account. Perhaps the ambivalence we find in attempting to provide a description of divine action is rooted in an intentional ambiguity in God's acts themselves.

In summary: I am proposing that the uniformity of nature is a divine artifact. God could produce a macroscopic world that behaved in much less regular ways by manipulating quantum events. However, there are two kinds of limits by which God abides. The first is respecting the inherent characteristics of created entities at both the quantum and higher levels—respecting their "natural rights." However, within the degrees of freedom still remaining many more strange things could happen than what we observe—billiard balls "going on a spree." So we must assume that God restricts extraordinary actions even further in order to maintain our ability to believe in an orderly and dependable natural environment.

4.4.2.1 *Chaos Theory: A Subsidiary Role*

The real value of chaos theory for an account of divine action is that it gives God a great deal of "room" in which to effect specific outcomes without destroying our ability to believe in the natural causal order. The room God needs is not space to work within a causally determined order—ontological room—but rather room to work within our perceptions of natural order—epistemological room.

It may be significant that two of Christians' most common subjects for prayer are health and weather. Weather patterns are clearly chaotic, so it is never possible to claim definitively that a prayer regarding the weather has or has not been answered. I suspect that because most bodily states are so finely tuned they too involve chaotic systems. Thus, the recovery from an illness and especially the timing of recovery cannot always be predicted. So do we pray for these things

---

[41]History, both in scripture and elsewhere, reports frequent miracles in ancient times; relatively few are reported today, and contemporary reports come more often from less-educated populations. Most commentators assume that we are seeing a decrease in gullibility. It is possible, though, that there are in fact fewer extraordinary events because, with our sharpened sense of the order of nature, with increased abilities to make measurements, our sense of the order of nature has become more *fragile*. As technological and scientific capabilities to test miracle claims have increased, so have our abilities to cast doubt upon causal regularity.

rather than others because we lack faith that God could "break a law of nature" or is it rather because of our long experience with a God who prefers to work on our behalf "under the cover of chaos"?

4.4.3 *Top-Down Causation: A Subsidiary Role*

The recognition of top-down considerations plays two vital roles in this proposal. In addition to the factor to be pursued below—God's top-down influence on the created order through human top-down agency—a second is as follows: I have argued that God's action at the sub-atomic level governs the behavior of nature's most basic constituents, but without violating their "natural rights." So in order to understand the limits of divine agency at that level and in all higher levels, we need to know what are the intrinsic capabilities of those entities. However, it appears that the behaviors that are "natural" to an entity (from any level of the hierarchy) are not simply given—entities can do more, have more degrees of freedom, when placed in a more complex environment. Studying the inherent powers of an entity in isolation will not tell us what it can do when incorporated into a higher regime. For example, humans can eat, move about, make noise, in total isolation from community, but our truly human capacities such as language only emerge in society. A solitary individual, or even an individual who is part of a herd rather than a society, cannot teach philosophy.

There are limits, of course, to the increased freedom that any particular regime can promote. For example, cats can be taught to play games and eat onions when incorporated into a household; pet rocks cannot, no matter how stimulating the company. However, this limitation applies to known regimes. As Polkinghorne has pointed out, we are not well-acquainted with the possibilities for either human life or natural events within a regime in which God's will is the dominant factor. Medieval Christians believed that the great chain of being was also a chain of command, broken by human (and angelic) disobedience. Saintly beings repaired the chain, and hence holy people such as St. Francis could command the animals. While this account does no more than Peacocke's to explain *how* divine influences are transmitted to sub-human beings, it does suggest that a phenomenon has been recognized throughout Christian history: natural beings and processes operate somewhat differently in the presence of people imbued with the presence of God.

4.5 *Divine Action in the Human Realm*

God has a number of ways to affect human beings by means of the spoken and written word. But this kind of communication is the transmission of effects via normal human processes, and we have to ask where these effects originated. How does original communication between the divine and human take place?

A theory consistent with the proposal of this paper is that God affects human consciousness by stimulation of neurons—much as a neurologist can affect conscious states by careful electrical stimulation of parts of the brain. God's action on the nervous system would not be from the outside, of course, but by means of bottom-up causation from within. Such stimulation would cause thoughts to be recalled to mind; presumably it could cause the occurrence of new thoughts by coordinated stimulation of several ideas, concepts, or images stored in memory. Such thoughts could occur in conjunction with emotions that suited the occasion. I suggest that concatenations of such phenomena that convey a message or attitude from God to the individual is what constitutes revelation.[42] I believe that this account fits with the phenomena of religious experience. It is interesting to note that medieval mystics placed a great deal of emphasis on the faculty of *memory*—taking it as an important means by which God made himself known to them.[43] It is also consistent with the extent to which revelation is formed of materials available in the person's culture.

The following account illustrates my suggestion: A student reported that the thought suddenly occurred to him that he should speak to a recent acquaintance about his drinking problem—even though he did not know that the person had such a problem. In conjunction with the thought, he had a sudden memory of his troubled relationship with his father due to alcohol, and felt an associated emotional impact from that memory. The conjunction of all of these experiences convinced him that he was receiving a message from God to approach the acquaintance and urge him to attend to the alcohol problem before it affected his relations with his children.

In short, religious experience is made up of the same materials of which ordinary experience is made. This is consistent with the view that God acts upon consciousness by stimulating and coordinating materials that are already stored in the subject's brain. So this is a top-down account of God acting upon our actions, since its explanation requires reference to God as the "environment" within which the person functions. However, it depends for its means of operation on a bottom-up account of God's affecting the brain. As stated above, I believe all top-down causes have to involve some point of contact between the larger whole and the affected part.

4.6 *Overview*

I have claimed that we need to distinguish among three different regimes for the purpose of devising a theory of divine action: the human, the quantum and, in between, the regime of law. However, these three regimes cannot be considered in isolation. God acts within the regime of law by actualizing, at chosen times, one or another of the built-in potentialities of each sub-atomic entity. Coordination of all such events generally produces the law-like behavior we observe—both statistical regularities of aggregates of quantum events and the law-like behavior of macro-entities and processes.

So when we consider the behavior of entities in the regime of law, God's ability to engineer desired outcomes is on the one hand limited by his decision to

---

[42]See George F.R. Ellis, "The Theology of the Anthropic Principle," in *Quantum Cosmology*, section 8.1.

[43]This notion originated with Augustine's Platonic epistemology, but there must have been some experiential correlate to keep the emphasis alive.

respect the "natural rights" of the entities with which he cooperates, and in this sense God's consequent freedom of action decreases as we go up the scale of complexity—more and more complex constraints are placed on divine outcomes by the more and more complex sets of entities with their "natural rights" to be allowed their characteristic limitations. Why this is so can be seen by considering an analogy—the increasing complexity of engineering an outcome in increasingly complex societies. Among a group of friends, exerting one's will is constrained only by one's own natural limitations and by the wills (rights) of the other individuals. However, if these individuals together constitute a social entity or institution with its own proper rights and responsibilities, further constraints are imposed (e.g., club rules require attendance at group activities, forbid certain activities). If this social entity is a part of a larger social entity (a national organization of local clubs) it will be further constrained by the character of that larger entity, and so forth.

On the other hand, as I have emphasized above, placing entities in more complex environments increases the scope of their inherent powers. To return to our analogy, there are things God can do with a national organization that could not be done with a collection of individuals. So, in some ways, this gives God more room to maneuver without violating their "natural rights." And finally, we have to reckon with the possibilities for interaction between God's top-down and bottom-up causation.

5 *Evaluation*

In this final section I shall attempt an evaluation of the position presented herein. First, I shall indicate the ways in which this proposal meets the theological and scientific criteria proposed in section 2. Second, I shall mention some of the objections that I expect will be raised and reply to them. Third, I shall mention advantages of this proposal over some others. Finally, I shall describe the issues that need further development.

5.1 *Theological Criteria*

This proposal meets the criteria set out in section 2 as follows:

5.1.1 *Doctrine*

The above account of divine action allows for God's cooperation with and governance of all events in a way that leaves (some) room for special (extraordinary) divine acts. It also emphasizes the non-coercive, freedom-respecting character of God's action in the human realm and extends these features to an account of divine action in the non-human realm as well.

5.1.2 *Knowledge of God's Actions*

I have proposed an account that for all practical purposes is observationally indistinguishable from a naturalistic or deistic account. The built-in ambiguity in distinguishing intentional action of God from natural events—the general hiddenness of God's action in random processes and chaotic systems—raises the question of how we could ever know of God's action. The answer is that we only see God's action by observing larger patterns of events. Consider this analogy: Rocks are arranged on a hillside to spell out "Jesus saves." It is obvious when looking at the whole that this has been done intentionally. However, investigation of the location of any single rock in the collection would not reveal that it had been intentionally placed. Similarly, in the student's account of his revelatory experience, the occurrence of the thought that he should speak to his acquaintance about a drinking problem, by itself, would merely be odd. It is only because of its occurrence in conjunction with the other experiences that he took it as possibly revelatory. Its revelatory status was confirmed when he acted on it: the acquaintance confessed to the problem and set out immediately to get help. So God's acts are recognized by the way particular events fit into a longer narrative, and ultimately into the great narrative from creation to the eschaton, from Genesis to Revelation.

In order to maintain a place for special acts of God, it is important to distinguish between two classes of events: those that would not have happened without causal input from God (and we are here assuming that *all* events fall into this category), and events that count as God's *actions*. In discussing human action we distinguish, from among all of the events that humans *cause*, the smaller class of those that express their intentions. Only the latter are described as *actions*. So all events are the result of God's causal influence; only some events express (to us) God's intentions. It is the latter that ought, strictly speaking, to be called God's actions.

5.1.3 *Prayer*

Petitionary prayer makes sense on this account, but more so for some kinds of events than others. Events that are recognized as possible yet unpredictable (i.e., the results of chaotic processes, unpredictable coincidences) are more to be expected than events that defy the law-like behavior of natural processes. However, prayers for the latter are not out of the question. One condition under which we might expect such prayers to be answered is when the divine act would serve a revelatory purpose, since, by hypothesis, God must occasionally act in extraordinary ways to make himself known.

It is clear that in cases where outcomes are not predictable (e.g., weather, healing), one of the most valuable conditions for recognizing the action of God is that it constitutes a meaningful complex of prayer and response. The prayer beforehand makes it possible for an *unpredictable* event—an event that "might have happened in any case"—to reveal the purposes of God. So while prayer might not be necessary to persuade God to act, it will be necessary for us to recognize the fact that God is acting.[42]

5.1.4 *Deism and Occasionalism*

The central goal of this paper was to present an account of divine action that steers a course between deism and occasionalism. I believe that this proposal does so. God's action in every event is guaranteed, and so is some measure of control over the course of events such that special, even extraordinary, acts are possible. At the same time, God's decision to cooperate with created entities rather than to override their natural characteristics means that entities above the quantum level, with their built-in capacities for action, are allowed by God to use them—natural causal relations are not denied. It is this "letting be" that provides an explanation for the fact that the universe does not appear to manifest the purposes of an all-wise and all-powerful God in all of its details.

5.2 *Scientific Criteria*

This proposal saves the appearance (within limits) of law-governed processes and justifies scientific research. However, the justification is theological: the universe can be expected to be intelligible since it is one of God's high purposes that it be so. In addition, it saves quantum physics from violation of the principle of sufficient reason.

5.3 *Objections and Answers*

The following are some of the objections that might be raised against this proposal. Others will be addressed in conjunction with the evaluation of competitors in section 5.4.

5.3.1 *Ad Hoc-ness*

One criticism of this position is that it appears *ad hoc*: God *can* make all sorts of wonderful things happen, but *almost never does so*. In defense, I claim that the apparent absence of divine action is *ethically* necessary.[43] First, unless and until we know more about how God's acts at the quantum level affect the macro-level, we really do not know what actions are possible for God without violating God's ethical principles.

Second, the intentional but metaphysically unnecessary decision on God's part to act openly only on rare occasions is necessary if God is to interact with humans without destroying their sense of the dependability of the natural order and

---

[42]There are surely other reasons for prayer, as well, such as building a relationship with God, and perhaps the praying itself in some way contributes to bringing about the desired effect.

[43]See Ellis, "Ordinary and Extraordinary Divine Action" (in this volume).

of their own responsibility. This not only answers the charge in question, but has the further advantage of answering the very troubling question raised for Christians who believe in providence: Why would God answer prayers for small things (cure of a cold), while apparently refusing to take actions that would prevent much suffering (an early death for Hitler)?

5.3.2 *Uncertainty of Science*

It may be said that this proposal is faulty because the science upon which it is based is controversial or likely to change. My reply is that the proposal is not *based* on the particularities of current quantum theory. Its real basis is ontological reductionism (a view that is not likely to be overturned so long as there is science as we know it) and on the theological claim that God works constantly in all creatures. I conclude that God therefore works constantly in the smallest or most basic of all creatures. This claim will stand, however those most basic constituents are described.

However, current theories in quantum physics do provide a valuable ingredient for this theory of divine action: the currently accepted supposition of indeterminacy at the quantum level provides a handy analogue for human freedom, and thus grounds for the claim that God's action is analogously *non-coercive* at the quantum level. I would be sorry to have to give up this element, but it is not essential to the proposal.

5.3.3 *Two Languages*

How shall we now speak of causes? For any event there will be at least two, usually three, necessary conditions: the prior state of the system, God's influence, and often influences from the environment. Our standard practice in answering questions about causes is to select from a set of necessary and jointly sufficient conditions the one that is most relevant to our purposes. On this account, every event can be considered from the point of view of science, and the natural conditions (previous state and environment) will then be cited as causes. Every event can likewise be considered from a religious point of view, where God's action is the relevant factor.

It might be said that this position is a version of a "two- language" solution, similar to some strategies for answering the problem of free will and determinism. However, the difference is that divine action and natural causation, on the view proposed here, are no longer opposing accounts (as are freedom and determinism), since neither the natural nor the divine condition for an event is assumed to be a sufficient condition.

I claim that this way of speaking about causes is not only consistent with our normal linguistic practice, but also reflects a common way of speaking of divine action in scripture. For example, Joseph says to his brothers: "And now do not be distressed, or angry with yourselves, because you sold me here; for God sent me here to preserve life" (Gen. 45:5). The full account of the event involves both human and divine agency. Joseph emphasizes God's providence while recognizing at the same time that his brothers can indeed be held accountable.

5.3.4 *God's Lack of Knowledge*

Peacocke claims that God's action at the quantum level is forestalled by the fact that particular events are as unpredictable to God as they are to us. My proposal evades this difficulty since by hypothesis these events are not random; they are manifestations of divine will.

5.4 *Advantages Over Previous Proposals*

The theorist in this area whose work comes closest to mine is W.G. Pollard, who suggested that God works through manipulation of all sub-atomic events.[44] Pollard has been criticized by both David J. Bartholomew and Barbour for providing an account whereby all events are *determined* by divine action. Such an account, they say, in incompatible with human freedom. My account avoids this problem, first, by qualifying bottom-up divine influences by means of top-down causation.[45] Second, and more importantly, my account of God's respect for the natural rights of all creatures leaves room for genuine human freedom.

Another criticism of Pollard is that he takes God's action at this level to be constrained within fixed statistical laws. However, I concur with Bartholomew, who claims that Pollard's work involves a misunderstanding of the very nature of statistical laws.[46]

The constraints upon God's action that I propose come instead from God's commitment to respect the innate characteristics with which he has endowed his creatures. This seems to leave some room for God to maneuver at the macro-level, but, as I mention below, the exact amount of room is difficult to ascertain. This same factor (constraint) allows me to answer a charge Bartholomew makes against Donald MacKay. MacKay claims that God is in detailed control of the behavior of all elementary particles.[47] But if this is the case, Bartholomew asks, why does God appear to act in such a capricious manner?[48] My answer: God's control is limited by his choice to cooperate with rather than over-ride created entities.

So we have returned to the issue of finding an account of *how* God acts that produces a result between two extremes: On the one hand, the account must not lead us to expect God to have total control of every outcome. If so, it would deny human freedom, clash with the apparent randomness and purposelessness manifested in some aspects of nature, and leave God entirely responsible for all of the evil in the world. On the other hand, such an account must not lead to the

---

[44]William Pollard, *Chance and Providence: God's Action in a World Governed by Scientific Law* (New York: Scribner, 1958).

[45]Barbour has already noted the need for this qualification in his discussion of Pollard's position in *Issues in Science and Religion*, 430.

[46]See David J. Bartholomew, *God of Chance* (London: SCM Press, 1984), 127-28.

[47]See Donald MacKay, *The Clockwork Image: A Christian Perspective on Science* (Downers Grove, IL: InterVarsity, 1974).

[48]Bartholomew, *God of Chance*, 25.

conclusion that God has no room within created processes intentionally to influence the course of events. I believe that my account successfully steers between these two extremes.[49]

5.5 *Unanswered Questions*

The most serious weakness of this paper is in describing the consequences of the theory of divine action at the quantum level for events at the macro-level. What, exactly, are the possibilities for God's determining the outcomes of events at the macro-level by governing the behavior of sub-atomic entities? What exactly are the limits placed upon God's determination of macro-events by his decision not to violate the natures of these entities? Is this a broad opening for divine action, or a very narrow window? The answer depends on sorting out issues in the relation of quantum physics to the rest of science. Polkinghorne states that:

> There is a particular difficulty in using quantum indeterminacy to describe divine action. Conventional quantum theory contains much continuity and determinism in addition to its well-known discontinuities and indeterminacies. The latter refer, not to all quantum events, but only to those particular events which qualify, by the irreversible registration of their effects in the macro-world, to be described as measurements. Occasions of measurement only occur from time to time and a God who acted through being their determinator would also only be acting from time to time. Such an episodic account of providential agency does not seem altogether theologically satisfactory.[50]

Polkinghorne would include among these possible instances of meaningful divine action, I believe, cases where sensitivity of chaotic systems to initial conditions involves changes so slight as to fall within the domain of quantum mechanics. The classic example of a macroscopic system that "measures" quantum events is Schrödinger's poor cat, whose life or death is made to depend on the status of one quantum event.

Against Polkinghorne's view, Robert Russell would argue that the important fact that has been overlooked here is the extent to which the general character of the entire macroscopic world is a function of the character of quantum events. Putting it playfully, he points out that *the whole cat is constituted by quantum events*!

We can imagine in a straightforward way God's effect on the quantum event that the experimental apparatus is designed to isolate; we cannot so easily imagine the cumulative effect of God's action on the innumerable quantum events that

---

[49]It also avoids the interventionist overtones of Bartholomew's suggestion that it might be better to assume that God leaves most quantum events to chance and only acts upon occasion to determine some specific outcome. See ibid., 130.

[48]Polkinghorne, "Metaphysics of Divine Action."

constitute the cat's existence. Yet this latter is equally the realm of divine action.[51] I have been assuming Russell's position throughout this paper. Yet even if Russell is correct, there still remains a question. Does the fact that God is affecting the whole of reality (the whole cat) *in a general way* by means of operation in the quantum range allow for the sort of special or extraordinary divine acts that I claim Christians need to account for? Or would such special acts be limited to the few sorts of instances that Polkinghorne envisions?[52]

A second open question comes from our lack of knowledge regarding the possibilities for top-down causation, and the role of "holist laws." In particular, we lack knowledge of the possibilities of divine top-down causation and of the possible behavior of natural entities within a regime constituted by the full presence and action of God. We have a glimpse of this regime in the resurrection of Jesus, and a hint from Paul that the whole cosmos awaits such a transformation. Are there states in between this final state, in which God will be all in all, and the present state of God's hiddenness in natural processes? Do the extraordinary events surrounding the lives of Jesus and the saints represent such an intermediate regime?

Finally, it has been the consistent teaching of the church that God respects the freedom and integrity of his *human* creatures. I have proposed as an axiom of my theory of divine action that God respects the "natural rights" of entities at the quantum level as well. Is it, then, the case that *all* created entities have intrinsic characters that God respects in his interaction with the world? And what does God do when the rights of creatures at different levels of the hierarchy come into conflict? The claim that God acts consistently throughout the hierarchy of complexity has consequences regarding what sort of thing God should and should not be expected to do with creatures within the intermediate realm between humans and quarks. For instance, it would be consistent with my proposal for God to cause Buridan's ass to eat, but not to cause Balaam's ass to speak. Does our experience of God's action in our lives bear out such a distinction, and does this distinction help explain why some prayers are answered and others not?

My hope is that despite these unanswered questions, the foregoing proposal provides insights that are worthy of further pursuit.[53]

---

[51]See Russell, "Quantum Physics in Philosophical and Theological Perspective," in *Physics, Philosophy, and Theology: A Common Quest for Understanding*, ed. Russell, William R. Stoeger., and George V. Coyne (Vatican City State: Vatican Observatory, 1988), 343-68.

[52]My hope is that this question can be addressed at a future conference.

[53]I thank all conference participants for their responses to this paper. Steve Happel and Bob Russell were especially diligent critics. Bob also needs to be thanked for the tremendous amount of editorial work he has done in putting together this volume and its predecessor.

ORDINARY AND EXTRAORDINARY DIVINE ACTION:
The Nexus of Interaction

George F.R. Ellis

0 *Prologue*

This paper touches controversial issues, and some of the possibilities discussed will undoubtedly make some readers uncomfortable. This is because it takes seriously in a particular way both the historic Christian message and a modern scientific perspective, emphasizing their cognitive claims as I understand them from a Quaker perspective. The reader may not share this double commitment. Nevertheless the argument is logically and epistemologically sound; the unease is at a theological and/or metaphysical level. This issue will be discussed briefly in the last main section. However, a full treatment cannot be given here; an in-depth justification for the view taken will be given in a forthcoming work.[1]

For the moment I make the initial claims that: (1) there are other types of knowledge besides that given by the "hard" sciences, for example, that given by philosophy, theology, humanistic, and artistic disciplines—the task is to find a viewpoint that does justice to these issues as well as to hard science, in a compatible way; (2) the hypothetico-deductive method used to support the viewpoint presented here is essentially the same as that underlying our acceptance of modern science; and (3) the main themes proposed, controversial as they are, are supported by as much or indeed more evidence (admittedly of a more general form than that used by physics alone) than many of the themes of modern theoretical physics.

The requirement in order to approach the material fairly is an open mind in looking at the various logically possible options, rather than simply selecting one particular metaphysical stance on an *a priori* basis. The important point is that we have to adopt some metaphysical position; we should do so here in a considered way.

1 *Introduction*

This paper is largely a response to Nancey Murphy's contribution to this volume, "Divine Action in the Natural Order: Buridan's Ass and Schrödinger's Cat." That

---

[1]Nancey Murphy and George F.R. Ellis, *On the Moral Nature of the Universe: Cosmology, Theology, and Ethics* (Minneapolis: Fortress Press, 1996) developing themes outlined in Ellis, *Before The Beginning: Cosmology Explained* (New York: Bowerdean/-Boyars, 1993).

paper is revolutionary because it represents a conservative interpretation of the Christian faith[2] which, unlike most other such interpretations, takes the content of modern science seriously as part of the task of constructive theology. The viewpoint here will be to basically agree with Murphy's paper, and comment on some specific issues raised by its thesis.

Accepting the main thesis of that paper, the themes I would like to discuss further are: (a) the issue of capricious action; (b) the issue of top-down causation through intention, and the particular causal nexus of the action; and © the issue of evidence for the position stated.

As regards (a), one of the main problems for the proposal is the charge of capriciousness in God's action, in terms of God deciding now and then to act contrary to the regular patterns of events but often deciding not to do so. One would like to have articulated some kind of criterion of choice underlying such decisions, and then an analysis given of how that criterion might work out in practice. This has to take very seriously indeed the issue of evil, pain, and suffering as experienced in the present-day world, of God's acceptance and allowance of horrors of all kinds, which one might *a priori* presume he/she could and would prevent if he/she so desired. If the usual Christian view is to make sense, there has to be a cast-iron reason why a merciful and loving God does not alleviate a lot more of the suffering in the world, if he/she has indeed the power to do so. This leads to the question of when divine action may be expected to take place, in either an "ordinary" or an "extraordinary" manner. Thus, one needs to characterize those concepts, and have some kind of criterion as to when each may be expected.

One possible approach to this range of issues is to emphasize the possibility of another domain of response of matter to life than usually encountered, as suggested by John Polkinghorne:[3] that matter might respond directly to God-centered minds through laws of causal behavior that are seldom tested (see section 4.4 below). Then the distinction between ordinary and extraordinary action becomes the question of whether we have entered this domain or not. This may provide a partial answer.

As regards (b), a central theme in Peacocke's writing,[4] the issue is what type of top-down causation might occur, and where the causal nexus could be whereby

---

[2]That is, it is in agreement with centuries-old aspects of the Christian tradition. See, e.g., *Dictionary of Christian Spirituality*, ed. Gordon Wakefield (London: SCM Press, 1989); and Denis Edwards, *Human Experience of God* (Ramsey, NJ: Paulist Press, 1983). However it is certainly not fundamentalist in its attitude; rather it is in agreement with the kind of modernizing approach advocated by Peter Berger in his superb book, *A Rumor of Angels: Modern Society and the Rediscovery of the Supernatural* (New York: Doubleday, 1969; reprint, New York: Anchor Books, 1990).

[3]See John Polkinghorne, "God's Action in the World," *CTNS Bulletin* 10, no. 2 (Spring 1990), 7; *idem, Science and Providence: God's Interaction with the World* (London: SPCK Press, 1989); and William Stoeger "Describing God's Action in the World in the Light of Scientific Knowledge of Reality" (in this volume).

[4]See Arthur Peacocke, "God's Interaction with the World" (in this volume).

this top-down action could be initiated in the physical world. I will particularly contrast general top-down influences which alter conditions over a wide range of events (as in many of the examples given by Bernd-Olaf Küppers[5]) with specific top-down actions, which are very focused in their aim and influence. I will argue that the latter is what is required for the Christian tradition to make sense, and that it requires something like the special action mentioned in Murphy's paper. This is probably related to the issue of free-will.[6]

Regarding the specific causal nexus, my view is in agreement with that of Robert Russell, William Pollard, and others, and recently supported and well-discussed in Thomas Tracy's paper "Particular Providence and the God of the Gaps."[7] The relevant points are: (b1) the need for some kind of "gap" in the strictly causal chain from physical cause to effect if specific divine action in the world is to be possible in a meaningful sense (I believe it may well be that one can make the same claim in respect to individual actions with a connotation of personal responsibility); (b2) the inability of deterministic chaos to provide a solution to this problem of causal gaps; and (b3) the fact that quantum uncertainty does indeed have this potential.

Overall these contentions are supportive of the argument in Murphy's paper.

As regards (c), while "proof" will not be available, one would like some broad brush-stroke defense of the position presented in terms of general lines of evidence.[8] The main point here is that, as emphasized in Murphy's present paper, one of the needs is to satisfy the Christian tradition in terms of doctrine and practice; but then the issue is, Whose doctrine? Whose practice? What is the foundation for choosing and supporting one particular brand of tradition?

Either one goes here for a rather inclusive, broad-stream interpretation which aims to be widely acceptable across the many varieties of Christian tradition, and therefore will inevitably be regarded as "weak" by many of them; or one aims to be more particular and detailed in terms of developing the view of some particular branch of that tradition in depth. But then the product becomes rather exclusive in its nature, and may be regarded as irrelevant by others. In either case the issue becomes that of validating what is claimed to be true by the chosen traditions or doctrines, in the light of manifest errors, in many cases, in what has been claimed in the past.

To cope with the issue of inclusivity, one can suggest that this defense should, first, have a broad base aimed at validating a religious worldview in

---

[5]Bernd-Olaf Küppers, "Understanding Complexity" (in this volume).

[6]See, e.g., Roger Penrose, *The Emperor's New Mind: Concerning Computers, Minds, and the Laws of Physics* (Oxford: Oxford University Press, 1989); and Euan Squires, *Conscious Mind in the Physical World* (Bristol: Adam Hilger, 1990).

[7]In this volume.

[8]Cf. Murphy, "Evidence of Design in the Fine-Tuning of the Universe," in *Quantum Cosmology and the Laws of Nature: Scientific Perspectives on Divine Action*, ed. Robert John Russell, Nancey Murphy, and C.J. Isham (Vatican City State: Vatican Observatory, 1993; Berkeley, CA: Center for Theology and the Natural Sciences, 1993).

general, strongly supported by widely acceptable evidence; second, support a more specifically Christian view developed as a second stage of the argument, refining its methods, detail, and evidence; and with support for a particular tradition developed in the third stage. I shall make some comments along these lines at the end. The proposal made here is that the idea of top-down causation, with different layers of description, effective laws, meaning, and evidence, is the best framework for understanding and testing the overall scheme suggested.

2 *Emergent Order and Top-Down Action*

As explained clearly by Küppers and Peacocke, in hierarchically structured complex systems we find both top-down action and emergent order.

First, the hierarchical structure introduces levels of emergent order, as described so ably by Arthur Peacocke:[9] irreducible concepts used to describe the higher levels of the hierarchical order are simply inapplicable at the lower levels of order. Thus, different levels of order and description are required, allowing new meanings to emerge at the higher levels of description. (Note that these are different-level descriptions of the same physical system, applicable at the same time.)

Second, this structure enables top-down action to take place whereby interactions at the lower levels cannot be predicted by looking at the structure at that level alone, for it depends on, and can only be understood in terms of, the structures at the higher levels.

In the specific case of biology, we find, beautifully described by Neil Campbell, a hierarchical structure as depicted in Figure 1 (*on following page*). As expressed by Campbell:

> With each upward step in the hierarchy of biological order, novel properties emerge that were not present at the simpler levels of organization. These emergent properties arise from interactions between the components. . . . Unique properties of organized matter arise from how the parts are arranged and interact . . . [consequently] we cannot fully explain a higher level of organization by breaking it down to its parts.[10]

Indeed not only are such different levels of description permitted, they are required in order to make sense of what is going on. This is true not only of biological systems: Küppers shows convincingly that such emergent properties are important even in a physical system such as a gas, being mediated by the system's structural

---

[9]Peacocke, "God's Interaction."

[10]Neil A. Campbell, *Biology* (Redwood City, CA: Benjamin Cummings, 1991), 2-3.

conditions and boundary conditions (as discussed further below).[11] Ian Barbour[12] and Peacocke[13] develop the theme of emergence in depth.

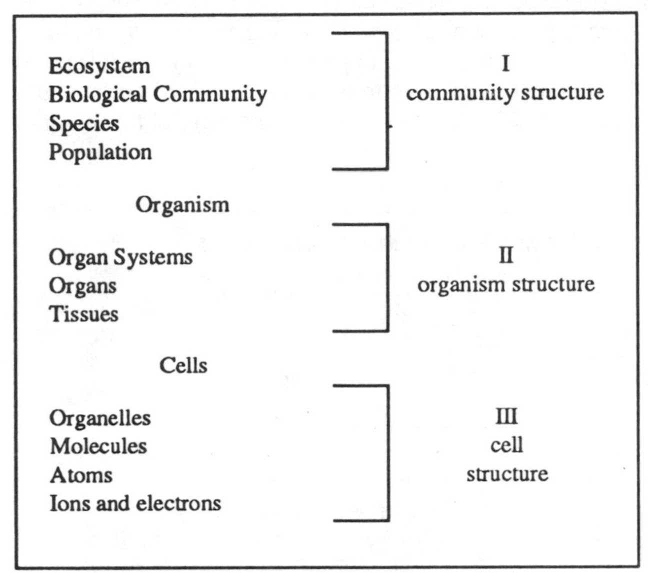

Figure 1: Biological levels of emergent order.

In a biological system, the two crucial levels of order are that of the cell and the individual organism; at each of these levels there is a higher level of autonomy of coherent action than at any of the other levels. A biologist regards individuals as the elementary components of a population, and cells as "elementary components" of the individual, while (broadly speaking) a microbiologist regards molecules and a biochemist, ions and electrons, as the elementary components. A physicist would continue down the hierarchical scale, reducing these to quarks, gluons, and electrons.

2.1 *Hierarchies of Software: Digital Computers*

A particularly clear example is given by modern digital computers, which operate through hierarchies of software: from the bottom up there are machine language

---

[11]Küppers, "Understanding Complexity."

[12]Ian Barbour, *Religion in an Age of Science*, The Gifford Lectures, 1989-1991, vol. 1 (San Francisco: HarperSanFrancisco, 1990).

[13]Peacocke, *Theology for a Scientific Age: Being and Becoming—Natural and Divine* (Oxford: Blackwell, 1987).

(expressed in binary digits), assembly language (expressed in hexadecimal), operating system and programming language (expressed in ASCII), and application package (e.g., word processor) levels of software. At every level there is a completely deterministic type of behavior described by algorithms applicable at that level. All of this is realized in terms of the motion of electrons flowing in the integrated circuits as determined by the laws of physics. This is where the actual action takes place, but it does so according to plans implemented at the higher levels of structure, and thereby effects actions that are meaningful at the higher levels.[14]

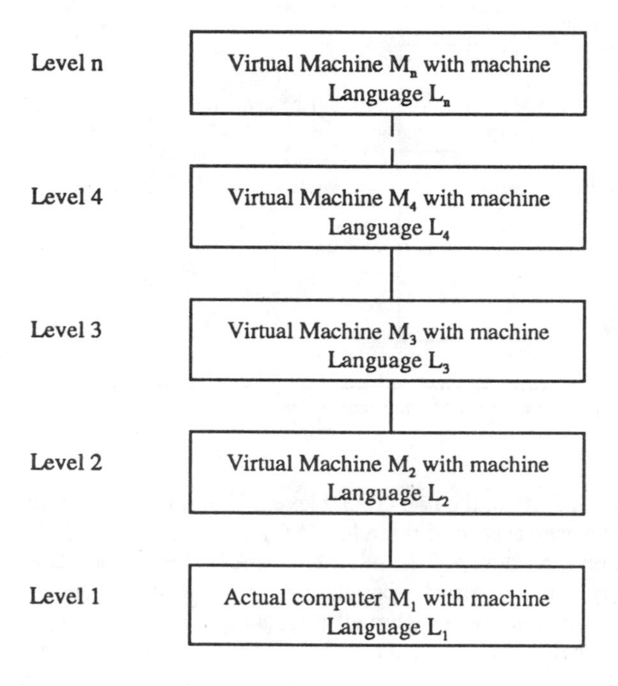

Figure 2: Generic Hierarchical Structure of a Computer.

Logically, a digital computer consists of a hierarchy of *n* different virtual machines M_n each with a different machine language L_n.[15] Its generic structure is expressed in Figure 2 (*above*).

---

[14]For a very clear exposition of the hierarchical structuring in modern digital computer systems, see Andrew S. Tannenbaum, *Structured Computer Organization* (Englewood Cliffs, NJ: Prentice Hall, 1990).

[15]Ibid., 2-3.

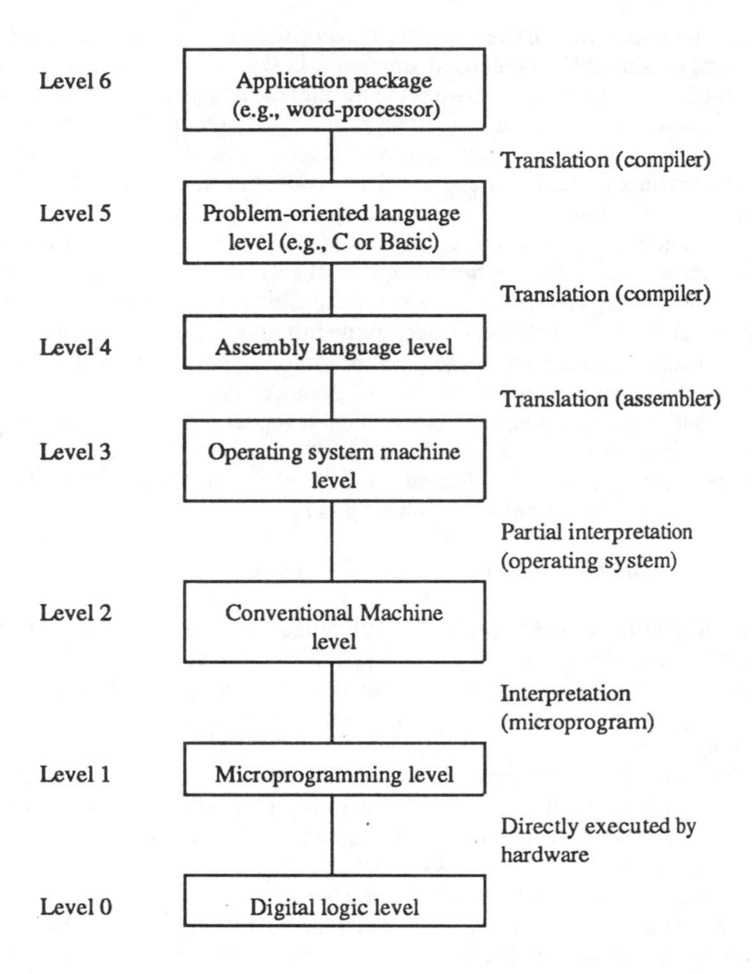

Figure 3: Typical Hierarchical Structure of a Digital Computer.

The physical computer M_1 does the actual calculation in machine language L_1; each virtual computer runs programs either by interpreting them in terms of the lower machine languages, or translating them into these lower languages (e.g., programs in L_2 are either interpreted by an interpreter running on L_1, or are translated to L_1). Each computer's machine language (at each level in the hierarchy) consists of all the instructions the computer can execute at that level. However, only programs written in language L_1 can be directly carried out by the electronic circuits, without the need for intervening translation or interpretation. In contemporary multilevel machines, the actual levels realized are shown in Figure 3 (*on preceding page*).[16]

The logical connections between the different levels in the computer, and the resulting machine languages at each level, are tightly controlled by the machine hardware and software. In particular, given the machine, the program loaded into particular memory locations, and the data resident in memory, each high-level instruction will result in a unique series of actions at the digital (hardware) level, which in turn will result in a unique series of consequences at each of the higher levels. Consequently the machine language at each level also has a tight logical structure with a very precise set of operations resulting from each statement in that language. The detailed relation of operations from high to low levels, and at each level, will depend on the actual memory locations used for the program and data; but the logical operation is independent of these details.[17]

In biological systems, with hierarchical levels as indicated above, the same kind of logical structure holds; however the "languages" at the higher levels are much less tightly structured than in the case of the computer,[18] and the links between different levels correspondingly less rigid.

2.2 *The Physical Mediation of Top-Down Action*

Consider now how the hierarchically structured action is designed to occur, in physical terms. We can represent this as follows: for a structured hierarchical physical system S, made up of physical particles interacting only through physical forces, top-down and bottom-up action are related as shown in Figure 4 (*on following page*).

The boundary B separates the system S from its environment E. Interaction with the outside world (the environment) takes place by information/energy/matter flow in or out through the boundary, and is determined by the boundary conditions at B. The structure of the system is determined by its structural conditions, which can be expressed as constitutive relations between the parts. I distinguish here structural conditions, fixed by the initial state of the physical system but then remaining constant in a stable physical system (e.g., the structure of a computer as determined by its manufacture), and initial conditions and boundary conditions as usually understood in physics (e.g., the initial state of motion of a fluid in a cell and temperature conditions imposed at the cell boundaries over a period of time).[19]

---

[16]Ibid., 4-7.

[17]These structures and their interconnections are described in considerable detail in Tannenbaum's book.

[18]The major aim of the AI (artificial intelligence) movement is to arrive at a correspondingly loose structure in the computer's higher-level languages.

[19]Küppers's concept of "boundary condition" conflates these three rather different concepts. See "Understanding Complexity."

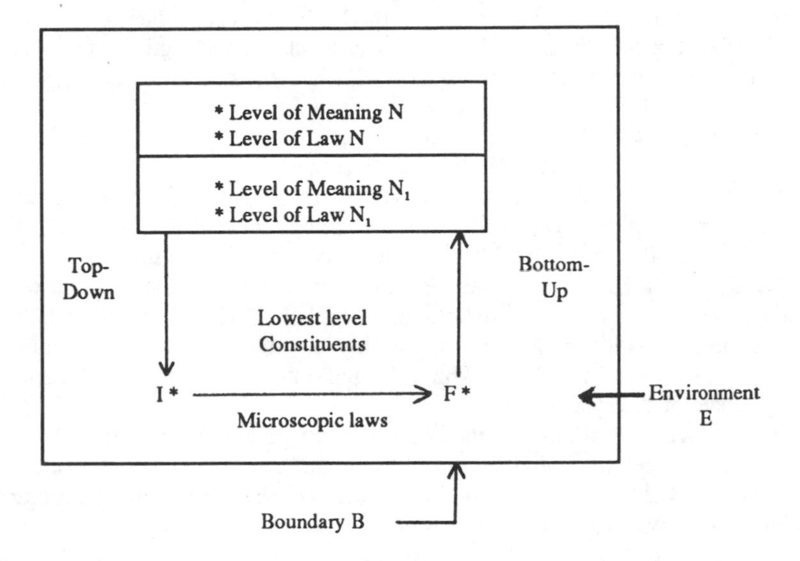

Figure 4: Top-Down and Bottom-Up Causal Interactions in a Generic Hierarchical System. *I* is the initial state of the constituents and *F* the final state; the microscopic laws of the system determine the movement from *I* to *F*. (For brevity only two of the levels of meaning and law have been shown; ordinarily there will be many.)

2.2.1 *Classical Physical Systems*

Examples: (1) the atmosphere; (2) an aircraft; (3) a digital computer.

In these systems, the action is strictly deterministic, though not necessarily predictable.[20] (This is ensured in designed systems such as (2) and (3), in order to obtain reliability; any quantum uncertainties are damped out, by design.)

A: Top-down action happens by means of states at the higher-level initiating coordinated action at the bottom level, which is governed by the basic causal relations underlying the system. The bottom-level components act on each other by regular physical laws, the resulting final state at the bottom level then determining conditions at the higher levels, because they define conditions at the higher levels through their aggregation (or "coarse-graining") properties. The last two steps are what is meant by 'bottom-up causation' in these contexts.

The coordination of action occurs through the structural arrangement and interconnection of lower-level entities (e.g., transistors, capacitors, etc.) to form higher-level entities (e.g., computers, television sets, etc.). Because the semantics

[20]If the computer output were predictable in any simpler way we would not need to run the computer program.

of the higher level are intrinsic to its nature, the language (vocabulary and syntax) at each level cannot be reduced to that at a lower level, even though what happens at each higher level is uniquely determined by the coordinated action taking place at the lower-levels, where it is fully described in terms of the lower-level languages. Thus, the whole structure shows emergence of new properties (at the higher levels) not reducible to those of the constituent parts.
Examples: (1) lowering the undercarriage in an aircraft (realized by gas particles exerting forces on a piston in a cylinder); and (2) a computer reading out a text file and printing it on the screen (realized by electrons impinging on the screen).

What happens to a given system is controlled by the initial state along with the boundary conditions. The system boundary is either: (i) closed (no information enters); or (ii) open (information enters; possibly also mass, energy, momentum). In the latter case we have to know what information enters in order to determine the future state of the system.

B: The final state attained at the bottom level is uniquely determined by the prior state at that level (the initial conditions) and the incoming information at that level—that is, by the "boundary conditions" (assuming a given system structure and given microlaws). This determines uniquely what happens at the higher levels. We assume that a unique lower-level state determines uniquely the higher-level states through appropriate coarse-graining. When this is not true, the system is either ill-defined (for example, because our description has omitted some "hidden" variables), or incoherent (because it does not really constitute a "system"). We exclude these cases. Note that the loss of information implied in the definition of entropy results because a particular higher-level state can correspond to a number of different lower-level states (each of which leads to that single higher-level state.).

Note 1: This statement does not contradict the idea of top-down causation. Any given macro-state at the top level will correspond to a restricted (perhaps even unique) set of conditions at the basic level. It is through determining a set of micro-states as initial conditions at the bottom level, corresponding to the initial macro-state, that the top-level situation controls the evolution of the system as a whole in the future. How uniquely it does so depends on how uniquely the top-level state determines a state at the bottom.[21]

Note 2: At the higher levels, the statement analogous to *B* may or may not be true (i.e., the system may or may not be causally determinate when regarded as a machine at a higher level); this depends on what micro-information is lost in forming the macro-variables at the higher levels, from the micro-variables.

Note 3: Although the bottom-level system is determinate, prediction of what will happen is in general not possible even at the bottom level, because of the possibility of chaotic behavior (sensitive dependence on initial conditions).

---

[21]Or, equivalently, it does so depending on how much information of the microstates is lost by giving only a top-level description—this information loss defining the entropy of the macroscopic state.

2.2.2 *Quantum Physical Systems*

In a system where quantum effects are significant,[22] we have a new element:

C: The bottom-level evolution is indeterminate in a quantum system, although the statistical properties of its evolution are determined. This lower-level indeterminacy may or may not result in significant higher-level indeterminacy, depending on the system structure. .

In many cases the properties at some higher level may be effectively determinate (the quantum uncertainties being washed out). However, this is not true when there is a sufficiently powerful amplifier in operation (e.g., photomultipliers in a telescope which allow detection of individual photons), or sufficiently sensitive dependence on initial conditions.[23] This theme has been developed by Russell.[24] He points out that quantum physics both produces the macroscopic world in all its properties and affects the macroscopic world (occasionally) through a single quantum event. Schrödinger's cat represents both aspects: it has bulk properties, such as volume, because of quantum statistics—based on the Pauli exclusion principle—and it lives or dies because of a single radioactive event. The latter (macroscopic effects due to a single quantum event) may well only happen during a "measurement"; the problem in deciding whether or not this is the case is that "measurement" in quantum theory proper is not yet a well defined concept. However, this effect is quite sufficient to allow the effects we have in mind in this paper.

2.2.3 *Simple Biological Systems*

By this we mean systems in the biological hierarchy at the level of an individual organism or lower.

Examples: (4) a mosquito; (5) a dog; and (6) a person.[25]

In these examples complex neural systems convey, route, and filter information in a hierarchically structured way so as to allow maximal local

---

[22]In this paper the prime quantum effect considered is that of indeterminacy (which is closely related to the problem of measurement). There are other equally important aspects of quantum theory—Fermi vs. Bose statistics, nonlocality, etc.; but they do not seem to bear directly on the argument at hand, except perhaps that of non-locality.

[23]These are really two ways of saying the same thing.

[24]See Robert John Russell, "Quantum Physics in Philosophical and Theological Perspective," in *Physics, Philosophy, and Theology: A Common Quest for Understanding*, Robert John Russell, William R. Stoeger, and George V. Coyne, eds. (Vatican City State: Vatican Observatory, 1988).

[25]In principle, the same kind of description applies to complex biological systems, e.g.: (7) people in a room; and (8) an ecosystem. But so many extra issues arise because of social, economic, and political interaction that it is better first to consider and understand the simpler examples.

autonomy and yet coordinate overall action,[26] the whole being coordinated by the extraordinarily complex structure of the brain.[27]

The fundamental point is that, despite this complexity, if in these systems what happens macroscopically is determined at the micro-level simply by the action of known physical laws, then the analysis is the same as in the case of the classical or quantum machines considered above. One can consider, for example, a moving human hand (realized by the motion of electrons and ions in muscles).

The analysis and examples given above lead to the following propositions about hierarchically-structured, physically-based systems, even given the high complexity of a living system:

Proposition 1: Top-down action underlies meaningful activity, for it enables lower levels to respond coherently to higher-level states, but does not by itself imply openness.

Proposition 2: Chaos generates unpredictability, but does not by itself underlie meaningful action.

Proposition 3: Quantum uncertainty allows openness (as it only makes probabilistic statements), which can be amplified to macro-levels.

In the latter case, the issue is the nature of this openness: is it truly indeterminate—representing a random process whose final state is not determined by the initial state—or is it in fact determinate, through some hidden variable presently inaccessible to us? We will return to this later. In any case the above analysis suggests the following speculation: Meaningful physical top-down action with openness in a hierarchical structure can occur only either (i) via injection of information from outside, that is, by manipulation of the boundary conditions (probably in a very directed manner, conveying specific information to specific sub-components); or (ii) through a process that resolves quantum uncertainty at the microscopic level by a choice of a particular outcome from all those that are possible according to quantum laws, thus resolving the uncertainties in a quantum mechanical prediction. This effect can then be amplified,[28] or it could be effective at the larger scale because it takes place in a coordinated way at the micro-level (as in superconductivity).

Note to (i): Bill Stoeger[29] has pointed out that it is essential to be clear about what is "inside" and what is "outside" the system considered— particularly when non-local effects occur. A more adequate characterization of a system to better

---

[26]See Stafford Beer, *Brain of the Firm: The Managerial Cybernetics of Organization*, 2d. ed. (New York: John Wiley & Sons, 1981), for an illuminating discussion.

[27]See, e.g., John C. Eccles, *The Human Mystery* (London: Routledge & Kegan Paul, 1984); or *idem*, *The Wonder of Being Human: Our Brain and Our Mind* (New York: Free Press, 1984); and Gerald M. Edelman, *Bright Air, Brilliant Fire: On the Matter of Mind* (New York: Basic Books, 1992), for contrasting views.

[28]See, e.g., I. Percival, "Schrödinger's Quantum Cat," *Nature* 351 (1991): 357ff. "DNA responds to quantum events, as when mutations are produced by single photons, with consequences that may be macroscopic—leukemia, for example."

[29]Private communication.

account for the observed phenomena may result in some of what was "outside" being brought "inside" the system. Our comment applies after such adjustments have been made.

Note to (ii): The basic point made here is that our present description of the quantum world is essentially causally incomplete,[30] as is clear from every discussion of the "measurement process" in standard quantum mechanics. Quantum theory determines the statistical properties of measurements, but does not determine the result of individual measurements where the initial state is not an eigenstate—a condition which includes almost all measurements. However, a specific final state does in fact result in each case. There is no known rule that leads uniquely from the initial state to the final state. Thus, the final state in almost every specific case is determined by some feature not described by present quantum theory, or is uncaused. The present view utilizes the first option.

Clearly this speculation contrasts with aspects of the views of Polkinghorne and Peacocke, but basically agrees with those of Russell, Murphy, and Tracy. I suggest that it is the logical outcome of any hierarchical structuring in which the bottom, low-level actions are governed by regular physical laws (i.e., we exclude a "vitalist" or "mentalist" interaction not mediated by physics).

3 *Ordinary Divine Action*

To set the scene, it is convenient to recapitulate some issues covered in many other essays in this volume. We need somehow to divide God's action in the world into ordinary and extraordinary action. This section concerns that which may be regarded as "ordinary," that is, those actions that are the result of the action of physical laws alone (God's effective action is secondary, through these laws, which are themselves established by his primary action). A theme at the conclusion of this section will be that it is reasonable and indeed in line with the religious worldview to characterize most "ordinary" action as revelatory and sacramental.

3.1 *Cosmological*

The first domain of action is the cosmological creative act:

Action 1: Creation of the universe, which has two aspects:

Action 1a: Initiation of the laws of physics and of the universe: creation of basic structure (setting up the regularities that underlie existence).

---

[30]This issue is separate from the further thorny problem of defining what a measurement is, in a fully quantum system, and when it will take place. See, e.g., M. A. Morrison, "Altered States: The Great Measurement Mystery," in *Understanding Quantum Physics: A User's Manual* (Englewood Cliffs, NJ: Prentice Hall, 1990).

Action 1b: Setting the boundary conditions for the universe: contingent choice from the possibilities compatible with the basic structure[31] followed by:[32]

Action 2: Sustaining the universe, through maintaining the sheer existence as well as the regularity of the universe (as described partially through the laws of nature we discover); thus underpinning existence in a reliable way.

The initial act of creation, if there was one (i.e., if there was a t=0), may properly be regarded as an extraordinary divine act, but in the past rather than the present; it has taken place, rather than being ongoing. The second (sustaining all events) is what underlies the predictable nature of the laws of physics, as is required for meaningful moral activity.[33]

Together these are the prerequisites and basis for ordinary divine action; that is, divine action carried out through the means of regular laws of behavior of the physical universe. The true creativity involved in these acts is in the selection of the laws of physics, and in choosing specific boundary conditions for them (whether in a single universe, or in an ensemble of universes) that enable the desired results to be attained (cf. the previous volume in this series).

3.2 *Functional*

The laws of physics in the existent universe provide the basis for the evolution and functioning of complex systems. They therefore allow ordinary divine action, which is "second-order" or "indirect" action. Its nature is fashioned by the laws of physics and the boundary conditions; it is understood specifically that divine input in such "ordinary" action—once the system is running—is to maintain the regular functioning of nature in such a way that it is describable by means of scientific laws, and therefore its results are largely determinate.[34]

In the relation of theoretical biology to fundamental physics, there are three main kinds of issue: the functioning of general living systems, evolution, and the issue of consciousness and free will. We will look at these in turn.[35]

---

[31]This may or may not imply a specific event at t=0. Cf. the discussions in *Physics, Philosophy and Theology*; and *Quantum Cosmology*.

[32]As seen from within the universe. Seen from outside, this may well be no different from Action 1.

[33]Ellis, "The Theology of the Anthropic Principle," in *Quantum Cosmology*; and Murphy, "Divine Action in the Natural Order: Buridan's Ass and Schrödinger's Cat" (in this volume).

[34]Quantum uncertainty and sensitive dependence on initial conditions to some extent limit predictability and allow for indeterminacy.

[35]The concerns of this section relate to the Anthropic Principle discussed in *Quantum Cosmology*; the point is that not every set of laws of physics will allow life to function.

Action 3: sustaining functioning of general living systems: this can be split into three parts:[36]

Action 3a: sustaining development: growth from a single cell to a complete organism; *Action 3b*: enabling physiological functioning of organisms; *Action 3c*: enabling community functionalism/ecology.

The basic mechanisms in all three cases are feedback controls operating in hierarchically-structured complex systems, made of matter functioning according to the fundamental laws of physics, and in the first two cases, organized according to digitally-coded information contained in DNA.

The first and second are highly controlled processes; it is very unlikely that chaotic mechanisms of any kind can play a significant role here. Indeed the whole purpose of feedback organization is to damp out any deviations from the desired developmental or physiological path; thus, these processes are usually of an anti-chaotic nature when properly functioning (they efficiently guide the system to a desired final state, despite errors in initial data or disturbances that may occur). What does occur here is self- organization, but based on very specific and highly controlled mechanisms (e.g., a reaction-diffusion equation with restricted boundary conditions, or cells moving over an extra-cellular matrix). Given the laws of physics, these mechanisms for the operation of life not only function but in some sense seem to be preferred solutions of the physical equations: experience seems to show that "physics prefers life" (e.g., simple organic molecules assemble themselves from an appropriate "primeval soup," providing the basis for more complex molecules to form). However, it seems a reasonable view that no special intervention is required to make all this happen; it is just the wonder of ordinary divine action (cf. the next subsection).

The third case, ecology, is less well-controlled (as is well known), and here chaotic effects may well happen. The most significant question (apart from learning how to cope with them) is whether this played any significant role in evolutionary processes, for example, by enhancing the range of the environments to which living beings were subjected. That will be a difficult question to answer; it may just as well have placed evolution in jeopardy as assisted it in creating more complex beings.

Action 4: evolution, shaping the nature of things as they are at present.

Here is where the issue of "design" arises, answered in conventional evolutionary theory by the statement that there is no design, only evolutionary selection, with evolution—an open-ended feedback process with the goal simply of survival—being adequate to describe the "design" of all living beings, including

---

[36]Cf. Ellis in *The Anthropic Principle: Proceedings of the Second Venice Conference on Cosmology and Philosophy*, ed. F. Bertola and U. Curi (Cambridge: Cambridge University Press, 1993).

humans.[37] The shift to cultural evolution implies a change in the nature of evolution, but still based on the same general process.[38]

One can, if somewhat diffidently, question whether the current orthodoxy is really adequate to explain all that we see—whether there is some degree of direction in the variations that take place (via genetic variation of DNA), or whether the variations are indeed totally random. Probabilistic calculations suggest that there may be a time problem if variations are indeed purely random (cf. the controversial claims by Fred Hoyle). The real issue that concerns me here is the question of evolutionary development of hardwired behavioral patterns of great complexity, despite the essential comment[39] that nothing experienced or learned can have any effect on the DNA passed on to offspring (these factors can effect whether DNA is passed on to offspring, but not the coding of that DNA). It may be claimed that the plasticity of the brain along with the Baldwin effect[40] are sufficient to explain development of all complex behavioral patterns.

However, I would like to see clear evidence that this is the case in such examples as bird migration, or the signals used for communication by the honey bee, let alone more complex examples in higher animals, where procreation is relatively rare and very complex interactions determine whether the animal survives long enough to procreate, so that any tendency to a specific behavioral pattern, determined by "hard-wiring" of the brain, is only one of many other features determining survival and procreation rates. Without being dogmatic about it, I would leave open a small question as to whether chance or fortuitous happenings alone in the evolutionary process are adequate in order that evolution succeed in the time scales available.[41]

Finally we come to the most difficult of the areas of "ordinary" action:

Action 5: enabling the functioning of the brain and mind: foundations of consciousness and free will; the foundation, in turn, of moral response.

Given some explanation of what happens in the mind, one can then envisage downward causation from intentions formed in the brain acting to enable specific events to happen in the body: bodily conditions alter, cells function in altered

---

[37]See, e.g., Richard Dawkins, *The Blind Watchmaker* (New York: Norton, 1986).

[38]See, e.g., Daniel C. Dennett, *Consciousness Explained* (London: Penguin Books, 1991).

[39]Cf. L. Wolpert, *The Triumph of the Embryo* (Oxford: Oxford University Press, 1990).

[40]Dennett, *Consciousness Explained*, 184-87. The basic point here is that if there is a high peak of suitability associated with some specific brain wiring state but not any nearby states, nevertheless nearby states will be more likely to survive because of brain plasticity. Their initial wirings will alter during their lifetime, because of plasticity of the brain connections, and will "explore" the region near where they start; all those ending up at (or passing through?) the highly preferred state will be more likely to survive than those that do not. But they will be more likely to end up there if they start nearby.

[41]See R.E. Lenski and J.E. Mittler, "The Directed Mutation Controversy and Neo-Darwinism," *Science* 259 (1993): 188ff., for a discussion refuting directed mutation.

environments, different currents flow and adjust electric potentials. Consequently muscles move, allowing limbs to fulfill the intent in the mind and alter conditions in the physical world. This clearly is a case of downward causation from the brain to events in the body.[42] The issue is how the mind relates to the brain,[43] a core question in terms of personal existence and meaning. An open-minded investigation must consider four features that might contribute (singly or together):

a) organized complexity,
b) chaotic motion ("openness"), .
c) quantum uncertainty,
d) mental fields.

Explanation via some combination of *a, b, and c* sees physics based on known fundamental (microscopic) laws as the basic answer, through allowing hierarchically structured organization and so emergence of higher levels of organization based on lower levels.

View I is that this complexity itself is sufficient; no invoking of chaos or quantum theory is necessary to attain consciousness and possibly not for free will, even though everything is fully deterministic. This includes the modern "mind is a computer" suggestion (supported by work on neural networks, and theoretical analyses such as that of Dennett,[44] but challenged by others such as Penrose.[45]

View II is the same as view I except that one needs the "openness" or unpredictability allowed by chaos theory, or the indeterminacy inherent in quantum theory (viewed as a purely random process), in order that consciousness can emerge. However, insofar as this is equivalent in the quantum case to adding a truly random variable to the equations, which is then amplified, this does not appear to help with the deeper issues.[46] Thus, views I and II perceive standard physics alone to be the total answer to the basis of consciousness. Mind is an emergent phenomenon, as are the other levels of organization in biology.[47] Mind is, in a sense, reducible to physics (the emergent order of biological systems, allowed by physics, is completely ruled by micro-level physics even though it entails and encodes higher levels of order). It is hard to see how free will and morality can be anything but an illusion on this notion (cf. the discussion of top-down causation above), particularly when we remember the development of the physical brain through the process of evolution governed by random mutation and selection through "survival of the fittest."[48]

---

[42]Cf. Küppers, "Understanding Complexity"; and Stoeger, "Describing God's Action."

[43]See, e.g., Squires, *Conscious Mind*; Dennett, *Consciousness Explained*; Edelman, *Bright Air, Brilliant Fire*; Eccles, *The Human Mystery*; and Penrose, *The Emperor's New Mind*.

[44]See, Dennett, *Consciousness Explained*.

[45]See, Penrose, *The Emperor's New Mind*.

[46]Cf. Tracy, "Particular Providence and the God of the Gaps" (in this volume).

[47]See Campbell, *Biology*.

[48]See Ellis, *Before the Beginning*, for further discussion.

View III is that quantum theory allows the uncertainty needed for the independent existence of mind, as it is an essentially incomplete theory of physical behavior. The conventional view[49] is that the specifics of what happens at the quantum level are uncaused: statistical behavior must be regular, but in each specific case what happens is purely random. Chance is treated as a causal explanation in itself, not relying on any other cause. Despite much propaganda for this viewpoint, it is essentially unsatisfactory; for chance does not cause anything, rather it is the name for a lack of cause. For example, Morrison states, after discussing the unsatisfactory state of the problem of measurement, that:

> Underlying the problem of measurement there is a deeper question. As a consequence of an ensemble measurement of an observable Q, the original state collapses into one of the eigenstates of Q. The question is, what mechanism determines which eigenstate a particular member collapses into? According to the conventional epistemology of quantum mechanics, the answer is that random chance governs what happens to each member of an ensemble. Many (your author included) consider this no answer at all.[50]

View III suggests rather that there is some cause: something not contained in our current physical descriptions of quantum theory determines the details of what happens in each specific case. This "something" may be related to mind in two ways. First, indeterminism is needed at the quantum level of nature if mind/ consciousness is to be effective in animals as in humankind,[51] and it extends the possibility of a non-algorithmic kind of activity that is essential in a full view of consciousness.[52] Second, mind/consciousness could be necessary to "collapse the wave function" and give a complete account of natural events, which quantum physics by itself cannot supply.

The suggestion is that the apparent randomness of quantum theory is not truly random but rather is a reflection of the operation of mind, intricately linked to the unsolved problems of the observer in quantum mechanics and the collapse of the wave function.[53] Imbedded in a complexly structured system, this provides the freedom for consciousness to function, "mind" being allowed to determine some of the uncertainty that quantum physics leaves open (thus being completely compatible with quantum physics, but allowing some other level of order to act in the physical world with openness). On this view one can maintain that information entry from mental to physical levels of nature is, for example, through the choice of when a quantum state will decay, which, because of quantum uncertainty, is not determined by known physical laws. This allows a transfer of information between levels of the world without an expenditure of energy or a violation of the known physical laws.

---

[49]See, e.g., Morrison, *Understanding Quantum Physics*, 70-73, 85-87, and 226-28.

[50]Ibid., 617-18.

[51]See, e.g., Squires, *Conscious Mind*; and Eccles, *The Human Mystery*.

[52]See Penrose, *The Emperor's New Mind*.

[53]Squires, *Conscious Mind*.

The basic physical question relating to quantum uncertainty is what, if anything, determines such apparently uncaused information selection (e.g., when an energetic state will decay)? It may be truly uncaused (nothing determines what occurs, it just happens capriciously—the standard dogma of quantum theory) or it may in fact be controlled (something determines what happens, we simply do not know what it is—in effect the hidden variable theory of quantum mechanics.) View III supports the latter, linking it to the reality of free will and morality (taken as solid experiential data about the real world),[54] and accepting as inevitable and even natural the consequent non-locality of causation—an important aspect of quantum theory.

View IV, based on *d*, entails something like vitalism: known physics, by itself, is not the answer. Some as yet undiscovered feature (thus, not quantum uncertainty, which is already known) links mind to brain and matter. Perhaps there is some as yet undiscovered force or field ("the mental field") underlying consciousness, which may eventually be discovered and studied as any other physical field. However, it is difficult to see how this will resolve the issues at stake unless its equations of motion are quite unlike any we have seen before.

In both latter views, something outside presently known physics acts and has material effects in the physical world (e.g., by fixing the time when a quantum decay takes place, which currently known quantum theory is unable to do). These views are probably consistent with proposals such as the radical dualist-interactionist theory of the brain and the self-conscious mind proposed by Eccles.[55] This kind of view is of course highly controversial. The challenge to those who disagree is to produce an alternative in which free will in a solely physics-based hierarchical system (the human brain) is not an epiphenomenon.

This discussion is relevant to the theme of this paper in two ways. First, when viewed from within the world, essentially the same issues arise in terms of special divine intervention (discussed below); we may well expect that an analysis of the two issues will be very much in parallel.[56] Indeed view III is basically consonant with the view of special divine action in Murphy's paper.[57]

Second, and related to the first point, what is at stake here is the closedness or openness of the physical world to other influences—not the rattle of a dice (as in view II) but the intervention of some purposeful consciousness that is not wholly bound into physical systems.[58] On the latter views, physics is not all that controls the functioning of the physical universe: at higher levels of organization, information is introduced that affects lower levels by top-down action. This theme

---

[54]Ellis, *Before the Beginning*.

[55]See Eccles, *The Human Mystery*.

[56]Although complex problems of dualism then arise: if our minds and God's can both influence what happens, how do they compete for such influence? This would have to be modeled on the basis of our understanding of the chosen mode of God's action. Cf. Ellis, "Theology of the Anthropic Principle."

[57]Murphy, "Divine Action."

[58]Tracy, "Particular Providence."

will be picked up again in the discussion of special divine action in the next section, and later sections will consider how that higher-level information could be inserted.

3.3 *The Divine in the Ordinary*

What is miraculous? The birth of a baby; the design and function of a flower or a tree; the everyday and the "ordinary":

> I like to walk alone on country paths, rice plants and wild grasses on both sides, putting each foot down on the earth in mindfulness, knowing that I walk on the wondrous earth. In such moments, existence is a miraculous and mysterious reality. People usually consider walking on water or in thin air a miracle. But I think the real miracle is not to walk on water or in thin air, but on earth. Every day we are engaged in a miracle which we don't even recognize: a blue sky, white clouds, green leaves, the black, curious eyes of a child—our own two eyes. All is miracle.[59]

While these features are "ordinary" given our laws of physics and the nature of our universe, which allows or even prefers these events to take place, they are not ordinary if one considers the range of all possible universes. This is where the "anthropic" arguments are relevant: most of these possibilities will probably not be actualized in most universes.[60]

Thus, we can be justified in regarding these everyday occurrences as extraordinary if we include in our range of concepts an ensemble of universes—real or imagined—in most of which life is not possible.[61] It is appropriate for the ordinary scientist to forget this while studying what happens within the given, taken-for-granted order of things in the universe-as-is. However, this issue cannot be forgotten in studying Cosmology in its broad sense;[62] remembering this frailty of life within the broader framework of possible universes gives a justification—even within a scientific framework—for a sense of awe and wonder at what we see around us, which is an essential part of many religious world views (the sense of the numinous). Indeed on many such views these "ordinary miracles" are evidence of design, albeit of design of the universe itself rather than direct design (through specific action) of the objects or beings involved.[63]

---

[59]Thich Nhat-Hanh, *The Miracle of Mindfulness: A Manual of Meditation* (New York: Random House, 1991), 12. This is also the standard viewpoint of nineteenth-century liberal Protestantism (cf. Friedrich Schleiermacher, *On Religion: Speeches to Its Cultured Despisers* [Cambridge: Cambridge University Press, 1988]), and continues in much of contemporary theology, in particular being part of the views of Peacocke and Barbour.

[60]I have to admit that it is almost impossible to make this statement precise and give it a watertight justification. It is, however, highly plausible.

[61]Cf. the anthropic discussion in *Quantum Cosmology*.

[62]By 'Cosmology' I intend to refer to a more complete account of reality than that provided by scientific cosmology. See Ellis, *Before the Beginning*.

[63]Murphy, "Evidence of Design."

The point is that our attitude to "ordinary" divine action, mediated through the laws of physics and their boundary conditions, can take into account both of these views: the ordinariness of this action, and also its "miraculous" nature, where this word reflects both on the probability of what has happened and on what is achieved by it.[64] The awe and wonder that attracts many people to a scientific career need not be totally lost as one immerses oneself in the details of scientific study.

4 *Extraordinary Divine Action*

We may define extraordinary divine action in the already existent universe, as that action which: (a) can reasonably be interpreted as expressing the intention of God, that is, it has a revelatory character; and (b) is not predictable through regular laws of behavior of matter; that is, the events concerned will not inevitably happen as a result of the laws of logic and physics.[65]

I identify two main themes here: revelatory insight and the possibility of miracles proper, and consider them in turn.

4.1 *Revelatory Insight*

The first aspect is:

Action 6: revelation as to the nature and meaning of reality. This may be taken as having two parts:

Action 6a: providing spiritual insight; *Action 6b*: providing moral insight.

4.1.1 *Spiritual Insight*

Whatever one's view may be of consciousness and free will in general, to make sense of the standpoint of Murphy's paper[66] and the broad Christian tradition, there must be a possibility of specifically revelatory processes being made accessible to the mind of the believer (and the unbeliever).[67]

The first point is that the existence of such a causal joint or communication channel is required as the foundation of Christian (and other) spirituality,[68] which we are taking to be a reality. This requirement underlies any theory of revelation whatever, for without some such causal nexus, an immanent God, despite his/her immanence, is powerless to affect the course of events in the world, but is simply

---

[64]Thus, all these events are subjectively special, in terms of the typology of modes of divine action presented in Russell's "Introduction" to this volume.

[65]In terms of the typology of modes of divine action in the "Introduction," they are objectively special.

[66]Murphy, "Divine Action."

[67]Cf. Ellis, "Theology of the Anthropic Principle."

[68]*Dictionary of Christian Spirituality*, ed. Wakefield.

a spectator watching the inevitable unfolding of these events. Such a God has no handle with which to alter in any way, in the minds of the faithful, the conclusion of that physical unfolding governed by the physical regularities (the "laws of nature") that he/she has called into being and is faithfully maintaining. Here I am rejecting the somewhat paradoxical notion of revelation without special divine acts.[69] While one can certainly envisage people who are unusually receptive or perceptive of God's action through natural processes, they cannot reach that stage of understanding without somehow knowing of the existence and nature of God. But this in turn requires some kind of specific revelatory act to convey those concepts, so that faith can be based in personal experience and knowledge rather than unsupported imagination, which could arrive at any conclusion whatever.

The second point is the use made of this capability by the creator. This is where various traditions diverge, and the position one obtains depends on one's view of revelation. It could in principle be used to convey information, images, emotions, instructions, or pre-conceptual intimations of the nature of reality to humanity. Which of these actually occurs depends on the nature of the revelatory process implemented by the creator, which must be compatible with his/her nature and the character of his/her action in the world. As a specific example, consider the theory of revelation proposed by Denis Edwards. He states:

> Only an adequate theology of experience can do justice to the Old and New Testament understandings that God breaks in on our individual lives, that the Spirit moves within us, that God's word is communicated to us, and that we live in God's presence. . . . It is possible to show that while we do not have access to God's inner being, and while God transcends our intellectual comprehension, yet we can and do experience the presence and activity of this Holy One in a pre-conceptual way.[70]

This experience is the reason why the kind of "causal joint" mentioned above is necessary; it could not plausibly be the result of the blind action of physical forces alone. How does this happen?

> When I speak of the experience of God I will always mean pre-conceptual experience . . . [this] allows us to speak of a real human awareness of God who yet remains always incomprehensible to our intellects. It is, I will argue, precisely as mystery that we experience God's presence and action.
>
> . . . experience of grace is experience of something that transcends us, which breaks in on our lives in a mysterious way, and which we experience as a gift given to us.[71]

This particular view is broadly in agreement, for example, with the Quaker view of the experience of the light of God within.[72] Thus, we may take it that the

---

[69]See, e.g., Maurice Wiles, "Religious Authority and Divine Action," in *God's Activity in the World*, ed. Owen Thomas (Chico, CA: Scholar's Press, 1983).

[70]Denis Edwards, *Human Experience of God* (New York: Paulist Press, 1983), 5.

[71]Ibid., 13; 28.

[72]See "Christian Faith and Practice in the Experience of the Society of Friends" (London: Yearly Meeting of the Religious Society of Friends, 1972).

envisaged channel of communication is used at least to convey pre-conceptual intimations of the nature of reality to humankind. It finds its expression in the profound insights of the mystics and saints, as well as the religious experiences in the lives of the countless faithful, those ordinary people who do their best to follow their understanding of a life of enlightenment.

The traditions diverge on the issue of whether more specific forms of spiritual revelations and insights are communicated to humanity (e.g., St. Paul on the road to Damascus; Jesus throughout his life, but specifically in the temptations in the desert and in the garden of Gethsemane). In the context of the present investigation, we can afford to be open-minded about this; once the existence of the causal link is established, it could be used for such purposes—if that was spiritually desirable.

Note 1: In either case, a process of discernment is required on the part of the receiver to test whether what appears to be some intimation or revelation is indeed so, or if it is a false (perhaps psychologically induced) manifestation. This is an area that has been considered by the spiritually aware for many centuries, and will not be discussed further here.[73]

Note 2: It must be emphasized that the idea of conveying such "information," where this word is used in the broadest possible sense as indicated above, does not in any way imply a coercive or monarchical use of that capability by the creator. Indeed it is fully compatible with a view of the universe based on self-sacrifice and kenosis.[74] Indeed without such a possibility for the flow of information, we cannot have any reliable idea of the nature of transcendent reality. Thus, it is precisely the availability of the intimations of reality through the envisaged link that enables us to conclude that this reality is better described by the theme of kenosis than by any other.

The supposition here is that these events proceed through the normal functioning of the brain but have an extra, non-inevitable character in the sense that they must—if they mean what they appear to mean on the Christian interpretation—convey information to the receiver that was not explicitly there initially (in an evolutionary perspective). This necessity supports view III above as to the functioning of the brain, as is discussed in the following section. This is then the foundation of Christian spirituality.

4.1.2 *Moral Insight*

However, the further need for meaningful human existence is for a more generally based understanding of the nature of ethics and morality, as a foundation for moral

---

[73]See, e.g., Edwards, *Human Experience of God*; and Murphy, *Theology in The Age of Scientific Reasoning* (Ithaca, NY: Cornell University Press, 1990), where this topic is discussed in depth.

[74]Cf. Peacocke, *Theology for a Scientific Age*; *idem*, "God's Interaction with the World"; Ellis, "Theology of the Anthropic Principle"; and Murphy and Ellis, *Cosmology, Theology, and Ethics*.

decisions and the search for meaning. It can be argued[75] that the deeper levels of ethics and morality also should come through this revelatory channel as intimations of reality and ethical rightness, rather than through some process based on evolution of the brain and culture, as envisaged in sociobiology. While the mechanism would be closely related to that by which consciousness and free will arise (cf. the previous section), this ethical understanding cannot—by its very nature, in order that it can have ethical meaning—be mandatory; that is, it cannot be supposed to follow inevitably from the operation of the laws of physics in the brain. In that case there would be a lack of the ability for free response, which is essential for ethical behavior to have meaning.[76] Thus, this should also be classified—according to the above definition—as extraordinary rather than ordinary divine action (but non-disruptive).

4.2 *Miracles Proper*

Finally, we come to the most controversial area of all—the possibility of:

Action 7: miracles: special actions of an exceptional kind, so that the physical outcome is altered from what it would otherwise have been. This could be either:

Action 7a: actions not based on ordinary laws of physics, indeed involving a suspension of those laws; or

Action 7b: actions affecting physical conditions directly,[77] based on a steering of what happens consistent with known laws; for example, through direction of quantum events, amplified by sensitive dependence on initial conditions to macroscopic effects.

This is where the traditions differ most, the modern liberal view denying their existence at all, in contrast to many more traditional views. They may or may not occur (or have occurred in the past); we will return to that issue in the next subsection. For the moment we simply consider this as a possibility in a non-committed, open-minded way. In doing so, we note that action 7a is the only possibility considered in this paper that does not respect the laws of physics;[78] all the rest do (they are all strictly compatible with the regularities of those laws).

Considering the first type of exception (7a), these certainly are possible, although there may be a problem of interface with the rest of the universe: If some exceptional interaction takes place in a space-time domain *U*, then in general these "illegitimate" effects will causally interact with events outside *U*, eventually spreading the consequences to a large part of the universe. Problems could arise at the interface of the region where the laws of physics hold and the region where they are violated; for example, how are energy, momentum, and entropy balances maintained there?

---

[75]See Ellis, *Before the Beginning*.

[76]See Murphy, "Divine Action"; and Ellis, "Theology of the Anthropic Principle."

[77]Apart from giving humans insight that leads to purposeful action, as in Action 6.

[78]In terms of the types of modes of divine action, these are objectively *special* interventionist events.

Leaving this technical issue aside, examples of what might conceivably occur range from the Resurrection to altering the weather or making someone well if they are ill. It is here that one needs to distinguish different strands of the Christian tradition, and the various ways they view the question of miracles. Some will take literally all the miracle stories in the Old and the New Testaments; others will explain away some, many, or even all of them. Supposing that they do occur, or have occurred, one has then to face the thorny questions: What is the criterion that justifies such special intervention? When would they indeed occur? These issues will be picked up in the next sub-section.

The second type (7b) is quite possible in principle too, the classic case being God affecting the weather through the "butterfly effect" but within the known laws of physics. In its effect this is similar to the previous possibility, but of course in practical terms this has to be seen through the eyes of faith: no physical investigation could ever detect the difference between such action and chance effect, even if it was clear that the desired rain had fallen just after a major prayer meeting called to petition God for an end to the drought. Thus, one has here the possibility of an "uncertainty effect" deliberately maintained in order that true faith be possible. Such intervention would never be scientifically provable. Whether we believe it takes place or not depends on our overall worldview and experience.[79]

4.3 *Capricious Action or Regular Criteria*

The problem of allowing miraculous intervention[80] to turn water into wine, to heal the sick, to raise the dead, or to alter the weather is that this involves either a suspension or alteration of the natural order.[81] Thus, the question arises as to why this happens so seldom. If this is allowed at all to achieve some good, why is it not allowed all the time, to assuage my toothache as well as the evils of Auschwitz?

Indeed when we look at the world around, seeing the anguish of Bosnia, Somalia, Mozambique, and so on, and seeing children dying of drought and famine in many parts of the world, we pray "God have mercy on us" and wonder what would induce him/her to do so: to relinquish for a minute the iron grip of physical law held there by his/her apparently pitiless will. After all, these laws hold in being the material in its inexorable course while it is used to destroy and torture humanity. Here one recalls the unspeakable horrors of "necklacing" in the townships of South Africa, or the materials used in previous times by clerics of

---

[79]Perhaps this corresponds to the non-basic objectively special events identified in the typology of divine action.

[80]It is not possible in the space available here to do justice to the debates on the enormous hermeneutical and historical problems concerning the miracles reportedly performed by Jesus, and their relation not only to enlightenment science but also to the problems of interpreting ancient, often contradictory, texts.

[81]Such an occurance is allowed and possible because the laws are the expression of the will of God, who could therefore suspend them if he/she wished. See Murphy, "Divine Action."

many theological persuasions who pitilessly tortured and burnt to death those of differing views. We even arrive at the extraordinary concept of God holding to their natural behavior and nature the nails and wood used in the cross at Calvary to crucify Jesus.

Thus, if the usual Christian view is to make sense, there has to be a cast-iron reason why a merciful and loving God does not alleviate a lot more of the suffering in the world, if he/she has indeed the power to do so, as envisaged in Murphy's paper. This reason has to be sufficient to outlaw any pity in all these cases, and to prevent the taking of that decision that would end the suffering. This is of course just the age-old problem of evil, brought to special focus by the claim that the laws that enable it to take place are the optional choice of God.

In broadest terms,[82] the solution has to be that greater good comes out of the arrangement we see, based on the unwavering imposition of regularities all or almost all the time, even though that conclusion may not be obvious from our immediate point of view. For example, death is not so important when life is considered in a full perspective that takes into account the promise of resurrection. More particularly the regularity and predictability gained by the laws of physics must be seen as the necessary path to create beings with independent existence incorporating freedom of will and the possibility of freely making a moral and loving response.[83] Pain and evil are the price to be paid both for the existence of the miracle of the ordinary (cf. the previous section) and for allowing the magnificent possibility of free, sacrificial response. But then—if miracles do occur—the issue is why on some occasions this apparently unchanging law should be broached; this would strongly suggest a capriciousness in God's action, in terms of sometimes deciding to "intervene" but mostly deciding not to do so.

What one would like here—if one is to make sense of the idea of miracles—is some kind of rock-solid criterion of choice underlying such decisions to act in a miraculous manner,[84] for if there is the necessity to hold to these laws during the times of the persecutions and Hitler's Final Solution, during famines and floods, in order that true morality be possible, then how can it be that sometimes this iron necessity can fade away and allow turning water to wine, or the raising of Lazarus? Here as before I am not going to deal directly with the enormous hermeneutical problems of interpreting texts on miracles. Instead I am asking a different kind of question. If we are to be able to make any sense whatsoever of these miracles, what one would like to have is some kind of almost inviolable rule that such exceptions

---

[82]Cf. Stoeger, "Describing God's Action."

[83]See John Hick, *Evil and the God of Love* (New York: Macmillan, 1977); Murphy, "Divine Action"; Stoeger, "Describing God's Action"; and Ellis, "Theology of the Anthropic Principle."

[84]These are criteria from our limited viewpoint, being applied to God's activity. Stoeger points out we must realize that in considering this, what appears to us to be intervention may not be so from God's viewpoint; and that while in some sense through revelation, God has given us access to his/her point of view, this is only a limited access.

shall not take place unless the most unusual circumstances arise—something like the following:

Assertion 1: Exceptional divine action (7) can take place only in the case of events that make a unique and vital difference to the future evolution of humanity as a whole, and/or its understanding of the action of God, significantly influencing the entire future of humankind.

This does not include making rain in a drought-stricken area, stopping slaughters, or saving children from starvation, but could include the Resurrection of Christ as one of the most important ways of God communicating with humanity about the nature of life here and after. It could just conceivably include some "steering" of biological evolution at vital junctures (cf. 4) in a way compatible with the laws of physics (cf. 7b), although in that case it would be impossible to prove that this steering had ever happened; believing this to be so would be an act of faith. The alternative, suggested in my previous paper,[85] is:

Assertion 1a: Exceptional divine action (7) never takes place, but action (6) does. Then extraordinary divine action must always be in the form of provision of pre-images of right action or of ultimate reality, as freely attested in the spiritual tradition, thereby guiding and assisting free agents as they struggle to understand the world; the "miraculous" option, although possible, would not be used. This view somewhat assuages the problem of evil in that the charge of capriciousness is removed: the same laws always hold—implemented in order that freedom and morality can exist. Regularity is always there, and the "rights of matter" are always respected.

I suggest that what is needed here is a testing and examination of such possible views, looking again, systematically, at the different kinds of claims about "miraculous" intervention and whether they would or would not be permitted by the criterion being considered, and what the moral and religious implications are (a centuries-old debate). As emphasized previously, this would be tantamount to choosing between various viewpoints on the nature of Christianity. My own present preference has been made clear above: I would exclude interventions 7a and 7b, because otherwise the charge of capriciousness becomes almost overwhelming.[86] Should one hold the opposite view, adopting a criterion something like that suggested, it is imperative to clarify what "essential" means in this context.

Stoeger raises the issue of a category of events which we call "miracle," which of itself does not necessarily have a determinative influence on the course of history, but seemingly involves abrogation or at least a transcending of the laws of nature, and functions as a sign of something deeper or more life-giving in a situation or in reality than is otherwise apparent. Most of Christ's miracles were in this category—as are other claimed healings. Their main purpose was not, or is not, the healing or transforming act itself, but rather the manifestation of the deeper level of reality which otherwise would be hidden (e.g., Jesus' cure of the paralytic

---

[85]Ellis, "Theology of the Anthropic Principle."

[86]Apart from a point made by Willem Drees about "respecting the integrity of science," relevant to 7a. See Willem Drees, "Gaps for God?" (in this volume).

as a sign of his power to forgive sins). My suggestion would be that insofar as these events actually happen—obviously an issue for debate—they belong to category 6 rather than 7.

The above criterion relates to "miraculous" events (7). Similar questions arise in regard to the provision for moral and spiritual visions to people (cf. 6): What determines when this is done and when not? It may perhaps be suggested here that these are always available to those willing to hear, who patiently wait on God. This is a partial answer, as one can suggest that sometimes compelling visions are indeed made available (cf. St. Paul or George Fox) that are not given at other times, and that a recurring feature of spiritual life are the "desert" times when such sustenance is not forthcoming. There may well be good spiritual reasons for this, but this too needs clarification. Thus, to complete the picture one would require some kind of criterion applicable in these cases too. This may be already implicit in the literature on Christian spirituality, but it needs to be drawn out and explicated in the present context.

4.4 *An Alternative Domain of Action*

There is one alternative way to avoid the charge of capriciousness. This is to consider the possibility that within the laws governing the behavior of matter, there is hidden another domain of response of matter to life than usually encountered: matter might respond directly to God-centered minds through laws of causal behavior, or there may be domains of response of matter encompassed in physical laws, but they are seldom tested because such God- centered minds are so seldom encountered.[87] Then the distinction between ordinary and extraordinary action becomes a question of whether or not we have entered this domain. What has been classified as "extraordinary" action above would be "ordinary" action but in a different set of circumstances leading to a different kind of response and behavior where God-centered thought dominates and matter responds. Thus, we have the possibility:

Action 7c: existence of a new order, a new regime of behavior of matter (cf. a phase transition), where apparently different rules apply (e.g., true top-down action of mind on matter), when the right "spiritual" conditions are fulfilled.[88]

Thus, the extraordinary would be incorporated within the regular behavior of matter, and neither the violation of the rights of matter[89] nor the overriding of the chosen laws of nature would occur. Thus, the laws and the nature of physics are respected. The charge of capriciousness would then fall away, in a way consistent with the views of Murphy's paper. This is related to collapsing the distinction between the natural and the supernatural, from God's point of view. An example could be Jesus' resurrection. Wolfhart Pannenberg suggests that this

---

[87]See Polkinghorne, *Science and Providence*; and Stoeger, "Describing God's Action."

[88]In Russell's terms, this is a "time-dependent miracle."

[89]Murphy, "Divine Action."

could be the first instance of the kind of transformation that awaits the entire cosmos.

Three new issues would arise. First, similar to criterion 1 above for exceptional divine action, one would want to have stated carefully something like:

Assertion 2: the condition requisite for such a change-of-phase in the operation of physical laws in a given situation is the presence of one or more people in that situation who have—as a consequence of God's initiative—handed over their lives to God so fully that they are able to act freely as channels of the divine will. This enabling feature then transforms the local functioning of physical law to a new domain.

This attempt at a criterion for what determines where such a phase transition takes place should not be taken too seriously in its details; it is intended rather to suggest the type of thing one might take into account in understanding this possibility. If it is indeed true that such a kind of transition can take place,[90] then characterizing its nature through a criterion of some kind, as suggested by the example above, would clearly be a description of one of the most fundamental features of the nature of the universe. Its clear characterization—even weakly—would be a major achievement.

Second, one would want to characterize the nature of what would be possible and impossible in this altered regime: what then are the laws of behavior of matter?

A third issue would be to give some kind of evidence that this intriguing but highly controversial possibility is realized. One could claim that there is existent evidence supporting this proposal, for example, in the historical record contained in the Bible. But apart from querying the status of the evidence itself (e.g., did miracles really take place?), it is not clear how uniquely those data can be taken as supporting this particular proposal (7c).

Could we give some other evidence that this kind of behavior does indeed take place? This seems very difficult but not entirely beyond the bounds of possibility; for example, careful scientific investigation into the claimed instances of faith-healing might be relevant. One would encounter here all the same kinds of problems that occur in scientific testing of para-normal phenomena. Perhaps the biggest problem would be the conceivably legitimate claim that the kind of skeptical watching involved in a scientific investigation is precisely one of the conditions preventing such an altered domain of behavior. Supporting such a claim would require some modification of criterion 2, so that it takes into account negative factors that might hinder the proposed change of state.

5 *Mind and Top-Down Causation*

As mentioned before, given the understanding attained so far, the further issues are what type of causation might occur whereby these intentions are made a reality, and where the causal nexus could be whereby this top-down action could be

---

[90]Polkinghorne, *Science and Providence*; Stoeger, "Describing God's Action."

initiated in the world (whatever interpretation one may give to the concept of special divine action).

It is essential here to distinguish two rather different kinds of downward causation. Firstly, there is generic downward causation: this influences a whole range of events through alteration of operational conditions in a region (e.g., variation in temperature or pressure or magnetic fields affects the way matter responds.) Most of the examples mentioned by Küppers are of this kind.[91] This kind of general top-down influence alters conditions over a wide range of events in a region, and affects them all.

By contrast, there is specific or directed downward causation, which influences very specific events as occurs, for example, in the human body or complex machinery and is essential to their functioning. Instances include brain action to move a specific muscle in the body, a command to a computer that activates a particular relay or sensor, or hitting a specific typewriter or organ key that effects the desired result. In each of these cases a very specific local change in environment (current flow, pH levels, etc.) is effected, which causes proximate events to proceed in a specific way that is very localized and directed. This is possible through specific communication channels (nerves in a human body, bus lines in a computer, wires in a telephone exchange, or fiber optics in an aircraft) conveying messages from the command center to the desired point of activity.[92]

The point here is that setting boundary conditions at the beginning of the universe can achieve generic downward action but not specific action. An event such as influencing a mental state requires specific acts, changing circumstances in a locally highly specific way, rather than an overall change in the boundary conditions (a change in temperature, for example). I reject the possibility of setting special initial conditions at the beginning of the universe (t=0) to make this happen. While this is theoretically possible, it would amount to solving the problems involved in a reversal of the arrow of time. It would require setting precisely coordinated initial conditions over a wide area of the universe so as to come together at the right time and place in such a way as to achieve the desired effect, despite all the interactions and interfering effects that will have taken place from the hot early universe, where the mean free path even for neutrinos is extremely small, up to the present day, where the possibility of agents acting with free will implies an essential unpredictability in the environment within which this distant effect will be propagating. This tuning would in fact be impossible to accomplish—with the usual arrow of time—for one highly specific event, let alone a whole series of such events, each to be accomplished independently. According to Oliver Penrose,[93] this feature is the essential foundation of the second law of thermodynamics, based on a lack of correlations in initial conditions in the past (in contrast to the existence of such correlations in the corresponding final conditions

---

[91]Küppers, "Understanding Complexity."

[92]Cf. Beer, *Brain of the Firm*.

[93]Oliver Penrose, "Foundations of Statistical Mechanics," *Reports on Progress in Physics* 42 (1979): 1937-2006.

in the future). This law can in principle be confounded; for example, one could reverse the motion of molecules from a fallen and broken glass to reassemble it. In practice, however, this is not possible[94]—or at least not without special directed intervention.

Thus, the specific top-down action needed requires either specifically directed lines of access to particular nerve cells (as in the physiology of the human body), or a universal presence with detailed and specific knowledge of and access to each atom (as conveyed by the idea of the immanent presence of God). The latter is what is required for the Christian tradition to make sense as envisaged in Murphy's essay in this volume. Thus, in order for any of the "special action" discussed in the previous section to be possible (and specifically the provision of pre-images of ultimate reality or notions of spirituality to a person's mind), the interaction must be such as to provide highly directed information and influence, rather than some broad, overall top-down influence.

5.1 *The Nexus of Interaction*

The point then is that the action envisaged will be top-down in the sense that it originates in some higher level of organization (the mind of a person, the mind of God) but is highly specific in the time and place of action (as discussed here in terms of the specificity of action). Unless we envisage a totally new form of interaction[95] it cannot be effected in terms of some broad overall interaction with the universe as a whole, or indeed at any higher level of organization. Rather it must be implemented in detailed local interactions at the atomic or particle level, where quantum uncertainty and non-locality are factors that cannot be neglected and can in fact conceivably provide a *modus operandi* without violation of any physical laws.[96] This must then be done in whatever coordinated way is required to effect the required results at the macroscopic scale.

Thus, the quantum mechanism identified in Murphy's and Tracy's essays will suffice, in principle, as the vehicle of intentional interaction by a transcendent being. This view requires the essential action of God who is ensuring that the "laws of physics" are obeyed and who acts in a hidden way in every classical realization of such action to determine its actual outcome. At the meso-scale this interaction would not be recognizable through any violation of physical laws; everything would proceed causally according to those laws. The supposition is that this quantum effect would be amplifiable through brain processes—similar perhaps to photon multipliers—to macroscopic levels where they could influence feelings or thoughts. This is a wide enough channel to convey to us all that is needed for revelation, and to be recognizable as such by those with eyes to see.

---

[94]See Penrose, *Emperor's New Mind.*

[95]This seems to be implied by Peacocke's proposals, but it raises problematic aspects in terms of its interaction with normal physics.

[96]See Murphy, "Divine Action"; Tracy, "Particular Providence"; and Squires, *Conscious Mind.*

Note that this would not mean that God in some sense calculates the effect of what would happen via specific neural stimulations and then delicately one by one acts in just the right way in each neuron; rather we must see how we act downwards on our own neurons. We think things, plan, imagine, and the delicate causal channels set up for that purpose convey these intentions in such a way that the appropriate neurons fire as required. On this analogy we would envisage God through the mode of transcendence planning certain pre-images, emotions, or whatever to be made available to us. The appropriate communication channels which are in place by means of divine immanence allow this intention to be communicated to the appropriate neurons, quantum uncertainty being the feature that allows this to happen at any desired place and time without violating known physical laws. Thus, we would envisage the conscious part of his/her intentional action being similar to ours: the intention is formed consciously, the details take care of themselves.

Now, this sounds very strange from the viewpoint of physics alone. However, that is not our starting point. Following Murphy, I am assuming one stream of thought within the variety of traditional Christian positions, and developing its logical implications. Thus, the underlying assumption is:

1: The immanent God is present everywhere and yet, as transcendent God, maintains the nature of physical entities, ensuring their regular, law-like behavior according to the description of local physical laws. In particular he/she causes quantum action to take place in a law-like way, according to the known nature of quantum physics.

Then the possibilities are:

2a: God determines the actual realization of quantum outcomes from the possible ones, choosing a specific result in each quantum measurement (which is undetermined by the imposed physical laws). There is an openness in the system, and God uses it to input the desired information. Or:

2b: These outcomes really are "uncaused," in that God chooses not to determine which of the possible outcomes eventuates. God rattles dice each time to determine the actual outcome from those that quantum theory allows, refraining from making a choice. There is an openness in the system, and God uses it to input random noise, or possibly a combination of these positions. In any of these cases, the issue is not whether there is divine action at the quantum level, (for effective immanence ensures that there is[97]), but rather, what use is made of this divine action at the quantum level?

Alternative 2a envisages coherent information input through this action, actualizing top-down action in a purposeful manner. Alternative 2b rejects this as a useful channel of action. The action still takes place, but is specifically structured so as not to be purposeful. In that case, it seems that the only channel for meaningful top-down action of the required kind[98] is through altering the boundary conditions of the system *S*. But this in turn has to happen through some physical

---

[97]Murphy, "Divine Action."

[98]Peacocke, "God's Interaction with the World."

means in the larger system $S' = S + E$, where the previous environment E is now included in the system to be explained. The whole problem recurs now for this larger system, with its new boundary E'.

The analysis supports the proposal of Murphy and Tracy that quantum uncertainty is a, perhaps even the only, vehicle through which special divine action (particularly as experienced in revelatory acts affecting human minds) can take place as required by many religious traditions. This provides an important part of the foundation of Christian spirituality. It also supports view III above as to the functioning of the brain,[99] which seems to be a closely related issue.

6 *Evidence*

The final topic I wish to discuss briefly is the issue of supporting evidence for these views.[100] It is clear from the nature of the argument that some aspects are compatible both with "chance" and with divine action; they will only be seen in the latter context through the eye of faith. But what then is the starting point for our discussion of the nature of faith? Furthermore, in view of conflicting standpoints, whose doctrine of faith and whose practice will one accept and why?

This is the whole issue of apologetics, which cannot be dealt with properly here.[101] However, some key points can be made. The suggestion will be that the "Christian Anthropic Principle"[102] selects a particular viewpoint based on the theme of self-sacrifice or kenosis,[103] which structures the argument and opts for specific Christian traditions from among the competitors. We can present the analysis in summary form by referring to the implied scheme of top-down action, with emergent layers of description and meaning,[104] that arises from that discussion.

The structure envisaged is one of layers of meaning and morality as shown in Figure 5 (*on following page*).

Top-down causation is active in this hierarchy in terms of action and meaning. The fundamental intention of the creator shapes the structure and brings into being the physical foundations. The interactions at the physical level are the basis for all the higher levels of order (through bottom-up action), enabling the existence of life through the fine-tuned nature of physical reality and allowing life to come into being through the processes of self-organization and evolution. Once conscious beings have come into existence, they create moral orders through psychological and sociological interactions. These orders then come into

---

[99]Cf. Squires, *Conscious Mind*; and Penrose, *Emperor's New Mind*.

[100]Cf. section 2 of Murphy's paper.

[101]A systematic presentation will be given in Murphy and Ellis, *Cosmology, Theology, and Ethics*.

[102]Ellis, "Theology of the Anthropic Principle," section 6.

[103]The different levels of kenosis are discussed in K.M. Cronin, *Kenosis: Emptying Self and the Path of Christian Service* (Rockport, MA: Element, 1992).

[104]Cf. section 2 above.

confrontation with the moral and spiritual order of ultimate reality, which exerts its influence on humans in a persuasive rather than coercive manner. Thus, while the upward and downward causal action is fairly rigid at the lower levels, it is not so at the higher ones. At the lower levels the interconnecting laws of action are those of physics, which are inviolate as long as special divine actions 7a are not taking place (and we have assumed they do not occur), while at the moral levels they are of the nature of persuasion and invitation, allowing choice and free response. This is their essential character.

Level 1: Spiritual/religious
Spiritual values: ← Data 1
kenosis in relation to transcendence

Level 2: Moral/ethical
Ethical values: ← Data 2
kenosis in relation to others: serving

Level 3: Social and ecological
Political, economic interactions: ← Data 3
community and ecosystem kenosis

Level 4: Personal/individual (psychological)
Consciousness, choice: free will ← Data 4
responsibility, kenosis/self

Level 5: Biological
Levels of biological organization: life ← Data 5
self-organization, evolution

Level 6: Physical
Level of physical entities and action ← Data 6
regularities of physical law

Figure 5: Hierarchical structuring of meaning and morality in the Universe. Top-down and bottom-up action occur as in the other hierarchically structured systems, leading to emergent meaning at the higher levels as indicated. The data at each level must be in terms of the kinds of concepts and meanings appropriate at that level.

In assessing this proposal relative to its competitors, there are separate data of different types appropriate to each level in the hierarchy. At each level the scheme suggested is indeed supported by a considerable volume of data, and provides an overall coherent scheme in agreement with those data (but not uniquely indicated by those at the physical level alone). However, choice of the whole structure is a metaphysical choice based on recognition of the appropriateness and rightness of what is presented, justified ultimately by the "good fruits" associated with this worldview. The further key element is dealing with apparent counter-evidence, for otherwise the proposal is vulnerable to the charge of being based on selected evidence only, ignoring awkward evidence pointing in other directions.[105]

A defense can be built on the lines indicated in Murphy's paper and my paper:[106] essentially, the overall scheme proposed is only possible, in terms of truly allowing free will and full moral choice, if the possibility of evil is allowed as well, with full acceptance of its consequences. This is both the "free-process" defense of Polkinghorne and the traditional free-will defense.

One's assessment of what has been suggested here will depend on one's prior assumptions. If one accepts that the traditional religious view (summarized above) is correct and, additionally, that the modern scientific view is correct, then one arrives fairly uniquely at the scheme suggested here (the essential element of choice is identifying the theme of kenosis as fundamental,[107] as against, for example, monarchical themes.). This leads to a holistic view, as sketched above, which accords with the data at all levels, once the apparent counter-evidence has been evaluated in the light of overall constraints on what is possible in view of God's ultimate aim in creating a universe where free moral response is possible. Two further points are of interest.

First, from this foundation I suggest that we arrive at the necessity for an upwards openness, in correspondence with the downwards openness fundamental to the proposal. The possibility of free moral choice requires "gaps" in the system, as discussed by Tracy. Indeed the downward causation is not rigid but involves persuasion rather than coercion, as mentioned above. Correspondingly, the upward causation must be open; this is required for the system to be consistent.[108]

Thus, on this view, rather than searching for the "gap" allowed by quantum uncertainty as a place where divine action can take place, we invert the argument: we demand that there must be such an openness in physical laws, in order that morality can be possible and that special divine action (as described above) can take place. That is, just as one demands certainty in physical processes at the macroscopic level, as discussed by Murphy, so that moral response is possible, additionally one demands causal gaps (as described by Tracy) at the microscopic level, so that top-down causation can lead to an openness in upward emergent properties and allow the kind of revelatory possibilities envisaged in this article. Thus, in a sense one predicts the necessity for an openness. While it may be that

---

[105]See Anthony N. Flew, *Thinking about Social Thinking: Escaping Deception, Resisting Self-Deception* (London: Harper Collins, 1991), for a discussion of the dangers of such selective choices.

[106]See Murphy, "Divine Action"; and Ellis, "Theology of the Anthropic Principle."

[107]Cf. Ellis, *Before the Beginning*; and *idem*, "Theology of the Anthropic Principle."

[108]This is really an aspect of W. Ross-Ashby's "law of requisite variety." See his *Introduction to Cybernetics* (London: Chapman & Hall, 1956); and Beer, *Brain of the Firm*.

this openness could occur otherwise than through the uncertainty inherent in quantum processes, my own analysis (in accord with Tracy and Murphy) is that it is indeed this openness which we should identify with that required for true morality to exist.

It follows then that there is no question of this proposal not "respecting" the randomness built into quantum physics, as if this has an independent ontological status. Rather this apparent randomness is just the openness required in physical reality in order that God's action can be effective without destroying the possibility of higher levels of order.

Second, because of the nature of any system of top-down causation and emergent order, it is clear that when considered in terms of the lower-level descriptions only, the meanings and concepts of the higher levels do not exist: they literally have no meaning. This is what worries those who view the proposal on the basis of the requirements of science alone: the scheme simply does not make sense when viewed from that perspective. The issue is what level of description is being used in one's analysis of reality; the proposal here only makes sense if one includes the highest levels of meaning.

7 *Conclusion*

The view of divine action presented in Murphy's paper[109] seems coherent and reasonable. It emphasizes first "ordinary action" in terms of the creation and preservation of the universe, providing the ground for the existence of the dependable physical systems that allow objects and people their independent existence and "rights," through the upwards emergence of physical properties based on physical laws. It also allows special divine action, particularly in terms of intimations of right action provided to those willing to see. God's action is then able to lead to action in the world through directed downwards causation in the body, and so to effective changes in the world.

Problems arise in terms of the possible choice to act specially in a miraculous manner as is certainly possible in this scheme of things. The issue then is how to avoid the charge of capriciousness and, in some sense, conniving with evil in those cases where such action is not taken. A clear-cut criterion controlling such interventions provides some kind of safeguard against such charges. This could be a partial answer, when taken in conjunction with a strong argument to the effect that the conditions leading to apparent evil are those required to create free will and independence.[110]

---

[109]I see Murphy's paper as being complementary to my own ("Theology of the Anthropic Principle"). I regard the two as being (in broad terms) in agreement with each other and with others in this volume, for example, that of Tracy.

[110]I am here avoiding an explicit reference to free evil spirits, e.g., a "Devil" operating independently of God, or to the Jungian alternative of a "dark side of God." This could be one of the areas where various Christian traditions differ strongly from each other, possibly

However, a different possibility is the existence of an alternative domain of action in the physical world, coming into effect in those cases where wills are in concert with God.[111] This preserves a fixed order of behavior in the universe without "miraculous" intervention, but allows "special action" to become commonplace where the conditions for this alternative order exist. This possibility needs further exploration to make clear the criteria that could govern such a "phase change" and to characterize some of the features of the new domain of action that could then arise. Experimental data relevant to this situation would appear to be rather few; the motivation for its acceptance on other grounds would then have to be compelling.

Clearly, the proposal that quantum uncertainty provides the necessary causal gap is highly controversial. However, if one takes into account the data as a whole and seriously attempts a holistic combination of both the religious and scientific views, this suggestion becomes less scandalous and, indeed, the necessity of microscopic uncertainty in physical laws virtually becomes a prediction of the understanding attained.[112]

---

leading to significant variations of the theme proposed in Murphy's paper.

[111]Polkinghorne, *Science and Providence*; Stoeger, "Describing God's Action."

[112]I thank all the members of the second Vatican Observatory/CTNS conference for the stimulating interaction with them that has led to the thoughts presented in this paper. I am particularly grateful to Bill Stoeger, Bob Russell, and Nancey Murphy for detailed comments on the manuscript, which have led to major improvements.

LIST OF CONTRIBUTING PARTICIPANTS

Willem Drees, Center for the Study of Science, Society, and Religion (Bezinningscentrum), Free University, Amsterdam, The Netherlands.

Denis Edwards, Lecturer in Systematic Theology, St. Francis Xavier Seminary, Adelaide College of Divinity, Flinders University, Adelaide, South Australia.

George F.R. Ellis, Professor of Applied Mathematics, University of Cape Town, Rondebosch, South Africa.

Langdon Gilkey, Professor Emeritus of Systematic Theology, The Divinity School, University of Chicago, Chicago, Illinois, USA.

Stephen Happel, Associate Professor of Religion and Culture, Chair of the Department of Religion and Religious Education, The Catholic University of America, Washington, D.C., USA.

Michael Heller, Professor of Philosophy, Pontifical Academy of Theology, Cracaw, Poland.

Bernd-Olaf Küppers, Professor of Philosophy, Institut für Philosophie, Friedrich-Schiller-Universität Jena, Germany.

Jürgen Moltmann, Professor Emeritus Evangelisch-theologische Fakultät der Eberhards-Karl-Universität, Tübingen, Germany.

Nancey Murphy, Associate Professor of Christian Philosophy, Fuller Theological Seminary, Pasadena, California, USA.

Arthur Peacocke, Director of the Ian Ramsey Centre, Oxford, England, Warden Emeritus of the Society of Ordained Scientists, and formerly Dean of Clare College, Cambridge, England.

John Charlton Polkinghorne, President, Queens' College, University of Cambridge, Cambridge, England.

Robert John Russell, Professor of Theology and Science In Residence, Graduate Theological Union, Founder and Director, The Center for Theology and the Natural Sciences, Berkeley, California, USA.

William R. Stoeger, S.J., Staff Astrophysicist and Adjunct Associate Professor, Vatican Observatory Research Group, Steward Observatory, University of Arizona, Tucson, Arizona, USA.

Thomas F. Tracy, Professor of Religion and Chair, Department of Philosophy and Religion, Bates College, Lewiston, Maine, USA.

INDEX

VATICAN PRESS